Service-Oriented Computing

Radi Doncheva

SERVICE-ORIENTED COMPUTING
Semantics, Processes, Agents

Munindar P. Singh
North Carolina State University, USA

Michael N. Huhns
University of South Carolina, USA

John Wiley & Sons, Ltd

Copyright © 2005   John Wiley & Sons Ltd,
                          The Atrium, Southern Gate, Chichester,
                          West Sussex PO19 8SQ, England

                          Telephone   (+44) 1243 779777

Email (for orders and customer service enquiries): cs-books@wiley.co.uk
Visit our Home Page on www.wileyeurope.com or www.wiley.com

Reprinted September 2005

Other Wiley Editorial Offices

John Wiley & Sons Inc., 111 River Street, Hoboken, NJ 07030, USA

Jossey-Bass, 989 Market Street, San Francisco, CA 94103-1741, USA

Wiley-VCH Verlag GmbH, Boschstr. 12, D-69469 Weinheim, Germany

John Wiley & Sons Australia Ltd, 33 Park Road, Milton, Queensland 4064, Australia

John Wiley & Sons (Asia) Pte Ltd, 2 Clementi Loop #02-01, Jin Xing Distripark, Singapore 129809

John Wiley & Sons Canada Ltd, 22 Worcester Road, Etobicoke, Ontario, Canada M9W 1L1

British Library Cataloguing in Publication Data

A catalogue record for this book is available from the British Library

ISBN 10: 0-470-09148-7 (H/B)
ISBN 13: 978-0-470-09148-7 (H/B)

Printed and bound in Great Britain by Antony Rowe Ltd, Chippenham, Wiltshire
This book is printed on acid-free paper responsibly manufactured from sustainable forestry
in which at least two trees are planted for each one used for paper production.

To the Singh and Huhns families, especially
To Mona on our fifteenth anniversary – Munindar
To Mary on our thirtieth anniversary – Mike

Contents

About the Authors

Munindar P. Singh is a Professor of Computer Science at North Carolina State University. From 1989 through 1995, he was with the Microelectronics and Computer Technology Corporation (better known as MCC). Munindar's research interests include multiagent systems and Web services. He focuses on applications in e-commerce and personal technologies. Munindar's 1994 book *Multiagent Systems*, was published by Springer-Verlag. He coedited *Readings in Agents*, which was published by Morgan Kaufmann in 1998. He has coedited several other books and authored several technical articles. Munindar's research has been recognized with awards and sponsorship from the National Science Foundation, DARPA, IBM, Cisco Systems, and Ericsson.

Munindar was the editor-in-chief of *IEEE Internet Computing* from 1999 to 2002 and continues to serve on its editorial board. He is a member of the editorial boards of the *Journal of Autonomous Agents and Multiagent Systems* and the *Journal of Web Semantics*. He serves on the steering committee for the *IEEE Transactions on Mobile Computing*.

Munindar received a B.Tech. in computer science and engineering from the Indian Institute of Technology, New Delhi, in 1986. He obtained a PhD in computer science from the University of Texas at Austin in 1993.

Michael N. Huhns is the NCR Professor of Computer Science and Engineering at the University of South Carolina, where he also directs the Center for Information Technology. Previously he was a Senior Member of the Research Division at the Microelectronics and Computer Technology Corporation. Prior to joining MCC in 1985, he was an Associate Professor of Electrical and Computer Engineering at the University of South Carolina, where he also directed the Center for Machine Intelligence.

Mike is a member of Sigma Xi, Tau Beta Pi, Eta Kappa Nu, ACM, IEEE, and AAAI. He is the author of over 180 technical papers in machine intelligence and an editor of the books *Distributed Artificial Intelligence*, Volumes I and II, and, with Munindar, *Readings in Agents*. His research interests are in the areas of multiagent systems, enterprise modeling and integration, and software engineering. From 1997 to 2003, he wrote a column *Agents on the Web* for IEEE Internet Computing.

Mike was an associate editor for *IEEE Expert* and the *ACM Transactions on Information Systems*. He is an associate editor for the *Journal of Autonomous Agents and Multiagent*

Systems. He is on the Editorial Boards of the *International Journal on Intelligent and Cooperative Information Systems*, the *Journal of Intelligent Manufacturing*, and *IEEE Internet Computing*. He was an advisor for the First International Conference on Multiagent Systems, 1995, and has been on the advisory boards for the International Workshops on Distributed Artificial Intelligence. He is a member of the board for the International Foundation for Multiagent Systems and the International Foundation on Cooperative Information Systems.

Mike received the BSEE degree in 1969 from the University of Michigan, Ann Arbor, and the MS and PhD degrees in electrical engineering in 1971 and 1975, respectively, from the University of Southern California, Los Angeles.

Contact information

Munindar P. Singh
Department of Computer Science
North Carolina State University
Raleigh, NC 27695-7535, USA

http://www.csc.ncsu.edu/faculty/mpsingh/
singh@ncsu.edu

Michael N. Huhns
Department of Computer Science & Engineering
University of South Carolina
Columbia, SC 29208, USA

http://www.cse.sc.edu/˜huhns/
huhns@sc.edu

Preface

The current World-Wide Web was intended to be used by people, but most experts, including the founder of the WWW, Tim Berners-Lee, agree that the WWW will have to evolve to include usage by computer systems. Moreover, the impact of computer usage will exceed that of human usage. The evolution is expected to occur through the design and deployment of Web services. The phrase *Web services* sometimes refers to services that employ a particular set of basic standards. Since these standards are all but incidental to the key concepts of services and services apply even in settings strictly different from the WWW, it is helpful to think of service-oriented computing as a more general topic.

The objective of this book is to explain the principles and practice of service-oriented computing, although most of its concepts are developed in the context of Web services. The book presents the concepts, architectures, theories, techniques, standards, and infrastructure necessary for employing services. It includes a comprehensive overview of the state of the art in Web services and related disciplines.

Services are a means for building distributed applications more effectively than with previous software approaches. For this reason, it is not useful to talk about services without talking about service-based applications, how they are built from services, and how services should be designed so they can feature as parts of serious applications. Indeed, the *raison d'etre* for services is that they be used for multiple purposes. And, services are used by putting them together or *composing* them—the rare case where a service is used without any contact with another service can be thought of as a trivial composition. For this reason, every aspect of services is designed to help them be composed: specifically, so they can be described, selected, engaged, collaborated with, and evaluated.

Many of the key techniques now being applied in building services and service-based applications were developed in the areas of databases, distributed computing, artificial intelligence, and multiagent systems. These are generally established bodies of work that can be readily adapted for service composition. Some additional techniques, although inspired by these areas, are being developed from scratch by practitioners and researchers in the context of service-oriented computing. These new techniques address the essential openness and scale of Web applications that previous work did not need to address. Both classes of key techniques should be incorporated into our best practices. In most cases, they can be applied on top of existing approaches.

which can be fleshed out with more powerful representations and techniques. These are established computer science approaches and serious practitioners are already using them, although they are omitted from most expositions of Web services. The rest of the book shows what these are and how they can be employed.

Description. Part II addresses techniques and methodologies for describing services. These techniques include ideas from conceptual modeling of database schemas and domain knowledge, and cover both representation and reasoning approaches. They lead naturally into some of the XML-based technologies gaining currency as the *Semantic Web*. The idea is that when services are described with sufficient richness, it is easier for providers to state what they offer and for consumers to specify what they need, leading to meaning-based interactions.

Engagement. Part III deals with how services may be engaged or executed so as to facilitate the simpler kinds of composition. Often, when services are described, there is an emphasis on invoking services as if they were no more than methods to be called remotely. We prefer the term *engagement*, because it more accurately reflects the power of the service metaphor for computing. Imagine going to a carpenter, a human service provider, to have some bookshelves built in your home. You will typically not instruct the carpenter in how to build the shelves. Instead, you will carry out a dialog with the carpenter in which you will explain your wishes and your expectations; the carpenter might take some measurements, check with some suppliers of hard-to-get materials (maybe some exotic woods), and suggest a few options; based on your preferences and your budgetary constraints, you might choose an option or get a second estimate from another carpenter.

Likewise, in computing, instead of merely invoking a particular method, you would engage a service. Here the relevant computational themes are peer-to-peer computing, messaging, transactions (traditional, nested, and extended or relaxed), workflow, business processes, and exception handling. A number of standards for services are emerging in these areas.

Collaboration. Part IV discusses advanced concepts that arise from a computational standpoint in composing services, where it is helpful to think of the services as collaborating with each other. Some of the key technologies that apply for collaboration include protocols, agents, contracts, service agreements, and negotiation techniques. The engagement techniques in Part III give us the basis for engaging services while considering various transactional properties. The techniques of collaboration in this part go several steps further in characterizing the interactions among the consumers and the providers of services, dealing with how they plan and enact service episodes, how they maintain consistency, negotiate, enter into and execute contracts and agreements, and carry out specified protocols. This part includes a discussion of monitoring compliance with contracts and service agreements.

Selection. Part V introduces concepts of service discovery and selection, and distributed trust. Service discovery in its simplest form involves registries where services can be registered and looked up. However, selecting desirable services in practice also involves accommodating notions of trust, endorsement, and reputation.

This part of the book also includes a discussion of how services can be evaluated by the parties using them. This is essential to complete the cycle of locating services, engaging the services, and then evaluating the services to determine if they were successful. Fair and accurate evaluation can enable the various parties to find, select, and engage the services that are superior in some way.

Engineering. Part VI focuses on the engineering of service-based applications. It discusses methodologies and techniques for building services in the context of some important classes of applications, especially knowledge management and e-business. This part also discusses the best practices for the main kinds of techniques described in the previous parts.

Directions. Part VII discusses some of the key trends in services and in architectures. It considers architectural policies, privacy, and personalization from the perspective of how services fit into the larger world. It also discusses more advanced philosophical notions, such as ethics and social mores, with a view to inspiring services that function in a manner that improves the network at large, not just optimizing the results for themselves.

Appendices. Part VIII has appendices on important background topics such as XML technologies and Web standards and protocols. The appendix on XML, in particular, is as extensive as a chapter and includes description of all the key XML technologies that you need to know in order to read this book. It also includes exercises for readers who wish to test their knowledge.

The organization of the book is designed to encourage the building of a series of projects beginning with the most basic applications of the most established standards, and ending with areas where technologies are still gelling.

Munindar P. Singh
Raleigh, North Carolina

Michael N. Huhns
Columbia, South Carolina

Note to the Reader

This book brings together a lot of interesting concepts that apply in service-oriented computing. Where possible, we have sought to describe the techniques that these concepts support—in other words, to make the concepts actionable. However, the concepts in many cases are subtle; you must master the concepts before you can be effective with the techniques.

We recommend that you read the text carefully and work through several of the theoretical exercises. If this book is successful, it will have piqued your interest about several topics. If you have a theoretical bent, you will want to pursue deeper results. Virtually all topics discussed here have a lot of depth, and a number of PhD problems lurk within. The book cites the key work for each topic, which will give you excellent starting points.

For those with an interest in practical implementations (and we recommend that even theoreticians work on a few, just to keep grounded), the book provides a fair amount of detail about how various techniques can be realized. It discusses virtually all the key emerging standards for service-oriented computing, and concludes with a discussion of engineering challenges and the emerging methodologies and best practices to address them. In this sense, it is a lot more practical than the typical advanced textbook. If you are reasonably experienced as a programmer, you can find the few necessary details to get started by downloading the latest versions of the tools from the Internet. The book's home page at http://www.csc.ncsu.edu/faculty/mpsingh/books/SOC/ includes useful pointers.

However, as you can well imagine, each standard and tool has a lot of nitty-gritty details. All too often these are not based on any theoretical concepts, but are accidents of history or practical concerns of implementations. This book does not describe such idiosyncrasies. In our own system development, we would learn such details by Internet searches or by trial and error, and promptly forget them as soon as we could! For the bleeding edge standards and tools, there may be no other way; for the older ones you can perhaps find detailed books. In either case, this book will help you master the concepts that will last a long time, rather than details that might be lost in the next release.

Lastly, please note that writing the first big book on a wide-ranging new topic is a daunting task. In our combined four decades of post doctorate experience, this is the largest intellectual venture that we have attempted. There will undoubtedly be lots of room for improvement. We welcome your suggestions!

Acknowledgments

A book such as this inevitably describes the work of others. Besides the authors cited within, we are indebted to our colleagues, students, teaching and research assistants, and the audiences of our tutorials at various conferences. In particular, we would like to thank Soydan Bilgin, Michael Maximilien, Yathi Udupi for help with some of the programming assignments, José Vidal for contributing several exercises and Paul Buhler for evaluating an early version of the text in his class. Amit Chopra gave extensive comments on some chapters. Several students, in particular Leena Wagle and Sameer Korrapati, gave useful suggestions on previous drafts. We thank Mona Singh and Hilary Huhns for their creative talents in drawing many figures.

We gratefully acknowledge support from the government and corporate sponsors of our research, which enabled us to develop the understanding of service-oriented computing that is described in this book. Munindar's research sponsors include Cisco, DARPA, Ericsson, IBM, and the National Science Foundation. Mike's research sponsors include the National Science Foundation, DARPA, the US Department of Agriculture, and NASA. Mike has greatly benefited from his association with Ray Emami, Alok Nigam, and their research team at Global Infotek, Inc.

As usual, we are deeply indebted to our families for their patience despite the onerous demands made by our writing.

Figures

Tables

Listings

Part I

Basics

Chapter 1

Computing with Services

When we browse the Web, fill a form, or make a purchase, we are participating in a distributed computation, whose other components we may know very little about. Web services are interesting, because they provide an approach for constructing and deploying such distributed computations in a manner that enhances the productivity of programmers, administrators, and users alike.

Most researchers and practitioners agree that today's Web, although successful in many ways, also has a number of limitations. Information on the Web is not organized. It can be inaccurate and inconsistent and, worse, incomprehensible. Current techniques for locating information offer low precision (including irrelevant results) and low recall (missing relevant information). Most of the information is static and Web sites typically do not exhibit well-structured programmatic behavior. The only kind of programmative behavior is when a form is posted to a script running at a specified URL, but these scripts have rigid interfaces and behaviors, and complicate the tasks of building and maintaining distributed applications.

This chapter reviews some key motivations for the emergence of Web services. The motivations are guiding some of the emerging standards and technologies for Web services and provide a basis for evaluating competing approaches.

1.1 Visions for the Web

Most plausible visions for the future of the Web are based on the following tenets, many of which have already begun to come about. The Web is and will be ubiquitous, have no central authority, and consist of "components" that are heterogeneous and autonomous. Today, the components are primarily Web pages, but increasingly they will be programs in general. In other words, the Web provides both content and services today, but there is an increasing emphasis on the latter. The Web is dynamic today in terms of its components being able to change arbitrarily. However, it is not fully dynamic in that the components can negotiate only a few limited aspects of their interactions: typically, the visual aspects of a page or whether cookies can be set and retrieved via a browser. The Web will begin to support *cooperative* peer-to-peer (P2P) interactions, while continuing to support client-server interactions.

1.1.1   Semantic Web

Tim Berners-Lee, the originator of the World-Wide Web, has described one such vision called the *Semantic Web*. Today's Web is geared for use by people. In other words, information is generally marked up for presentation and is displayed accordingly by a browser. Humans can usually interpret the content of the information because they have the necessary background knowledge, which they share with the creators of the given page. Unless programs are created to represent and exploit such knowledge, they are limited to processing in a hard-coded manner, which does not suit a dynamic setting where the details can easily change.

For example, we can write a so-called screen-scraping program that extracts the price of a book from a search-results page on amazon.com. This program would, of course, rely upon the syntax of the Web page being encoded in some sort of a formal grammar. Intuitively, such a program may be instructed to read the price from the results page by parsing it appropriately. Depending on the structure of the page at the given site, these instructions could be heavily *ad hoc*, for example, get the seventh cell from the third row of the second table in the fifth frame. Although tools exist to simplify such parsing and extraction, this task still requires painstaking effort by programmers. Moreover, the program would fail or behave erroneously when some seemingly irrelevant change is made to the structure of any of the Web pages that it reads.

In the Semantic Web, the page would be marked up not only with presentation details, but also with a separate representation of the meaning of its contents. In other words, the results page of our example would say what the price was. A program instructed to extract the price would find the price even if the layout of the page was changed. In other words, the markup on the Web would progress from the merely syntactic—capturing just the structure of the information, to the semantic—capturing the meaning of the information.

1.1.2   Peer-to-Peer Computing

Another important trend relevant to our topic is peer-to-peer computing (P2P). The Web today is used for interactions in which most of the information resides on one side (server) and most of the intelligence on the other (client). The asymmetry between the interacting parties means that the information tends to be aggregated in large servers, which become significant to the functioning of the entire system. If the servers fail, then the whole system is adversely affected; if they are compromised, then security in the system may be violated. The key idea behind P2P is that the different components are peers or equals of each other. Each has aspects of being a server and a client—for this reason, the peers are sometimes termed "servents."

Under P2P, the Web would consist not of passive pages to be accessed by programs, but of active programs that can communicate with one another. In principle, such programs may carry out negotiations with one another and proactively offer suggestions to one another.

However, current P2P approaches lack semantics, meaning that the applications must hard-code how they interact, precluding flexible negotiation. For this reason, they have been instantiated in simple applications, such as file sharing, where humans provide the semantics.

1.1.3 Processes and Protocols

The Web today is by and large static and passive, although, as noted above, it is possible to invoke programs (e.g., servlets) over it. However, the challenge is to find and correctly invoke the programs. This challenge is addressed by first-generation service approaches, which provide a means by which the parameters and outputs of programs can be specified, usually in a notation based on the *eXtensible Markup Language* (XML). Further, these standards provide a means to locate the services one might wish to invoke and provide support for how to invoke those services over the Web, using the *Hypertext Transfer Protocol* (HTTP), the same protocol as used by Web browsers.

Services are invoked to carry out business processes, which also involve high-level interactions among various parties. In contrast to service invocations, which are single-shot two-party interactions, processes are typically long-lived multiparty interactions. Except for toy examples such as looking up the weather or converting currencies, services are usually employed as parts of larger processes. For this reason, a process-oriented view of services is gaining prominence. Standards for modeling business processes are being developed as close extensions of standards for Web services.

1.1.4 Pragmatic Web

When we put the above together, we get what may be termed the *Pragmatic Web*. The term *pragmatic* here means that the Web is understood as being used for processes in context. The ramifications of this term are made clear by the rest of this book. Because this book describes the technologies of Web services from the perspective of how they could be used to build large, open information systems, it may be thought of as describing the emerging Pragmatic Web.

What is important about the Pragmatic Web is that it begins to put together the enabling technologies for negotiation. Negotiation is common in real life. Yet, in today's computer applications, most negotiations are handled ahead of time by humans; only the most trivial negotiations are carried out automatically. This is changing with the emergence of online markets that support bidding and bartering. Market algorithms and strategies for optimal bidding are not central to service-oriented computing.

However, participating successfully in a negotiation depends upon modeling the semantics of what is being negotiated, interacting as a peer, and carrying out long-lived interactions as part of realistic business processes. These aspects are central to the thesis of this book. This book does not seek to solve the problems of negotiation that arise in various applications, because the specific techniques would depend on the application in question. However, it seeks to provide the concepts and techniques of Web services that would be an integral part of the solutions in those applications.

In our vision, the Web will become usable by machines as well as by people, and will become active rather than passive. The following summarizes the main trends regarding the Web:

- *Automation*. Human $\Rightarrow$ Machine.

- *Richer markup*. The move from HTML to XML has regularized the syntax of documents and data structures found over the Web. However, XML is not adequate for capturing the semantics or formal meaning of these documents and data structures. Richer representations are needed. These are distinct layers above XML, although they might be given an XML syntax as well.

- *Richer activities*. Passive $\Rightarrow$ Active; Services $\Rightarrow$ Processes.

- *Greater interaction*. Client-Server $\Rightarrow$ P2P $\Rightarrow$ Cooperative.

- *Accommodating context*. Semantics $\Rightarrow$ Mutual Understanding $\Rightarrow$ Pragmatics.

The above trends bring us well beyond the Semantic Web as currently understood, and add pragmatics to the picture. For this reason, it is best to term these the Pragmatic Web.

1.2 Precursors

Let us quickly review the history of information technology from the perspectives of distributed computing and information modeling. As is well known, computing has evolved from centralized systems to time-shared ones, to client-server computing, and now on to peer-to-peer computing:

- First-generation information systems provide centralized processing that is controlled and accessed from simple terminals that have no data-processing capabilities of their own.

- Second-generation information systems are organized into servers that provide general-purpose processing, data, files, and applications and clients that interact with the servers and provide special-purpose processing, inputs, and outputs.

- Third-generation information systems, which include those termed peer-to-peer, enable each node of a distributed set of processors to behave as both a client and a server; they may still use some servers.

- Emerging next-generation information systems are cooperative, where autonomous, active, heterogeneous components enable the components collectively to provide solutions.

Already, systems are emerging that include aspects of cooperation wherein the components not only can deal with each other as peers, but also can understand each other at a higher level. Here we take the term *cooperative* to include both cases of intelligent help and intelligent competition.

In the early days of computing, data and applications were inextricably intertwined, leading to poor maintainability and upgradeability. By separating the data from the applications, database management systems enabled each to exist independently of the other. The hope was that data could be used for applications other than those for which it was intended and that applications could access databases other than those they were designed to access. This hope was partially realized in that, with sufficient effort, one could build new applications that access old data and feed new data into old applications. However, in its fullest form, this hope was dashed, because although access methods became standardized on the Structured Query Language (SQL) for relational databases, the semantics of the data remained as *ad hoc* as ever. For example, two relations with the column employee may or may not refer to the same concept.

The same problem occurred for the Internet. In the early days, applications and data formats were *ad hoc*. The standardization on HTML enabled the advent of browsers through which people could access information anywhere on the Web. This is fine as long as humans are engaged in understanding and processing the information, but does not lend itself well to automation. So it is natural that semantics will draw increasing attention on the Web as well.

The Web is simply the culmination of this trend to heterogeneity. There is an interesting dilemma here: the more distributed and independently managed that resources on the Web become, the greater is their potential value, but the harder it is to extract that value. Web services, by being usable across independently designed systems, facilitate creating systems of high value whose value can be extracted with greater ease than previously.

1.3   Open Environments

The above trends in information systems toward increasing distribution, decoupling, local intelligence, and collaboration have been accompanied by a similar evolution in networking, from proprietary local networks to wide-area private networks, such as *extranets* and *virtual private networks* (VPNs), to the public Internet. The result is that information systems have components that cross organizational boundaries, i.e., are open. The term *open* implies that the components involved are autonomous and heterogeneous, and system configurations can change dynamically.

Often, we would want to constrain the design and behavior of components, thus limiting the openness of the information system. However, the system would still have to deal with the rest of the world, which would remain open. For example, a company might develop an enterprise integration system that is wholly within the enterprise. Yet, this system would have to deal with external parties, e.g., to handle supply and production chains. In other words, the system would still need to function in an open environment. For this reason, it is helpful to think in terms of such environments. We now review some of the key characteristics of open information environments.

Let's begin with a review of the concepts of autonomy, heterogeneity, and dynamism as they relate to open information environments. A simple way to understand and distinguish these concepts is to associate them with the independence of users, designers, and administrators, respectively.

1.3.1   Autonomy

Autonomy means that the components in an environment function solely under their own control. Imagine dealing with an e-commerce site. It may or may not remove some items from its catalog. It may or may not even deliver the goods it promised. Of course, one might seek legal recourse if a contract is violated! In fact, the autonomy of the components is the reason that contracts and compliance are so important for open environments. We will return to these topics in Chapters 15 and 18.

Simply put, software components are autonomous because they reflect the autonomy of the human and corporate interests that they represent on the Web. In other words, there are sociopolitical reasons for autonomy. Resources are owned and controlled by autonomous entities and that is why they behave autonomously.

There are also a couple of technical reasons for autonomy. The simplest one is that a component that behaves unexpectedly might be doing so because of error, i.e., a mistaken requirement or a faulty implementation. If we can handle such components, then our system will be robust. A more subtle reason is that sometimes components are designed so as to be externally opaque in certain respects. For example, a well-encapsulated data type implementation hides its internal structures; therefore, the behavior of instances of this data type would not be controllable with respect to the hidden aspects. Consider a dictionary data type implemented as a hash table, with supported methods for inserting elements to the dictionary and iterating over all elements of the dictionary. The size of the hash table and the hash function are not revealed. To a programmer using this implementation, the ordering of the elements in the iteration would appear as uncontrollable. This ordering could even change across successive invocations if, in the interim, the hash table is resized because of internal considerations, perhaps motivated by space and efficiency.

A major practical example of this occurs in legacy enterprise systems wherein database management systems might be designed or configured to decide unilaterally (based on internal considerations) whether to allow a transaction to complete. To other components, their decision on whether a transaction may complete or not appears as purely autonomous. Lastly, certain instances of autonomy reflect the possibility of errors. For example, if a file system can fail, a Web site on which you submit a form may fail to record your changes, thus appearing to have unilaterally decided to discard your submission.

A consequence of autonomy is that updates can occur only under local control. In other words, you can request another party to do something, but you cannot force them to do it. This simple point illustrates a limitation of object-oriented computing. We can invoke methods on objects and, if we have the handle for an object, the object performs the method so invoked.

By contrast, for open environments, there is another layer of reasoning so that a component that is requested to perform a method may decide whether or not to accept the request. An advantage of service-oriented computing over object-oriented computing is that it respects autonomy.

1.3.2   Heterogeneity

Heterogeneity means that the various components of a given system are different in their design and construction. Just as for autonomy, there are both sociopolitical and technical reasons for heterogeneity. Component designers and architects might wish to construct their components in different ways, e.g., to satisfy different performance requirements. Often, the reasons are historical: components fielded today may have arisen out of legacy systems that were initially constructed for different narrow uses, but eventually expanded in their scopes to participate in the same system.

Heterogeneity can arise at a variety of levels in a system, such as networking protocols, encodings of information, and data formats. Clearly, standardization at each level reduces heterogeneity and can improve productivity through enhanced interoperability. This is the reason that standards such as the Internet Protocol (IP), HTTP, Universal Character Set (UCS), UCS Transformation Format (UTF-8), and XML have gained currency. Standards always evolve and different software components may lag behind or overtake standards in various respects. In general, it is easier to establish and comply with lower-level standards.

Heterogeneity also arises at the level of semantics and usage, where it may be hardest to resolve and sometimes even to detect. For example, a payroll system and a benefits system might both deal with employees. Yet, the payroll system might treat employees as those being paid on a regular basis, whereas the benefits system might treat employees as those receiving health benefits on a regular basis. Under some cases, the systems might happen to work correctly and be mutually consistent. A manager may obtain information aggregated from the two systems and meaningfully calculate, for instance, the average monthly expenses per employee. But a real-world event might cause their inherent heterogeneity to lead to differences in the behavior. Consider what happens when Anne, a paid employee with benefits, retires. If the organization continues to pay benefits for its retirees for the first year of their retirement, Anne would appear to be an employee in the benefits system but not in the payroll system. The aggregated data would not quite be as meaningful any more.

Heterogeneity can cause complications for the functioning of a component, because it means that less can be assumed about the other components. However, there is an excellent reason why heterogeneity emerges and should be allowed to persist. To remove heterogeneity would involve redesigning and reimplementing the various components. Even if the different designers are willing to bear the associated costs, removing heterogeneity is difficult, because doing so assumes that we can come up with a conceptually integrated design. However, integration is not easy and is fragile. This means that the conceptually integrated system, if one can be built, would tend to be unreliable. Most importantly, such a system would be fragile: as the components evolve because of changing local requirements, we would have to keep reintegrating them.

Therefore, it is more pragmatic to let the components be heterogeneous, but to impose various kinds of weak requirements on their interactions. After all, this is the reason why we have standardized protocols, such as TCP/IP and HTTP.

1.3.3 Dynamism

An open environment can exhibit dynamism in two main respects. First, because of autonomy, its participants can behave arbitrarily. In particular, they can change their behavior because of how they happen to be configured. Second, they may also join or leave an open environment at whim. It is worth separating out this aspect as a reflection of the independence of the system administrators. A large-scale open system would of necessity be designed so as to accommodate the arrival, departure, temporary absence, modification, and substitution of its components.

With regard to the first type of dynamism, individual components can change dynamically in their behavior, architecture and implementation, and interactions. That is, there might be changes in their externally observed behavior, how they achieve or produce their behavior, and how they interact with other components.

1.3.4 Challenges

Open environments pose significant technical challenges. In particular, we must develop approaches that can cope with the scale of the number of participants and respect the autonomy and accommodate the heterogeneity of the various participants, while maintaining coordination. Specifically, because of the scale, we cannot count on knowing all the available resources in terms of their functionality, reliability, trustworthiness, and so on. This means that discovering the required resources, deciding how to use them, engaging them, and checking their compliance are all significant challenges.

As a consequence of their autonomy and heterogeneity, the components must be treated in a *local* manner. In other words, each component must locally decide how to proceed in its interactions with others. This is in tension with assembling global information. Some level of global information is essential for ensuring that the different parties are coordinated. Yet, the presence of global information creates the possibilities of inconsistencies and causes potential difficulty for maintenance. Whereas the components in an open environment may often have some interdependencies in practice, they would have only a few such interdependencies if they are designed in a correct manner that preserves the autonomy and heterogeneity of the components.

An argument for preserving the autonomy and heterogeneity of the components is that it forces us to design simple or narrow interaction protocols, thereby eliminating any unnecessary dependencies among the components.

An argument for dynamism is that, if assumed, it greatly simplifies the challenge of configuring and administering a system. Moreover, it makes the system resilient to certain kinds of failures and enables the exploitation of certain emerging opportunities. For example, if a component fails, a system that was designed with dynamism in mind would take the failure in its stride: it could easily patch in a replacement component. Likewise, if a better component than one being used becomes available or a better deal comes along, such a system could switch components to better meet the overall business objectives.

Also, in practical settings, it is often appropriate to relax the constraints among the various components. Thus, global information is obtained or aggregated only when needed. More importantly, it is often OK to let inconsistencies emerge provided they can be corrected quickly enough (depending, of course, on the specific application at hand). The corrective actions in many cases will have a global basis, but can be applied locally. For example, an e-commerce transaction can complete correctly only if the goods are received by the purchaser and the payment is received by the vendor. It would be nearly impossible to synchronize these events perfectly, but it is possible to use a reliable payment mechanism, such as a credit card and a reliable delivery service. If the vendor fails to ship because of an unexpected shortfall, it can cancel the debit to the credit card. Section 5.4.2 considers a more detailed example. In general, there are many good examples of relaxed constraints in the way people have done business before the advent of computers. Some of these examples can be readily adapted for online settings.

1.4   Services Introduced

Just like *objects* a generation ago, *services* is now the killer buzzword of our era. And wherever you turn, some vendor or analyst is promoting services. They are like motherhood and apple-pie in modern computing. But unlike motherhood, services mean different things to different people. Web services have been defined as:

- A piece of business logic accessible via the Internet using open standards (Microsoft).

- Encapsulated, loosely coupled, contracted software functions, offered via standard protocols over the Web (DestiCorp).

- Loosely coupled software components that interact with one another dynamically via standard Internet technologies (Gartner).

- A software application identified by a URI, whose interfaces and binding are capable of being defined, described, and discovered by XML artifacts and supports direct interactions with other software applications using XML-based messages via Internet-based protocols (W3C).

Although our emphasis is on Web services, it is instructive to review how different communities conceive of services, reflecting their backgrounds and concerns.

- *Networking:* a service is characterized by bandwidth, availability, error rate, and similar properties.

- *Telecommunications:* a service is considered to be either a specific telephony feature, such as caller ID or call forwarding, or a basic connection service, such as narrowband versus broadband (itself of a few varieties).

- *Systems:* a service is for billing, storage, and other key operational functions. These functions are often parceled up into the so-called operation-support systems.

- *Web applications:* a service corresponds to Web pages, especially those with forms or a programmatic interface thereto.

- *Wireless:* in wireless versions of the Web, a service includes messaging, as in the popular short message service (SMS).

If there is agreement here, it is that a service is a capability that is provided and exploited, often but not always remotely. Accordingly, our working definition of a Web service is that it is functionality that can be *engaged* over the Web. Later sections explain the ramifications of engagement fully, but the essence is that engagement goes beyond mere invocation of services. However, the above answers provide a litmus test for judging what role one expects a "service" to play in a distributed system.

1.5   Using Services

Services provide a programming metaphor that supports the right kinds of programming models for open, distributed systems. Service architectures are modular, because each service inherently offers a certain provider–subscriber interface. This interface enables much flexibility, for instance, by allowing proxy agents transparently to provide new services based on old services and to compose services as appropriate. In general, composability is a powerful property for engineering software and more than sufficient justification for all the current interest in services.

Although services must be invoked, their invocations will often be implicit. For example, many of the networking and telecom services are not invoked as such; they are merely variants of other, more basic services that are invoked. That is, you might invoke a packet delivery service to send a series of packets over a network; with the same programming interface and depending on what underlying service is provisioned, you might obtain different guarantees as to the packet delivery in terms of, say, jitter. In telecom, the definition of a service is of regulatory (and hence economic) interest. For example, looking up a phone number is a standalone service, whereas call forwarding is a feature of telephony. To get a feel for these distinctions, see the Federal Communications Commission's ruling on reverse phone number lookup [FCC, 1996]. Telecom might not be of direct interest to Web specialists, but similar considerations and even regulations might begin to apply to Web services, either because they involve telephony or because the increasing economic importance of Web services attracts the attention of legislators.

1.6 The Evolving Web

The Web is, at first look, ubiquitous and so uniformly accessible that it is easy to begin thinking of it as a single large system. Its single distinguishing purpose seems to be the exchange of marked-up documents and its single distinguishing characteristic seems to be the hyperlinks among those documents.

However, the Web truly is many things to many people. Although the Web's many uses have similarities, we would be best off understanding their main variants so we can program accordingly. To this end, it is helpful to review a classification proposed by Bill Joy, former Chief Scientist of Sun Microsystems. Joy's classification consists of four main kinds of "Web," which he distinguishes based on the modalities of the interface as experienced by a user [Joy, 2000]:

- *Near Web*: conventional mouse-keyboard-monitor interaction with a personal computer, typically for purposes such as surfing the Web.

- *Far Web*: interaction with a computer from across a room as with a TV remote control, typically for entertainment, such as listening to music or viewing a movie.

- *Here Web*: interaction with a mobile device, with narrow bandwidths for input and output.

- *Weird Web*: interaction through emerging interface technologies, such as voice and wearable computing.

Joy defined two additional webs where there are no direct user interactions. These are the business-to-business (B2B) Web, dealing with the supply networks of business-to-business electronic commerce, and the pervasive Web, dealing with device-to-device interactions.

It is helpful to place the characteristics of services over the Web into a historical perspective, as shown in Table 1.1.

Table 1.1: A historical view of services over the Web

Generation	Scope	Technology	Example
First	All	Browser	Any HTML page
Second	Programmatic	Screen scraper	Systematically generated HTML content
Third	Standardized	Web services	Formally described service
Fourth	Semantic	Semantic Web services	Semantically described service

Systematically generated HTML content refers to data-driven Web sites that have a well-defined visual structure. Commercial Web sites such as amazon.com are examples. These can be automatically parsed, although the grammar through which they are parsed may be *ad hoc* and difficult to maintain as the structure of the content is not explicit and reflects the visual structure of a page. Formally described services are those that are described via current Web services standards (as introduced in Chapter 2). In current practice, these are being released with specialized toolkits by leading vendors. Exercises 1.1, 1.2, and 1.3 ask you to review three of these toolkits. Semantically described services are those that go beyond current Web services to explicitly encode the meanings of the services.

1.7 Standards Bodies

Since services involve serious work and interactions among the implementations and systems of diverse entities, it is only natural that several technologies related to services would be standardized. As in much of computer science, standardization in services often proceeds in a *de facto* manner where a standard is established merely by fact of being adopted by a large number of vendors and users. However, standards bodies play an important role. Sometimes they take the lead in coming up with *de jure* standards. At other times, they clean up and formalize emerging *de facto* standards, and lend some semblance of order to the marketplace.

The following are the most important standards bodies and initiatives for services. This book will refer to their specific contributions numerous times.

IETF. The Internet Engineering Task Force is charged with the creation and dissemination of standards dealing with Internet technologies. Besides the TCP/IP suite and URIs, the IETF is also responsible for HTTP and other protocols of interest to services, such as Session Initiation Protocol (SIP) and SMTP.

OMG. The Object Management Group has been developing standards for modeling, interoperating, and enacting distributed object systems. Its most popular standards include the Unified Modeling Language (UML) and Common Object Request Broker Architecture (CORBA). OMG has recently proposed the Model-Driven Architecture (MDA).

W3C. The World-Wide Web Consortium is an organization that promotes standards dealing with Web technologies. The W3C has mostly emphasized the representational aspects of the Web, deferring to other bodies for networking and other computational standards, e.g., those involving transactions. The W3C's main standards of interest for services include XML, XML Schema, WSDL, SOAP, and WSCI.

OASIS. The Organization for the Advancement of Structured Information Standards standardizes a number of protocols and methodologies relevant to Web services, including the Universal Business Language (UBL), UDDI, the Business Process Execution Language for Web Services (BPEL4WS), and, in collaboration with UN/CEFACT, ebXML.

UN/CEFACT. The United Nations Center for Trade Facilitation and Electronic Business focuses on the facilitation of international transactions, through the simplification and harmonization of procedures and information flows. Its mission is to improve the ability of business, trade, and administrative organizations, from developed and developing economies, to exchange products and services effectively, and so contribute to the growth of global commerce. One of UN/CEFACT's most important developments is the specification for the Electronic Business eXtensible Mark-up Language (ebXML), which is a framework for the global use of electronic business information.

WS-I. The Web Services Interoperability Organization is an open, industry organization chartered to promote the interoperability of Web services across platforms, operating systems, and programming languages. It creates and supports generic protocols for the interoperable exchange of messages between services. Its primary contribution to date is Basic Profile version 1.0 (BP 1.0). BP 1.0 is a consistent specification for basic Web services comprising SOAP 1.1, HTTP 1.1, XML 1.0, XML Schema Parts 1 and 2, UDDI Version 2, and WSDL 1.1.

BPMI.org. The Business Process Management Initiative is working to standardize the management of business processes that span multiple applications, corporate departments, and business partners. Microsoft based XLANG on the pi calculus, IBM used Petri Nets for WSFL, and BPMI.org unified the two approaches with the Business Process Modeling Language (BPML). In this regard, BPML 1.0 is similar to BPEL4WS.

WfMC. The Workflow Management Coalition develops standardized models for workflows and workflow engines, as well as protocols for monitoring and controlling workflows.

FIPA. The Foundation for Intelligent Physical Agents promotes technologies and specifications that facilitate the end-to-end interoperation of intelligent agent systems for industrial applications [FIPA]. FIPA's standards include agent management technologies and agent communication languages.

1.8 Overview of this Book

This book is organized according to two different, but complementary schemes. The first organization scheme is based on the major levels of abstraction for service-oriented computing, ranging from raw messages to individual services to conversations to choreography to sophisticated forms of orchestration supported by high-level contracts among teams of interacting, autonomous participants. This scheme is depicted in Figure 1.1, which illustrates the levels of abstraction, their relationships, and the aspects of Web services and service-oriented computing being addressed. The second scheme tracks the development of Web services and service-oriented computing from their heritage to their current incarnation, and where they either are heading or ought to head. The general themes of this organization cover basic connection, quality of service (QoS), and enterprise interoperation.

There have been several major efforts to standardize services and service protocols, particularly for electronic business. One of these is electronic business XML (ebXML), which has produced the rightmost stack shown in Figure 1.1. The leftmost stack is the result of development efforts by the Semantic Web research community in conjunction with the W3C. The central stack is primarily the result of standards efforts led by IBM, Microsoft, BEA, HP, and Sun Microsystems. By and large, these have been separate from standards bodies, but will be ratified eventually by one or more appropriate such bodies.

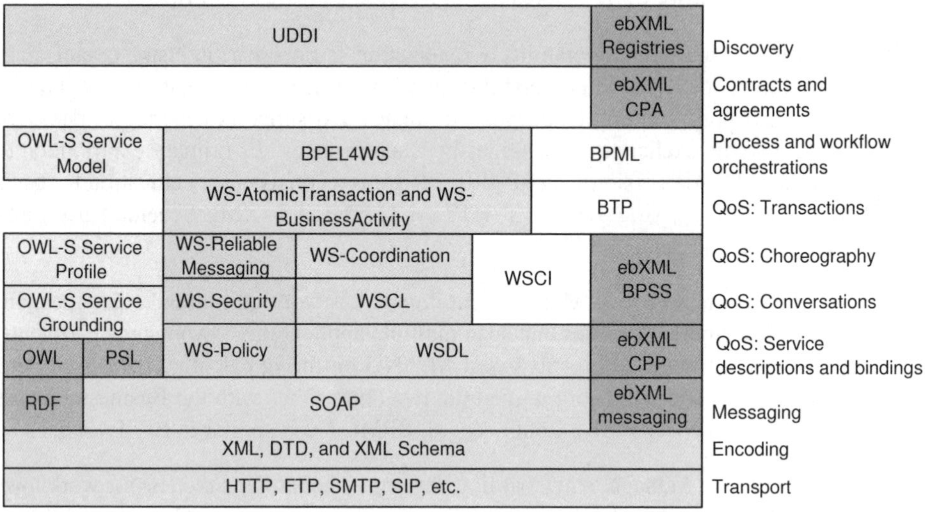

UDDI					ebXML Registries	Discovery
					ebXML CPA	Contracts and agreements
OWL-S Service Model	BPEL4WS				BPML	Process and workflow orchestrations
	WS-AtomicTransaction and WS-BusinessActivity				BTP	QoS: Transactions
OWL-S Service Profile	WS-Reliable Messaging	WS-Coordination	WSCI	ebXML BPSS		QoS: Choreography
OWL-S Service Grounding	WS-Security	WSCL				QoS: Conversations
OWL	PSL	WS-Policy	WSDL		ebXML CPP	QoS: Service descriptions and bindings
RDF		SOAP			ebXML messaging	Messaging
XML, DTD, and XML Schema						Encoding
HTTP, FTP, SMTP, SIP, etc.						Transport

Figure 1.1: The relationship of the different proposed standards and methodologies for automating electronic business

Each stack makes use of the following abstraction levels:

- The *transport* layer provides the fundamental protocols for communicating information among the components in a distributed system of services.

- The eXtensible Markup Language, *XML*, is the foundation for interoperation among enterprises and for the envisioned Semantic Web. Standards at this level describe the grammars for syntactically well formed data and documents, and how the well formedness can be validated.

- The messaging layer describes the formats using which documents and service invocations are communicated.

- The *service descriptions and bindings* layer describes the functionality of Web services in terms of their implementations, interfaces, and results.

- A *Conversation* is an instance of a protocol of interactions among services, describing the sequences of documents and invocations exchanged by an individual service.

- *Choreography* protocols coordinate collections of Web services into patterns that provide a desired outcome. Choreography is used *across* a domain of control to ensure harmony and interoperability.

- *Transaction protocols* specify not only the behavioral commitments of the autonomous components, but also the means to rectify the problems that arise when exceptions and commitment failures occur.

- The *orchestration* layer has protocols for workflows and business processes, which are composed out of more primitive services and components. They specify the control flows and data flows needed for the processes to be executed correctly. Orchestration implies a centralized control mechanism (e.g., the conductor in an orchestra), whereas choreography does not (e.g., the dancers on a stage). Orchestration is typically used *within* a domain of control.

- *Contracts* and *agreements* formalize commitments among autonomous components in order to automate electronic business and provide outcomes that have legal force and consequences.

- The *discovery* layer specifies the protocols and languages needed for services to advertise their capabilities and for clients that need such capabilities to locate and use the services.

1.9 Notes

XMethods provides a number of services for simple tasks such as currency conversion, and so on.

Microsoft MapPoint (http://www.microsoft.com/mappoint/webservice/) is a Web service for searching city maps (it is based on the old mapblast.com technology that was acquired by Microsoft).

1.10 Exercises

1.1. Study the amazon.com Web service toolkit licensing and evaluate their offering (with examples) in terms of the concepts of autonomy, heterogeneity, and dynamism. Next, imagine what additional forms of autonomy, heterogeneity, and dynamism you might experience from this toolkit, even if they are not explicitly documented.

1.2. Repeat Exercise 1.1 but for the Google Web service toolkit.

1.3. Repeat Exercise 1.1 but with respect to the Microsoft MapPoint Web service toolkit.

1.4. Which (zero or more) of the following techniques would preserve autonomy among the participating components:

- Message-passing via TCP/IP sockets between component A and component B?
- Remote procedure call (RPC) between component A and component B?
- Remote method invocation (such as Java RMI) between component A and component B?
- Email using SMTP between component A and component B?

1.5. Using a Web programming approach such as Java Server Pages (JSP) or Active Server Pages (ASP), build a Web page that lets a user search a local database. Consider the bookstore domain, where information is stored about authors, titles, publishers, years of publication, and prices. For simplicity, the database may be implemented as a flat file. This exercise is a precursor to implementing a Web service, thus helping you to learn the nuances of your local installation.

Chapter 2

Basic Standards for Web Services

Although Web services in the sense of current standards are only now emerging, the idea of providing services over the Web is quite old. If we checked, most of us would probably find that we have been providing and using Web services for years. For example, mail reflectors and their associated mailing lists are services to which we can subscribe. There are even on-line catalogs of the mailing lists and the particular topics to which they are devoted. The primary difference between older service offerings and contemporary Web services is that human intervention was previously required.

The modern view of services goes beyond the above, however, in terms of accommodating the openness of Web systems. This means describing services in a standard manner, arranging for them to be discovered in a standard manner, and invoking them, also in a standard manner. The general architectural model for Web services is shown in Figure 2.1. It consists of three types of participants:

1. Service providers, who create Web services and advertise them to potential users by registering the Web services with service brokers.

2. Service brokers, who maintain a registry of advertised (published) services and might introduce service providers to service requesters.

3. Service requesters, who search the registries of service brokers for suitable service providers, and then contact a service provider to use its services.

The architecture for Web services is founded on principles and standards for *connection*, *communication*, *description*, and *discovery*. For providers and requesters of services to be connected and exchange information, there must be a common language. This is provided by the eXtensible Markup Language (XML). A common protocol is required for systems to communicate with each other, so that they can request services, such as to schedule appointments, order parts, and deliver information. The Simple Object Access Protocol (now known just by its acronym, SOAP) [Box et al., 2000] currently provides the common communication

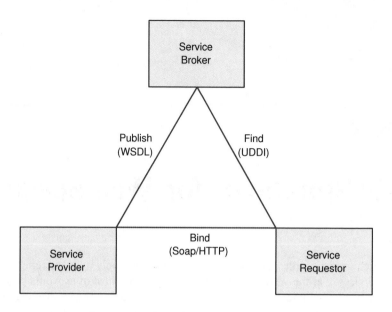

Figure 2.1: The general architectural model for Web services

protocol. The services must be described in a machine-readable form, where the names of functions, their required parameters, and their results can be specified. This is provided by the Web Services Description Language (WSDL), sometimes pronounced "whiz-dull." Finally, clients—users and businesses—need a way to find the services they need. This is provided by Universal Description, Discovery, and Integration (UDDI), which specifies a registry or "yellow pages" of services. Besides standards for XML, SOAP, WSDL, and UDDI, there is a need for broad agreement on the semantics of specific domains. In fact, that is where the deeper challenges lie. Chapter 6 introduces approaches for representing the semantics of services.

This chapter next introduces the key elements of the three main standards for Web services. Many of the details of these standards are of greatest interest to those implementing tools, not those implementing applications and systems using the tools. That is, application programmers will rarely need to operate at the level of the standards themselves. Just as they do not commonly use HTTP (much less TCP/IP) directly but through programming interfaces, they will use programming interfaces and tools that provide a layer of abstraction above the standards. Still, it is useful to develop an understanding of the standards themselves.

2.1 XML

The most fundamental of the above languages is XML. XML tags convey information about the meaning of data, not just how the data should appear, as is the case for HTML. XML regularizes the syntax of HTML, so that it is easier to parse and process. Either of these can provide definitions of tag structures and syntaxes that could be industry-wide and standard, or idiosyncratic and local. XML has the advantages in that, first, it can be extended for new applications and, second, it supports both documents (unstructured data) and structured data (e.g., from databases). Third, XML enables structure-sensitive queries, meaning that we can query an XML document based on its structure and its values in specific places or fields. Fourth, XML-tagged data can be mechanically validated.

XML provides a data format for documents and structured data, but does not specify the semantics of the format. To share information and knowledge among different applications (i.e., for interoperability), a shared set of terms describing the application domain with a common understanding is needed. Chapter 8 introduces a formal language for describing the application domain, which is also expressible in XML, but is conceptually at a higher level than XML. Appendix A provides a further description of XML and some relevant techniques.

2.2 SOAP

SOAP was originally intended to provide networked computers with remote-procedure call (RPC) services written in XML. It has since become a simple and lightweight protocol for exchanging XML messages over the Web using HTTP, SMTP, and SIP—the Session Initiation Protocol for Internet telephony and other peer-to-peer applications [Schulzrinne, 2004]. In practice, however, HTTP is the most common transport for SOAP and is the option that is included in interoperability standards, such as BP 1.0. Chapter 3 returns to this point.

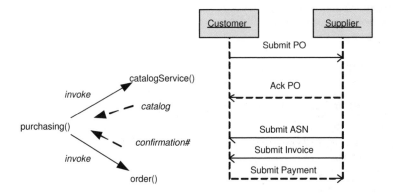

Figure 2.2: A customer can order goods from a supplier directly by method invocation (left) or indirectly by sending a purchase order message (right)

Figure 2.2 shows examples of SOAP interactions modeled as methods and as messages, respectively. In these examples, a manufacturing company (say, Dell) might directly invoke a supplier's (say, Intel's) catalog and order functions, or might send a purchase order message to the supplier. When services are modeled as methods, then service composition is achieved via scripts that invoke methods.

SOAP defines an envelope for transmitting messages and rules for representing remote-procedure calls. Listing 2.1 shows an example SOAP message packaged in an HTTP POST. HTTP POST is used rather than HTTP GET due to its ability to carry a payload. The message, sent to www.socweather.com, requests that the Web service perform the *GetTemp* procedure using "Honolulu" as the value of the *City* parameter and "now" as the value of the *When* parameter. The SOAP response message carries the resultant temperature as the parameter *DegreesCelsius* with a value of "30".

Listing 2.1: A SOAP request

```
POST /temp HTTP/1.1
Host: www.socweather.com
Content-Type: text/xml; charset="utf-8"
Content-Length: xxx
SOAPAction: "http://www.socweather.com/temp"

<!-- The above are HTTP header fields -->
<?xml version="1.0"?>
<env:Envelope
 xmlns:env="http://schemas.xmlsoap.org/soap/envelope/"
 env:encodingStyle=
   "http://schemas.xmlsoap.org/soap/encoding/"/>
 <env:Body>
  <m:GetTemp
   xmlns:m="http://www.socweather.com/temp.xsd">
    <m:City>Honolulu</m:City>
    <m:When>now</m:When>
  </m:GetTemp>
 </env:Body>
</env:Envelope>
```

Listing 2.2: A SOAP response corresponding to the request of Listing 2.1

```
HTTP/1.1 200 OK
Content-Type: text/xml; charset="utf-8"
Content-Length: xxx
SOAPAction: "http://www.socweather.com/temp"

<?xml version="1.0"?>
<env:Envelope
 xmlns:env="http://schemas.xmlsoap.org/soap/envelope/"
```

```
env:encodingStyle=
  "http://schemas.xmlsoap.org/soap/encoding/"/>
<env:Body>
  <m:GetTempResponse
    xmlns:m="http://www.socweather.com/temp.xsd">
    <DegreesCelcius>30</DegreesCelcius>
  </m:GetTempResponse>
</env:Body>
</env:Envelope>
```

The SOAP body, i.e., for SOAP RPC (see Section 5.2.2), describes method calls and responses. It must contain the target object's URI, the method name, and the parameters of the message. The parameters are specified in a struct that contains an accessor for each parameter in the method's invocation and an accessor for the return value. Note that the types of the parameters are not specified—this is left for WSDL, which is described in Section 2.3. The EncodingStyle attribute specifies rules for deserializing the SOAP message.

2.2.1   Processing

SOAP messages are assumed to be routed from one recipient to the next until they arrive at the final recipient. Thus SOAP is understood in terms of three kinds of *nodes*: sender, intermediary, or ultimate recipient. The SOAP model involves two main roles: next and ultimateReceiver, and the trivial role none. The role of next is played by any intermediary and by the final recipient; the role of ultimateReceiver may only be played by the final recipient. No participant can play the role none. In effect, each participant must play the role of next, unless someone sends a message and never wishes to receive a reply. Each of these roles is defined via a URI to ensure its uniqueness. However, the semantics of the roles is not formally expressed.

2.2.2   Body and Header

Each SOAP message must include a body, which is generally interpreted by the ultimateReceiver. However, an intermediate node may also interpret the body. Such behavior is considered to be against the spirit of the specification, although there is no way to ensure compliance. The processing semantics is left to the application, meaning that it must be negotiated through some out-of-band communication.

Both Listings 2.1 and 2.2 demonstrate simple SOAP messages. The main element of a message is the envelope, whose contents are processed as appropriate by the application, i.e., the interpretation of the contents is not part of the SOAP specification.

As shown in Listings 2.1 and 2.2, an envelope contains a mandatory body. In addition, a SOAP envelope may optionally contain a header. The header provides a control channel for passing additional directives and information that would influence a recipient's treatment of the body. A SOAP message may travel from a sender to a receiver by passing different

endpoints along the message path. SOAP header blocks contain information that might be
intended for SOAP intermediate endpoints, as well as the ultimate endpoint. The SOAP actor
attribute is used to address a header element to a particular endpoint. A header partitions into
blocks, which have their own namespace and describe some logically and computationally
related aspect of the processing of the body. As for the body, the semantics of the header
blocks is left to the application.

Listing 2.3: Example SOAP header

```
<!ENTITY SOAPENV
    http://www.w3.org/2003/05/soap-envelope>
<env:Header>
  <c:converse
    xmlns:c='http://www.socweather.com/conversation.xsd'
    env:role='&SOAPENV;/role/next'>
  <c:msgID>
    uuid:0123456789-0123456789-0123456789
  </c:msgID>
  </c:converse>
</env:Header>
```

The intended role for processing a header block can be identified via the role attribute; by
default, it means that the ultimateReceiver should process the given block. A header block that
is marked with a true value for the mustUnderstand attribute means that any node processing
the message must process the given block or throw a fault. The header may incorporate
information for transaction management and authentication, without an *a priori* agreement
between the interacting parties.

2.2.3 Faults

Exceptions are inevitable in computations in open environments. To accommodate exceptions, SOAP supports an element known as fault. A SOAP body may include up to one fault
element. The fault element would have the subelements faultcode, faultstring, faultactor, and
detail. Of these, faultcode is most precisely defined in the standard. The following are the
legal values of fault code:

- *Client*. The fault claims that the client formulated its request incorrectly. The server
  will not be able to entertain this request again and so the client should not repeat it.

- *Server*. The fault indicates that the server encountered internal problems. It may be
  able to entertain the same request later, and the client is free to retry.

- *VersionMismatch*. The fault asserts a mismatch between the server and the request.
  The request should use the soapenv namespace (as explained in Appendix C).

- *MustUnderstand*. The fault claims that a SOAP role failed to process a header that was
  marked mustUnderstand for that role.

2.2.4   Message Exchange

SOAP provides a means for communicating information (specified as XML) from a sender to a receiver. Just a single transfer of information would rarely be adequate. Therefore, richer message exchange patterns are essential. One simple and well-known pattern is the remote procedure call (RPC). A SOAP RPC is based on the following information:

- the address of the target SOAP node, which will be ultimateReceiver;

- the name of the method to be invoked;

- the arguments and return value, if any;

- additional header blocks.

One way to achieve the effect of richer message exchanges in SOAP is to use a header block to carry some sort of a conversation identifier possibly along with a message identifier. These would give the application sufficient information to correlate different messages, i.e., to recognize them as part of the same conversation. Further, the message identifiers can be used to encode other information such as a replyTo attribute.

SOAP intermediaries can be of two kinds. A *forwarding intermediary* can modify the header blocks of a SOAP message and decide where to forward it. The modifications and forwarding decision are made based on the contents of the given message as well as potentially the ongoing message exchange pattern. An *active intermediary* that can process an incoming SOAP message could act in a manner that need not be specified in the message. Examples are policy actions such as encryption or adding a new header with a timestamp or creating an audit trail. Thus SOAP intermediaries can be used to implement process flows.

Although the above approach seems quite powerful, it throws the burden of programming the right decisions upon the application designers. Consequently, this approach is fraught with risk from the standpoint of productivity and correctness. However, if tools were available to generate the correct intermediaries, then the ability to retrofit or modify a process would enable great flexibility.

2.2.5   Limitations

Like HTTP, SOAP is a character-based, rather than a binary protocol, making it easier to secure, i.e., encrypt and decrypt. Programmers can easily examine and comprehend the contents of SOAP messages and tools are easier to build. SOAP is also popular because, as a consequence of its riding on established protocols such as HTTP, it readily works through firewalls, and thus is able to form the basis for e-commerce over the Web. However, SOAP is inefficient for many applications, because data are transmitted in character, not binary form. Moreover, SOAP headers are large and in some cases the header size overshadows the payload size.

Since the original work on SOAP predates the XML Schema standard, SOAP has ended up with two syntaxes for representing the data. One is the so-called *Section 5 Encoding*, described in the eponymous section of the SOAP 1.1 specification; the other is XML Schema. The latter is the preferred approach.

In conceptual terms, SOAP is a stateless protocol. Although you can, of course, add and interpret conversation identifiers to lend some statefulness to the interaction, each SOAP message is unrelated to any other message. Hence, SOAP does not describe bidirectional or multiparty interactions. One can use conversation identifiers at the application level to build a conversation with an appropriate message pattern. However, this is not supported by the protocol itself. Thus, SOAP implementations would provide no support for any such enhanced message patterns and any standardization would have to be through a separate process. Conversations relate to business protocols and are an important theme, however, that are revisited in Chapter 13 and Chapter 18.

The SOAP specification is continuing to be revised. It does not yet describe bidirectional or multiparty communication, which would be useful for composing Web services from multiple providers. Also, there is no way to transfer transaction semantics across a SOAP call. At the present time, there is no standardized way to pass security credentials, although this problem should be solved soon due to the work on security standards that is currently underway. SOAP is effective for simple interoperability between single clients and servers, but for more complex interoperability among heterogeneous systems a message-queuing component should be used by each participant to provide transaction and security support.

Exercises 2.11 and 2.12 ask you to propose enhancements to SOAP to address some of the main shortcomings of its current incarnation.

2.3 WSDL

The architectural model for Web services presupposes that services can be found and used. This in turn presupposes accurate descriptions of services. The Web Services Description Language (WSDL) is an XML language for describing a programmatic interface to a Web service [Christensen et al., 2001]. The description includes definitions of data types, input and output message formats, the operations provided by the service (such as *GetTemp*), network addresses, and protocol bindings. WSDL can best be understood in terms of an example, as described via the code shown in Listing 2.4.

Listing 2.4: A WSDL example

```
<?xml version="1.0"?>
<!-- the root element, wsdl:definitions, defines a set of -->
<!-- related services -->
<wsdl:definitions name="Temperature"
  targetNamespace="http://www.socweather.com/schema"
  xmlns:ts="http://www.socweather.com/TempSvc.wsdl"
  xmlns:tsxsd="http://schemas.socweather.com/TempSvc.xsd"
  xmlns:soap="http://schemas.xmlsoap.org/wsdl/soap/"
  xmlns:wsdl="http://schemas.xmlsoap.org/wsdl/">
```

```
<!— wsdl:types encapsulates schema definitions of —>
<!— communication types; here using xsd —>
 <wsdl:types>
<!— all type declarations are expressed in xsd —>
  <xsd:schema
   targetNamespace="http://namespaces.socweather.com"
   xmlns:xsd="http://www.w3.org/1999/XMLSchema">

<!— xsd def: GetTemp [City string, When string] —>
   <xsd:element name="GetTemp">
    <xsd:complexType>
     <xsd:sequence>
      <xsd:element name="City" type="string"/>
      <xsd:element name="When" type="string"/>
     </xsd:sequence>
    </xsd:complexType>
   </xsd:element>

<!— xsd def: GetTempResponse [DegreesCelsius integer] —>
   <xsd:element name="GetTempResponse">
    <!— XML Schema entry as above —>
   </xsd:element>

<!— xsd def: GetTempFault [errorMessage string] —>
   <xsd:element name="GetTempFault">
    <!— XML Schema entry as above —>
   </xsd:element>
  </xsd:schema>
 </wsdl:types>

<!— wsdl:message elements describe potential transactions —>
<!— Most messages, as here, have only one part. Multiple —>
<!— parts provide a way to aggregate complex messages —>

<!— request GetTempRequest is of type GetTemp —>
 <wsdl:message name="GetTempRequest">
  <wsdl:part name="body" element="tsxsd:GetTemp"/>
 </wsdl:message>

<!— response GetTempResponse is of type GetTempResponse —>
 <wsdl:message name="GetTempResponse">
   <wsdl:part name="body" element="tsxsd:GetTempResponse"/>
 </wsdl:message>

<!— wsdl:portType describes messages in an operation —>
```

- *Request-response.* Receive a request and emit a correlated response.

- *Solicit-response.* Emit a request and receive a correlated response.

The paired operation types could be based on the unidirectional types, but they are kept because they identify important design patterns. For example, being the server in RPC corresponds to request–response and being the client in RPC corresponds to solicit–response. These operation types, thus, anticipate SOAP's message exchange patterns. Of the above types, one-way and request–response are the only ones that are commonly employed. These are readily supported by HTTP and by common object-oriented programming approaches.

WSDL 2.0 offers a richer set of primitives than the above. These primitives include receiving or sending multiple responses to a single query. The details are not of great significance to the overall goals of this chapter. Richer message patterns are discussed in the context of conversation modeling later.

2.3.3   Creating WSDL Models

It helps to separate out a WSDL specification into two main components: the interface and the implementation. Splitting the WSDL specification in this manner improves modularity and separates the service interface, which is reusable, and may have multiple implementations.

The WSDL interface is the more abstract component. It describes a service by fleshing out the definition element in terms of the types, import, message, portType, and binding subelements. An interface may import other interfaces.

The WSDL service implementation considers the specifics of binding a service. Its definition element must include an import element to import at least one WSDL interface document and includes a service element, which includes port elements. The import element specifies an identifier for the namespace being imported as well as its location.

2.4   Directory Services

The purpose of a directory service is for components and participants to be able to locate each other, where the components and participants might be applications, agents, Web service providers, Web service requesters, people, objects, and procedures. Directories collect and organize location and description information and make it available to any clients that might need it. Directories also function as the primary supporting mechanism for dynamism, as defined in Section 1.3, because they are the repository for information about changes that have occurred to their entries.

There are two general types of directories, determined by how entries are located in the directory: (1) name servers or "white pages," where entries are found by their name, and (2) "yellow pages," where entries are found by their characteristics and capabilities.

The implementation of a basic directory is a simple database-like mechanism that allows participants to insert descriptions of the services they offer and query for services offered by other participants. A more advanced directory might be more active than others, in that it might provide not only a search service, but also a brokering or facilitating service. For example, a participant might request a brokerage service to recruit one or more agents who can answer a query. The brokerage service would use knowledge about the requirements and capabilities of registered service providers to determine the appropriate providers to which a query could be forwarded. It would then send the query to those providers, relay their answers back to the original requester, and learn about the properties of the responses it passes on (e.g., the brokerage service might determine that advertised results from provider X are incomplete, and so seek out a substitute for provider X).

Two major standards for directories are emerging: ebXML registries and UDDI registries. Unfortunately, neither supports semantic descriptions, and thus neither supports semantic searching on functionality. Searches, as a result, can only be based on keywords, such as a service's name, provider, location, or business category. ebXML registries have an advantage over UDDI registries in that they allow SQL-based queries on keywords. As described in the next section, UDDI provides white-pages, yellow-pages, and green-pages services.

2.5 UDDI

The Universal Description, Discovery, and Integration (UDDI) specification [UDDI, 2000] describes a mechanism for registering and locating Web services. It defines an online registry where organizations, i.e., service providers, can describe their organization and register their Web services. The registry can then be used by service requesters and users to locate the services they need. For our purposes, UDDI makes it possible for providers to relate their services to each other and for a requester to discover services, a prerequisite for composing them.

2.5.1 Conceptual Model

UDDI *white pages* consist of the following information fields:

- Business name.

- Text description: a list of multilanguage text strings.

- Contact information: names, phone numbers, fax numbers, and Web sites.

- Identifiers that a business may be known by, such as D-U-N-S (also known as the "DUNS number") and Thomas Register.

The *yellow pages* consist of business categories organized as the following three major taxonomies:

- Industry: North American Industry Classification System (NAICS), a six-digit code maintained by the US Government for classifying companies.

- Products and services: Ecma International (for classifying information and communication technology systems) and United Nations Standard Products and Services Code (UNSPSC).

- Geographical location: ISO 3166 for country and region codes.

The yellow pages are implemented as name–value pairs to allow any valid taxonomy identifier to be attached to the white page for a business. Searches of a yellow pages can be performed to locate businesses that service a particular industry or product category, or are located in a particular geographic region.

The *green pages* consist of the information businesses use to describe how other businesses can conduct electronic commerce with them. Green-page information is a nested model comprising business processes, service descriptions, and binding information. The information is neutral as to language, platform, and implementation. The services can also be categorized.

UDDI is itself a Web service that is based on XML and SOAP. For example, a business registration is an XML document. A client uses a set of predefined SOAP interfaces to search the registry for a desired Web service. Providers use SOAP interfaces to register two types of information: (1) *technical models* (tModel), which are abstract service protocols that describe an individual Web service's behavior, and (2) *business entities* (businessEntity), which describe a service implementation and provide descriptions of the specifications of multiple tModels. Note that each distinct specification, transport, protocol, or namespace is represented by a tModel. However, a UDDI registry does not actually store the specification and such details. A UDDI tModel simply contains the addresses (URLs) where those technical documents can be found, metadata about the documents, and a key that identifies that tModel.

Figure 2.3 shows the yellow, white, and green pages for a business. A businessEntity is the top-level structure for all of the information related to a business. This is shown more formally in Figure 2.4. The core components of a UDDI businessEntity and the relationships among them are shown in Figure 2.5.

For our purposes, we will be interested mostly in registering Web services, so we will want to map WSDL descriptions of Web services to UDDI service descriptions. Figure 2.6 shows the correspondence between the fields of a WSDL description and the fields of a UDDI businessService.

2.5.2   UDDI APIs

UDDI specifies two APIs for programmatic access to a UDDI registry: the *Inquiry API* for retrieving information from a registry and the *Publish API* for storing information there. The Publish API requires authenticated access—which is particular to a registry and not specified by UDDI—but the Inquiry API does not. The APIs currently support 28 SOAP messages, the most important of which are the following:

- Inquiry API

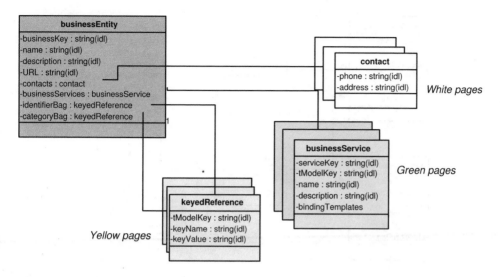

Figure 2.3: The yellow, green, and white pages representing a business entity in a UDDI registry

- Finding a business or its service and their characteristics
    * find_business, returns ⟨businessList⟩
    * find_service, returns ⟨serviceList⟩
    * find_binding, returns ⟨bindingDetail⟩
    * find_tModel, returns ⟨tModelList⟩
- Retrieving the details needed to interact with a business
    * get_businessDetail, returns ⟨businessDetail⟩
    * get_serviceDetail, returns ⟨serviceDetail⟩
    * get_bindingDetail, returns ⟨bindingDetail⟩
    * get_tModelDetail, returns ⟨tModelDetail⟩
- Publishing API
    - Saving information about a business or its services
        * save_business, returns ⟨businessDetail⟩
        * save_service, returns ⟨serviceDetail⟩
        * save_binding, returns ⟨bindingDetail⟩
        * save_tModel, returns ⟨tModelDetail⟩
    - Delete things
        * delete_business, returns ⟨dispositionReport⟩

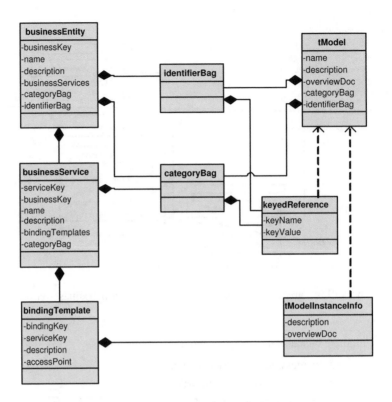

Figure 2.4: The UML information model for a business entity in a UDDI registry

- * delete_service, returns ⟨dispositionReport⟩
- * delete_binding, returns ⟨dispositionReport⟩
- * delete_tModel, returns ⟨dispositionReport⟩
- Security
  - * get_authToken, returns ⟨authToken⟩
  - * discard_authToken, returns ⟨dispositionReport⟩

2.5.2.1 Registering and Publishing a Service

Now that we understand the basic components of a UDDI entry, let's look at an example registration from WeatherService, Inc. and how its service for reporting current temperatures might be discovered and then used by a client. WeatherService, Inc. would first exchange two SOAP messages with a UDDI registry (possibly the registry maintained by IBM at https://uddi.ibm.com/ubr). The first SOAP message would invoke the operation get_authToken to establish authentication. The second, shown in Listing 2.5, would register WeatherService, Inc. as a business entity.

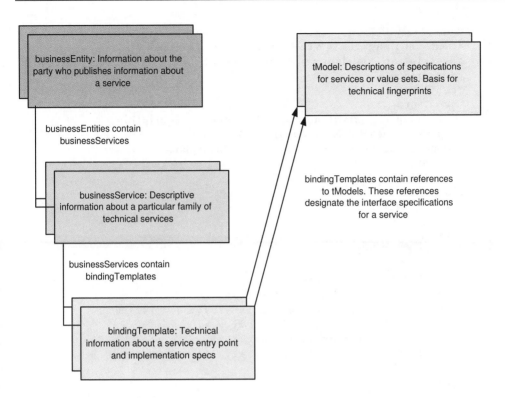

Figure 2.5: The core data structures and the relationships among them for a UDDI business entity

Listing 2.5: SOAP body of an example UDDI registration of a business entity

```
POST / HTTP/1.1
Host: www.socweather.com
Content-Type: text/xml; charset="utf-8"
Content-Length: nnnn
SOAPAction: ""

<?xml version="1.0" encoding="UTF-8" ?>
<env:Envelope xmlns:env="http://schemas.xmlsoap.org/soap/envelope/">
<env:Body>
<save_business xmlns="urn:uddi-org:api_v3">
 <businessDetail truncated="false">
  <businessEntity businessKey="...K1...">
   <discoveryURLs>
    <discoveryURL useType="homepage">
      http://www.socweather.com/WeatherService.html
    </discoveryURL>
```

```
  </discoveryURLs>
  <name xml:lang="en">WeatherService Inc.</name>
  <description xml:lang="en">Provider of temperature services
  </description>
  <contacts>
   <contact>
    <description xml:lang="en">President</description>
    <personName>Hot N. Cold</personName>
    <phone useType="Voice">803-555-1234</phone>
   </contact>
  </contacts>
  <identifierBag>
   <keyedReference
    tModelKey="uddi:uddi.org:ubr:identifier:dnb.com:d-u-n-s"
    keyName="DUNS:_WS_Inc." keyValue="12-123-1234"/>
  </identifierBag>
  <categoryBag>
   <!-- NAICS Classification -->
   <keyedReference tModelKey="uuid:K6"
    keyName="Meterological_services" keyValue="541990"/>
   <!-- ISO 3166 Geographic Taxonomy -->
   <keyedReference tModelKey="uuid:K7"
    keyName="North_Carolina,_USA" keyValue="US-NC"/>
  </categoryBag>
  <businessServices>
   <businessService serviceKey="...K2..." businessKey="...K1...">
    <name xml:lang="en">Temperature Service</name>
    <description xml:lang="en">
     Given a time and city, it returns a temperature
    </description>
    <bindingTemplates>
     <bindingTemplate bindingKey="...K3..." serviceKey="...K2...">
      <description xml:lang="en">
       This service uses a SOAP/RPC encoded endpoint
      </description>
      <accessPoint URLType="http">
       http://www.socweather.com/TempSvc
      </accessPoint>
      <tModelInstanceDetails>
       <tModelInstanceInfo tModelKey="uuid:...K4..."/>
      </tModelInstanceDetails>
     </bindingTemplate>
    </bindingTemplates>
   </businessService>
  </businessServices>
 </businessEntity>
```

```
</businessDetail>
<tModelDetail truncated="false">
 <tModel tModelKey="uuid:...K4...">
  <name>TempSvc Specification</name>
  <description xml:lang="en">tModel for service interface definition
  </description>
  <overviewDoc>
   <overviewURL>http://www.socweather.com/TempSvc.wsdl
   </overviewURL>
  </overviewDoc>
  <categoryBag>
   <keyedReference tModelKey="uuid:...K6...."
    keyName="uddi-org:types" keyValue="wsdlSpec"/>
  </categoryBag>
 </tModel>
</tModelDetail>
</save_business>
</env:Body>
</env:Envelope>
```

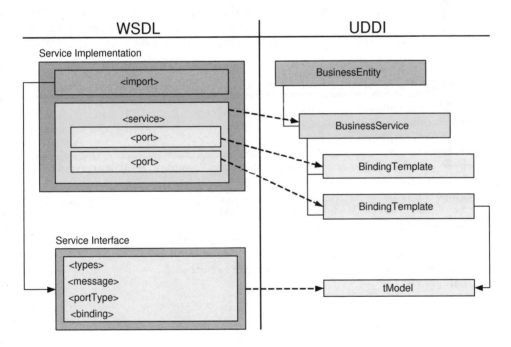

Figure 2.6: Correspondences between a WSDL document and a UDDI registration document

Let's consider some of the fields in this registration for WeatherService, Inc. First, we are sending to the registry the command save_business, whose attribute specifies that we are using version 3 of UDDI. A save_business message contains a businessDetail and any number of tModels. The businessDetail states where we can find the homepage for WeatherService, Inc., how we could call its president, what its D-U-N-S number is, and how WeatherService, Inc. is classified according to NAICS and ISO 3166 (that is, what sort of business it is and where it is located).

Continuing with our description of the registration, only one of the services provided by WeatherService, Inc. is listed: *TemperatureService*. Its details are given in a bindingTemplate that has two essential parts: (1) the precise URL where the service can be accessed and (2) the tModel that provides the access information. The tModel in this case has an identification key pointing to the tModel that is given as part of the registration, but the tModel might have been provided in a separate message to the registry or might even be part of another business's registration. tModels in this regard are like global user-definable and reusable data types.

The tModel in our registration message specifies that our *TempSvc* is defined in WSDL and provides a pointer to the appropriate WSDL document.

Next, WeatherService, Inc. sends three SOAP messages to the UDDI registry to register the following three tModels describing the behavior of its temperature services. The behavior is described in terms of the portType for the service and its protocol bindings. The overview-Doc element in the first two messages (Listings 2.6 and 2.7) contains an overviewURL element, which contains the URI for the WSDL interface of the service being published. These listings describe the registration of tModels that can be used as part of other services, not just the ones for WeatherService, Inc. Note again that the registry does not store the WSDL document, just a pointer to it.

Listing 2.6: The first of three tModels for WeatherService, Inc. specifying its port information

```
<save_tModel xmlns="urn:uddi-org:api_v3">
 <tModel tModelKey="uuid:...KA..." >
  <name>GetTempPortType</name>
  <overviewDoc>
   <overviewURL>http://www.socweather.com/TempSvc.wsdl</overviewURL>
  </overviewDoc>
  <categoryBag>
   <keyedReference tModelKey="uuid:...KB..."
    keyName="portType_namespace"
    keyValue="http://www.socweather.com/TempSvc"/>
   <keyedReference tModelKey="uuid:...KC..."
    keyName="WSDL_type" keyValue="portType"/>
  </categoryBag>
 </tModel>
</save_tModel>
```

Listing 2.7: The second of three tModels for WeatherService, Inc. specifying its protocol bindings

```
<save_tModel xmlns="urn:uddi-org:api_v3">
 <tModel tModelKey="uuid:...KD...">
  <name>TempSvcSoapBinding</name>
  <overviewDoc>
   <overviewURL>http://www.socweather.com/TempSvc.wsdl</overviewURL>
  </overviewDoc>
  <categoryBag>
   <keyedReference tModelKey="uuid:...KE..."
    keyName="binding_namespace"
    keyValue="http://www.socweather.com/TempSvc"/>
   <keyedReference tModelKey="uuid:...KF..."
    keyName="WSDL_type" keyValue="binding"/>
   <keyedReference tModelKey="uuid:...KG..."
    keyName="portType_reference" keyValue="uuid:...KH..."/>
   <keyedReference tModelKey="uuid:...KI..."
    keyName="SOAP_protocol" keyValue="uuid:...KJ..."/>
   <keyedReference tModelKey="uuid:...KK..."
    keyName="HTTP_transport" keyValue="uuid:...KL..."/>
   <keyedReference tModelKey="uuid:...KM..."
    keyName="uddi-org:types" keyValue="wsdlSpec"/>
  </categoryBag>
 </tModel>
</save_tModel>
```

Listing 2.8 is an additional description of WeatherService, Inc.'s businessService, containing further details about the port for the Temperature Service.

Listing 2.8: The third of three tModels for WeatherService, Inc. updating the service it provides

```
<save_service xmlns="urn:uddi-org:api_v3">
 <businessService serviceKey="...K2..." businessKey="...K1...">
  <name>Temperature Service</name>
  <bindingTemplates>
   <bindingTemplate bindingKey="...KP..." serviceKey="...KN...">
    <accessPoint URLType="http"> http://www.socweather.com/TempSvc
    </accessPoint>
    <tModelInstanceDetails>
     <tModelInstanceInfo tModelKey="uuid:...KQ...">
      <description xml:lang="en">The wsdl:binding the wsdl:port
         implements; instanceParms specifies the port local name.
      </description>
      <instanceDetails>
       <instanceParms>GetTempPort</instanceParms>
      </instanceDetails>
     </tModelInstanceInfo>
     <tModelInstanceInfo tModelKey="uuid:...KR...">
```

2.6   Notes

The subject of this book is the development of open information systems. It therefore deals with abstractions for applying services. Consequently, our emphasis is on studying the associated standards from the perspective of how they can be used, rather than how they can be implemented over the infrastructure. This is, in particular, the case with SOAP, whose implementation involves considerations of encoding and of the functioning of the underlying protocols. SOAP supports MIME attachments, which enables SOAP to be used for exchanging binary data of arbitrary form. Similarly, implementing UDDI involves considerations of databases and directory services, which are outside the scope of this book.

ISO 11179 is a standard for registering data elements.

2.7   Exercises

2.1. Providing the temperature using the World Wide Web is a service. What is the essential difference between a Web site such as www.weather.com where one can type in a zip code and find out the temperature at that location, and a GetTemperature Web service such as discussed in this chapter?

2.2. The "Object" in SOAP refers to which one of the following?

- Nothing
- The communication object
- The objects instantiated by both ends of the conversation
- The object to be accessed at the server, for which the client receives a reference
- The delivery object

2.3. Which one of the following would be the value of the actor attribute in a SOAP message?

- A role
- An agent name
- A URI to a WSDL file
- The URI of an agent's SOAP binding
- An unconstrained xsd:string

2.4. SOAP handles exceptions via which one of the following?

- The fault element
- The exception element
- The throws element

- Nothing, because SOAP does not handle exceptions

- A "500" return code

2.5. When a node receives a SOAP message, which of the following should it do first?

- Process all header blocks that are targeted to the node.

- Process the body.

- Process the body, but only if the node is the final recipient.

- Create an instance of the SOAP object.

- Find and obey the *mustUnderstand* attributes.

2.6. True or False? A node must understand only those blocks in a SOAP message that have their actor set to a role the node can play.

2.7. The following SOAP envelope:

```
<!ENTITY SOAPENC
         "http://www.w3.org/2001/06/soap-encoding">
<s:Envelope xmlns:s="...">
 <s:Body>
  <order xmlns="..."
    s:encodingStyle=&SOAPENC;>
   <partName xsi:type="string">valve</partName>
   <quantity xsi:type="string">12</quantity>
  </order>
 </s:Body>
</s:Envelope>
```

corresponds to which Java method signature(s)?

- String order(String partName, String quantity)

- void order(String quantity, String partName)

- void order(String quantity, Integer partName)

- void partName(String s); void quantity(Integer n)

- None of the above; in this case, specify whatever parts you can of a correct signature.

2.8. Develop a WSDL document describing a stock quote service. Define the following message types: loginRequest, logReply, stockQuoteRequest, stockQuoteResponse, and logoutRequest, and the following operations: QuoteToUser, LogIn, ProvideQuote, LogOut, and QueryNYSE.

2.9. Consider two main kinds of message exchange patterns in SOAP (and operation types in WSDL): (1) request-response and (2) one-way. How would you implement one-way messages over HTTP, which is a request-response protocol? How would you implement request-response messages over SMTP, which is a one-way protocol (do not assume any special receipt notification functionality, which some mailers support)?

2.10. Extend your solution to Exercise 2.9 to accommodate the other two of the WSDL operation types: (1) solicit-response and (2) notification. How would you implement these over HTTP? How would you implement these over SMTP?

2.11. How would you enhance SOAP so that its payload could be compressed? The idea is not simply to propose a new syntax but to show how the processing would be affected. A design requirement is to work over the existing infrastructure without any changes. For example, you cannot reasonably assume that you will be able to design and launch successfully a new version of HTTP, which would make your task much easier. Ensure the feasibility of the processing required in your proposed approach.

2.12. How would you enhance SOAP to accommodate some elements of security? Specifically, how would you accommodate authenticating the sender of a SOAP request from the perspective of a recipient and authenticating the responder from the perspective of the requester. Consider the use of simple credentials such as digital certificates or a login ID and password.

2.13. In a WSDL file, which one of the following most closely corresponds to a method or function name in an imperative programming language?

- operation
- portType
- message
- service
- type

Map the others to their closest analogs in such programming languages.

2.14. A WSDL message consists of parts that:

- each have a type from some type system;
- have a sender and receiver;
- are free-form;
- can themselves be messages;
- must be declared in the SOAP header.

2.15. Which one of the following is a WSDL transmission primitive that cannot be supported by an endpoint?

- multicast
- one-way
- request–response
- solicit–response
- notification

2.16. Which one of the following lists the top-level elements of a WSDL document?

- types, message, portType, binding, service
- portType, binding, service
- types, message, operator, portType, binding, service
- binding, service
- header, body

2.17. Produce the WSDL description for a phone book service that supports the operations *GetPhoneNumber* and *SetPhoneNumber*. The operation *GetPhoneNumber* accepts a name parameter of type String and returns a phone number of type String, whereas *SetPhoneNumber* accepts a name parameter of type String and a phone number parameter of type String and returns nothing.

2.18. The UDDI protocol is used for which one of the following?

- Finding SOAP services that implement a given interface.
- Describing the interface of a SOAP service.
- Communicating between SOAP and .NET.
- Describing the communication protocol SOAP service.
- Describing how a SOAP service is deployed on a Web server.

2.19. A UDDI registry holds descriptions of (choose one):

- the business, service, and bindings;
- the UDDI clients;
- the interfaces implemented by the registered Web services;
- the encoding mechanism;
- the users' preferences.

2.20. A UDDI tModel is:

- A way of describing the various business, service, and template structures stored within the UDDI registry.

- A technical description of the Web services represented by the business service structure.
- The English description of the business.
- A request-response message pattern definition.
- A transitional model.

2.21. A set of related UDDI registries are deployed using:

- a federated model;
- a hierarchical model;
- a master-slave model;
- an n-tier model;
- a centralized model.

2.22. A UDDI registry can be accessed using which SOAP interface(s)?

- InquireSOAP and PublishSOAP.
- UDDISOAPInterface.
- SOAP with HTTP.
- Publish and Query.
- uddiSOAP.

2.23. The first step in translating a WSDL file to be used with UDDI is to:

- split it into an interface file and an implementation description file.
- register the WSDL file as a UDDI tModel.
- change all the types from XML Schema to UDDI Schema.
- send it to a UDDI registry and gather the error messages.
- rewrite it using the UDDI Schema.

2.24. Imagine that the Web services described in Exercise 2.17 are offered by NSCU Phone, Inc. Using the UDDI Publish API, write the XML files that would be sent to a registry to register NSCU Phone, Inc. and its services.

2.25. Using the UDDI files you developed as part of Exercise 2.24, register your business at the UDDI test registry maintained by IBM at https://uddi.ibm.com/testregistry/registry.html. To do this, first obtain a username and password for the test registry. Second, download and install a tool such as UDDI4J available at http://uddi4j.org/. Third, if you choose the UDDI4J tool, recompile it with an appropriate SOAP transport implementation, such as Apache Axis or SOAP 2.2, which can be found at xml.apache.org, or HP SOAP, which can be found at http://hp.com/go/webservices. Fourth, create and run your Java application.

2.26. Imagine you are in charge of a UDDI registry, named MyUDDI.com and located at http://www.MyUDDI.com. When someone, for example, sends a find_tModel in a SOAP message, your registry responds with the appropriate tModel in a SOAP response. Your registry is providing a service, which can be described by a WSDL document. Write the WSDL description of the find_tModel service for your registry.

2.27. Web services, as currently implemented by using WSDL and SOAP, work well when a requestor wants a single instance of a service that can be had via one interaction. That is, there is a single request and a single response. Often, however, the interaction is more complicated, as when a buyer (requestor) is purchasing a service from a seller (provider). In this case, the buyer will ask the seller for a price quote. After receiving a quote, the buyer will issue a purchase order. The seller will acknowledge the purchase order. The buyer can then access the service from the seller. The interactions form an extended conversation. How can the UDDI, WSDL, SOAP, and XML-based Web service infrastructure be used or changed or adapted to support interactions of this sort?

cation by selecting links (state transitions), resulting in the next page (representing the next state of the application) being transferred to the user's browser and rendered appropriately.

REST is an architectural style for designing Web services, not a standard. It attempts to capture the characteristics that have made the Web successful and are guiding its evolution. While not a standard, REST does use the standards for HTTP, URI, resource representations such as XML, HTML, GIF, and JPEG, and MIME types such as text/xml, text/html, image/gif, and image/jpeg.

Familiar Web services, such as those for ordering books or for searching catalogs, are typically REST-based, even if they were not explicitly constructed with REST in mind. However, let's use REST to construct an example Web service.

3.2   A RESTful Example

The NorthSouth Carolina University has deployed a Web service to enable its students to:

- get a list of courses;

- get detailed information about a particular course;

- register for a course.

By making the following URL available, this Web service enables a client application to get the course list:

```
http://www.nscu.edu/courses
```

Note that *how* the Web service generates the course list is not apparent to a client. All the client knows is that if it submits the above URL, then a document containing the list of courses is returned. NorthSouth Carolina University is thus free to modify the underlying implementation of this resource (provided of course that it preserves their meaning and behavior) without affecting the functioning of its clients, which is an extremely convenient characteristic to have.

Here is the document that the client application receives, assuming that the application can handle XML:

```xml
<?xml version="1.0"?>
<!ENTITY NSCU 'http://www.nscu.edu'>
<p:Courses xmlns:p="&NSCU;"
           xmlns:xlink="http://www.w3.org/1999/xlink">
   <Course id="CS101" xlink:href="&NSCU;/courses/CS1"/>
   <Course id="CS102" xlink:href="&NSCU;/courses/CS2"/>
   <Course id="CS201" xlink:href="&NSCU;/courses/DataStruc"/>
   <Course id="CS202" xlink:href="&NSCU;/courses/ProgLang"/>
</p:Courses>
```

Note that the document incorporates links that can be used by the client to obtain detailed information about each course. This is a key feature of REST. The client transfers from one state to the next by examining and choosing from among the alternative URIs in the response document. For example, the client could get detailed information about the ProgLang course by issuing the request

```
http://www.nscu.edu/courses/ProgLang
```

This results in the following document being sent to the client:

```
<?xml version="1.0"?>
<!ENTITY NSCU 'http://www.nscu.edu'>
<p:Course xmlns:p="&NSCU;"
          xmlns:xlink="http://www.w3.org/1999/xlink">
 <Course-Num>CS202</Course-Num>
 <Name>Programming Languages</Name>
 <Requirement>Required for CS majors</Requirement>
 <Syllabus xlink:href="&NSCU;/courses/ProgLang/syllabus"/>
 <CreditHours type="semester">3</CreditHours>
 <Prerequisite xlink:href="&NSCU;/courses/DataStruc"/>
</p:Course>
```

Observe how this document is linked to still more documents—the syllabus for this course may be found by traversing the hyperlink. Each response document allows the client to get more detailed or related information.

The Web service makes available a URL for course registration. The client creates a registration instance document, Reg.xml, that conforms to the registration schema that NorthSouth Carolina University has provided in a WSDL document and published in a UDDI registry. The client submits Reg.xml as the payload of an HTTP POST.

The registration service responds to the HTTP POST with a URL to the submitted Reg.xml document. The client can retrieve this document any time thereafter to update or edit it. Reg.xml has become an item of information that is shared between the client and the server. By giving Reg.xml a URI, the server has, in essence, exposed it as a Web service (although not necessarily a standard Web service in the sense of Chapter 2 or BP 1.0). A resource is a conceptual entity, which is given a representation, i.e., a concrete manifestation, in REST. REST does not generally place constraints on resources, e.g., the following URL

```
http://www.nscu.edu/courses/ProgLang
```

is a logical one, not a physical one. Thus there does not need to be, for example, a static HTML page for this course. In fact, if there were a thousand courses, then a thousand static HTML pages would not be a very good design. As a better design, NorthSouth Carolina University could implement the service that returns detailed data about a particular course by (as an example approach) carrying out the following steps:

- employing an application server that parses the string after the host name and invokes an appropriate servlet based on one or more tokens;

- having the servlet parse the argument string;

- using the course number to query a course database;

- formulating the database tuples as an XML document;

- returning the XML document as the payload of the HTTP response.

As a matter of style URLs should not reveal the implementation technique used. Servers need to be free to change their implementation without affecting clients. URLs that informally refer to the implementation would either restrict the server or be misleading.

To summarize, here are the main characteristics of REST:

Client-Server. By assuming client-server interactions, REST separates interface concerns from data-storage concerns, enabling them to evolve independently.

Statelessness. Each request from a client to a server must contain all the information necessary to understand the request, and cannot take advantage of any stored context on the server.

Caching. To improve network efficiency, responses can be labeled as cacheable, enabling a client to store and reuse a cacheable response rather than requesting it again later.

Uniform interface. All resources are accessed via a uniform interface based on the following four constraints:

1. Identification of resources through URIs, where a resource corresponds to the semantics of what the author intends to identify, rather than the value corresponding to those semantics at the time the reference is created.

2. Manipulation of resources through their representations, where the representations of the resources are interconnected via URLs, thereby enabling a client to progress from one state to another.

3. Self-descriptive messages, which include their own metadata, but maintain the uniformity of the interface by limiting the scope to one of an evolving set of standard data types selected dynamically.

4. Hypermedia as the engine for the application state.

Layered components. Intermediaries, such as proxy servers, cache servers, and gateways, can be inserted between clients and resources to support additional properties such as performance and security.

Code-on-demand. Optionally, clients can be extended dynamically by downloading and executing code in the form of applets or scripts.

The following are the basic principles of REST Web service design:

1. The key to creating services in a REST network (i.e., the Web) is to identify all of the conceptual entities that you wish to expose. Above are some examples of resources: a course list, detailed information about a course, and a registration document.

2. Create a URL for each resource. The resources should correspond to nouns, not verbs. For example, do not use this:

> http://www.nscu.edu/courses/getCourse?id=CS101

Note the verb, getCourse, which indicates a particular process for the implementation. Instead, use a noun:

> http://www.nscu.edu/courses/CS1

3. Categorize your resources according to whether clients can just receive a representation of the resource, or can modify (add to) the resource. For the former, make those resources accessible using an HTTP GET. For the latter, make those resources accessible using HTTP POST, HTTP PUT, or HTTP DELETE.

4. Make all resources that are accessible via HTTP GET free of side effects, so that invoking the resource does not modify it.

5. Include hyperlinks in your resource representations that enable clients to obtain more detailed or related information. Do not try to put all information in a single response document.

6. Specify the format of response data using a schema.

7. Describe how your services are to be invoked using either a WSDL document or, simply, an HTML document.

REST is a set of architectural constraints that attempts to minimize latency and network communication, while at the same time maximizing the independence and scalability of component implementations. This is achieved by placing constraints on connector semantics, instead of on component semantics, which has been the focus of other architectural styles. REST enables the caching and reuse of interactions, dynamic substitutability of components, and processing of actions by intermediaries, thereby supporting a world-wide hypermedia system.

REST is the architectural style of the Web, and describes what makes the Web work well. Adhering to REST principles will make your services work well in the context of the Web.

3.3 SOAP and REST

SOAP has received some criticism because of its apparent violation of some of the principles that REST espouses. In particular, SOAP 1.1 required an HTTP POST binding and thus hid the identity of the Web resource being accessed within the body of the message. Recall from the above that REST requires Web resources to be identified in a manner that clearly separates the identification information from any data and control information. REST advocates the use of HTTP GET for accessing resources. In other words, it advocates that methods that satisfy the following two properties be specified as HTTP GETs, where the URI completely specifies the target object to be retrieved:

Safety. This holds for methods that are free of side effects on the given object. Query methods would be the canonical safe methods.

Idempotency. Loosely following the database transaction terminology, idempotent methods are those whose repeated occurrences have no additional side-effects beyond the first occurrence. In the database sense, an idempotent method would be resilient to restarts (or, in the Web context) reloads. That is, if the method were not completed when it had to be aborted and restarted, then assuming it eventually completed successfully, its ultimate effect would be the same as if it had completed successfully on the first attempt.

The association of GET with safe and idempotent methods is recorded in the HTTP protocol specification. Notice, however, that the protocol can offer no means of ensuring compliance with these guidelines. However, the specification does claim that when a GET is executed, it can be presumed that the user acting through the *user agent* (browser) is interested in retrieving information but not in changing it.

In acknowledgment of the above objections to SOAP 1.1, SOAP 1.2 supports a HTTP binding, which uses HTTP GET and places the resource URI in the HTTP header—the same as any other HTTP GET. Listing 3.1 illustrates this usage.

Listing 3.1: RESTful SOAP: Example of HTTP GET binding

```
GET /www.socweather.com/temp?city=Honolulu&when=now HTTP/1.1
Host: www.socweather.com
Accept: text/html, application/soap+xml
```

Notice that the HTTP GET formulation does not allow a request body. This means there can be no SOAP content in the request: that is, no headers and no arguments.

3.4 Developing and Using Web Services

Services do not exist in a vacuum. Often the business logic that a service presents would exist in some application programs, possibly already deployed. Therefore, a popular way to go about developing services is as follows.

On the server-side, programmers can take the internal code implementing the given business logic and generate service descriptions from that business logic. WSDL specifications can be readily generated from popular languages such as Java and C#, but also from other languages. The service can be made available for invocation via SOAP. At the same time, the WSDL specifications can be made available to prospective clients.

On the client-side, programmers would take the service descriptions and apply generic tools to map WSDL into interfaces of their desired programming language. They would then create their application using these interfaces and finally execute the application using SOAP to invoke the services provided by the server.

To enable dynamic binding, the server would use a programming interface to publish services just as the client would use a programming interface to find the services it needs.

3.4.1 Programming WSDL

WSDL appears complex, but is conceptually simple. A simple overview of the components in a WSDL file is shown in Figure 3.1. It is designed neither for readability nor succinctness, but for computers to process. As a result, it is straightforward for tools to generate WSDL automatically from source code, especially if it is object-oriented. It is easiest if the source code is in an object-oriented language, although even languages such as Cobol can be used. Tools such as Microsoft's Visual Studio .NET and Oracle Developer provide this functionality. As programmers implement or modify their implementations, they can generate WSDL specifications automatically. Clearly the efficiency helps or seems to.

However, there is a point of caution. The above kinds of tools end up exposing the implementation details of the underlying object-oriented framework, which would prove misguided in many settings [Vinoski, 2002]. In particular, automatically exposing business objects externally is risky, because those objects would generally have been designed and previously employed for limited internal purposes. Exposing such interfaces outside of the organization in which they were designed to function means that their behavior might not be adequate and may interfere in strange ways with their internal functioning. Also, the moment internal details are exposed externally we end up with dependencies that limit the evolution of the software components. Sometimes there can be mismatches caused by the differences in how the objects are interpreted by service consumers and providers. For example, session-based or stateful objects may not cohere with a pure invocation-based approach. It would be safer from the professional software engineering standpoint to formulate the service interface carefully and then to develop systems to implement it, and not to expose any more details than are explicitly called for by the interface.

3.4.2 Java for Web Services

Several tools for Web services now exist. The open-source Apache eXtensible Interaction System (Axis) tool from the Apache Project is a SOAP engine, which includes important functionality for WSDL as well.

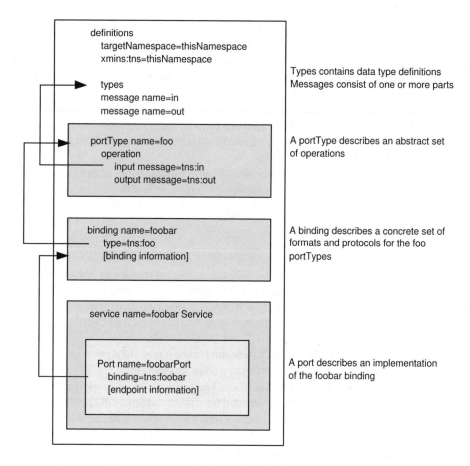

Figure 3.1: A simple view of the WSDL data model

3.4.2.1   JAX-RPC and SAAJ

The Java API for XML-Based RPC (JAX-RPC) and SOAP with Attachments API for Java (SAAJ) provide Java application programmer interfaces for processing SOAP messages. JAX-RPC is the higher-level of the two and builds on top of SAAJ. JAX-RPC handles conversions between Java objects and XML and performs type-checking on the conversion. JAX-RPC also includes tools to generate WSDL documents from Java code and Java code from WSDL documents.

For SOAP, SAAJ itself builds on JAXP and provides a simpler API geared toward SOAP. For instance, the API includes methods for managing SOAP connections, composing SOAP messages, extracting appropriate contents (headers, body) from them, and handling responses.

The Java API for XML Messaging (JAXM) provides APIs for creating and processing SOAP messages. JAXM is lower level than JAX-RPC and has been superseded by SAAJ.

3.4.2.2 Web Services Invocation Framework

The Apache Project's Web Services Invocation Framework (WSIF) is an approach for invoking WSDL-based services. WSIF takes a client perspective. However, it is based on WSDL descriptions of services and, in principle, is independent of the binding. Naturally enough, a binding for SOAP is available and is an important one, but a WSIF client could easily be ported to another binding.

3.4.2.3 JAXR

The Java API for XML Registries (JAXR) is a Java-based approach for accessing many different kinds of registries, including ISO 11179, OASIS, ebXML, and UDDI. It is most useful for accessing a UDDI or ebXML registry (discussed in Section 2.4) to advertise or discover a service.

3.4.2.4 JAXP

JAXP is an API for processing XML documents. One of JAXP's components is a parser based on the Document Object Model (DOM), which views a well-formed XML document in terms of the parse tree to which it corresponds. DOM provides a conceptually simple means to traverse the parse tree of a document via recursive-descent processing. An alternative parser based on the Simple API for XML (SAX) is also included. Further, JAXP has support for the XSL Transformations (XSLT).

3.4.3 .NET

Microsoft's .NET has tools that support essentially the same functionality as the Java family of tools. For instance, .NET includes wsdl.exe, which generates stubs from WSDL documents and generates WSDL documents from code.

3.5 Web Services Interoperability

The Web Services Interoperability Organization (WS-I) is an industry group that promotes interoperability at a level that is above the standards proper [WSI, 2004]. WS-I members include some of the leading Web service vendors, such as IBM, Microsoft, BEA, and Sun. WS-I makes recommendations about standards that in essence package the standards into compatible sets. These recommendations are termed *profiles*.

Currently, the WS-I has developed a profile known as the *Basic Profile 1.0*. This profile bundles SOAP 1.1, WSDL 1.1, XML 1.0, XML Schema, and HTTP 1.1. Further, the WS-I Basic Profile 1.0 imposes the following restrictions:

- SOAP should be used only with its HTTP POST binding.

- The SOAPAction header in the HTTP POST should be a quoted string.

- A SOAP recipient should return an HTTP response immediately upon receiving an HTTP request, and this response should be an HTTP success code (200 or 202) or an HTTP error code. A success code does not mean that the request was processed or even that it was well-formed (as a SOAP request). The HTTP response must not contain a SOAP envelope.

- A SOAP requestor must ignore any SOAP envelope that may be returned by a SOAP recipient (which would be in violation of the above restriction anyway).

- The WSDL message patterns are limited to request-response and one-way.

- Only XML Schema encodings are recognized, not SOAP Section 5.

WS-I identifies several other points of potential disagreement or ambiguity and seeks to resolve them, so that it can guarantee that if the various parties obey its recommendations, they will be able to interoperate. As explained above, application programmers should not be dealing with the raw protocols; instead, they should be exercising Web services through suitable tool suites and programmer interfaces. Consequently, standards such as WS-I's Basic Profile 1.0 are of greatest direct interest to tool developers.

3.6   Notes

Leading vendors such as HP, IBM, and Microsoft provide public UDDI registries for programmers to deploy their Web services. If you wish to provide your own UDDI registry, jUDDI found at http://www.juddi.org/, is an open source Java-based implementation of a UDDI registry. It includes a toolkit that enables developers to build access to UDDI registries within their own applications. jUDDI can act as the UDDI front-end on top of existing directories and databases. jUDDI-enabled applications can look up services in the UDDI registry and then proceed to invoke those services directly. Similar to jUDDI, UDDI4J (available at http://uddi4j.org/) is a Java class library that provides an API to interact with a UDDI registry.

3.7   Exercises

3.1. Consider a leading e-commerce Web site such as amazon.com from the consumer perspective (in the case of amazon.com, this would offer more functionality than their explicit Web service interfaces). Consider the basic steps of registering, signing in, searching a catalog, selecting some goods for purchase, providing shipping and payment information, and paying to conclude the deal.

   - Produce a state transition diagram corresponding to the various steps, possibly showing the various screens that you encounter during your interactions. Label the states with the choices available and the transitions with the choices taken by the client.

- Evaluate the above design with respect to the characteristics of REST.

3.2. Repeat Exercise 3.1 for an auction site, such as ebay.com from the perspective of buyers, who can (among other things) view the prices bid on a given item and create bids themselves. How might you model the closing time for an auction? Notice that interactions here are potentially longer lived. Would there be any significant change if you add sellers to your model explicitly?

3.3. Browser-Based Information System Using REST: Implement an intelligent vending machine, described as follows:

An intelligent vending machine accepts credit cards from customers. When a customer selects an item, the machine issues the item to the customer and issues to the credit card company a debit entry to the customer's account. The machine also maintains a list of suppliers, one supplier for each type of item. When the machine's inventory for an item falls below a threshold, the machine automatically reorders that item from the appropriate supplier.

Your implementation should proceed according to the following steps:

- Construct a model in UML for the application using Visio or Rational Rose:
  - Construct use-case diagrams for the vending machine and for the credit card company.
  - Construct a class diagram for the vending machine application, showing classes for Customer, Supplier, Vending Machine, Item, and any needed subclasses.
  - Construct a sequence (interaction) diagram for the application.
  - Construct a state-machine diagram for the vending machine as it goes from Idle to Getting Credit Card to Accepting Selection to Processing Credit Card to Dispensing Item to . . .
  - Construct a class diagram for the credit card company, showing their cardholders and card numbers. Make the classes persistent so that you can use Rational Rose to generate the Data Description Language (DDL) statements for the database you will need below.
- Construct your implementation in C# or as a Java applet. It should include the following:
  - A menu-based selection for items (allow at least 4 choices). To be RESTful, the choices should be URIs.
  - A user input (text-based) for entering a credit card number.
  - A check of the customer's credit card number with the credit card company's database of valid card numbers. Create the database using Microsoft Access or MySQL from the DDL created above (or generate your own DDL) and connect it to your application using ODBC or JDBC.

- A simulation of issuing the item to the customer.
- A simulation of sending a debit entry to the credit card company.
- A simulation of reordering items.

Suggestion: Use fixed (or pop-up) windows to display the progress of the simulated processes, and to enable user interactions with them.

Chapter 4

Enterprise Architectures

Several implementation architectures are competing for dominance in the Web service market. This chapter describes some of the architectures and indicates their differences and relative advantages. For the most part, it turns out that the various approaches are conceptually similar if not indistinguishable for our purposes. This is comforting because it means that the techniques described in this book can be realized with equal ease over different implementation architectures.

4.1 Enterprise Integration

An insidious terminological confusion is prevalent in a lot of the current literature. Often, the terms *integration* and *interoperation* are used interchangeably. However, the term "integration" indicates that the given components or resources are pulled together into one logical resource with a single schema, whereas the terms "interoperation" means that the given components work together. Usually, integration is not appropriate because it would violate the autonomy of the participants. Moreover, reasons of productivity and maintainability make integration undesirable and interoperation of autonomous entities more sensible. In practice when "integration" is realized, it is actually interoperation.

However, even in such settings, "integration" merely refers to a particular programming approach based on invoking functionality remotely. That is, the approach emphasizes imperative distributed programming using application programmer interfaces. A better approach would involve protocols that express the arms' length relationships among the interacting parties. However, in deference to common buzzwords, we use the term integration in this section. We revisit this point in Section 6.2.

The earlier work on addressing the challenges of heterogeneity involved integrating different information resources. This body of work was termed *enterprise information integration* (EII). More recently there has been work on *enterprise application integration* (EAI) where the business logics of the various applications are suitably interrelated and information flows from one to the other. Thus, whereas EII was about information modeling, EAI emphasizes environments where applications can be hosted. EAI more often involves considerations of performance and reliability. EII and EAI taken together can be termed *enterprise integration* (EI).

The two major modern approaches to EAI are based on Microsoft's .NET and Sun's Java 2 Platform, Enterprise Edition (J2EE). .NET emphasizes Web services, as does J2EE. Several J2EE server vendors are adding Web service capabilities to their platforms.

A common component of an EI architecture is a *metadata registry*, which records the identities and locations of various resources in the enterprise. A particularly common example of a registry is a directory service, typically based on the Lightweight Directory Access Protocol (LDAP). Directories were originally used to record information about people, such as their phone numbers and email addresses, as well as organizational information, such as who reports to whom in the given enterprise. Directories have since expanded to record the locations of shared resources in general. These resources can be databases, email servers, application servers, other directories, OLAP (on-line analytical processing) tools, and so on.

Metadata registries are used in modern information architectures to provide a simplified means of configuring complex systems. They facilitate the publishing and discovery of fairly fine-grained information, such as particular XML schemas, information models, and services. Interestingly, the technology behind enterprise metadata registries (e.g., LDAP) is commonly used to implement service registries such as those based on UDDI.

Metadata registries can be given a narrow or private scope where they are meant to hold information that is of interest only to a few applications and systems. For example, there could be separate registries for the billing and payroll departments of an enterprise. These registries would be used to configure applications within such departments. One reason to confine their access to particular departments would be security; another reason would be limiting external dependencies to the resources that are being made available to the rest of the enterprise. Wider-scope registries would be used by an enterprise to expose some information externally to its customers and partners, or to the public. The wider-scope registries can facilitate semantic mediation between components by providing a common handle to the models and semantic descriptions that are shared by the parties concerned.

Metadata registries are an excellent response to the challenges of dynamism, as defined in Section 1.3. Dynamism arises not only in open settings, but also within enterprises. It poses challenge for administering a system whose components may not be fixed and may change their functionality. The larger the system the more dynamic the setting. The more dynamic the setting the harder to manage the system. Because metadata registries lend structure to such a system, they simplify accommodating dynamism. If a registry can be kept up to date, it can be used to configure and reconfigure a system as needed.

4.2 J2EE

The Java 2 Platform, Enterprise Edition (J2EE) platform builds on the Java programming language by providing a framework for developing and deploying Java applications centered around an application server. Following the general principles of multitier architectures, J2EE separates into three main layers:

1. Presentation, for interacting with users, usually via client applications.

2. Business objects, for capturing the business logic that is the essence of the Java application being supported. This is the core part of J2EE and consists primarily of *Enterprise Java Beans* (EJBs), which are distributed objects defined according to rules that make them self-describing in terms of the methods that can be invoked on them. Different types of EJBs correspond to user sessions and to information entities. They can be serialized and materialized. Some can be made persistent on external databases. Yet other EJBs, called *Message-Driven Beans* (MDBs) listen for and respond to messages from a message queue.

   EJBs facilitate the development of maintainable applications because they capture the business logic of an application in a manner that is independent of its deployment. In other words, when an application is designed using EJBs, the decision about how to allocate resources to the EJBs can be delayed. If a certain EJB sees a large demand, it can be deployed using additional resources, for example, a larger number of threads.

3. Backend, which handles interactions with databases, ERP systems, and other systems (even those that are not in J2EE). This layer is associated primarily with the J2EE Connector Architecture (JCA), which supports both synchronous and asynchronous communications among the systems.

A J2EE application server consists of a Web container and an EJB container. The former hosts servlets and JSP pages. It invokes enterprise beans in the EJB container. The application server also interacts with the external world through message queues and databases. The EJB container provides system-level services including connectivity, security, and transactions. This is a productivity benefit during application development and management.

J2EE includes support for message queues through the *Java Message Service* (JMS). Messages can be sent and received synchronously by clients and EJBs. MDBs, mentioned above, consume messages asynchronously. Messages sent and received can be part of a distributed transaction.

The *Java Naming and Directory Interface* (JNDI) is a programming interface for metadata registries. A J2EE application or EJB can locate resources and other components on the fly, thus separating configuration details from the implementation. *Java Database Connectivity* (JDBC) provides libraries to open and maintain connections with relational databases, retrieve relational metadata, prepare and submit SQL statements, and process result sets.

J2EE is illustrated in Figure 4.1. J2EE provides the following components and their associated functionalities:

- Session, Entity, Message Driven Beans (EJB);

- Transaction Management (JTA/JTS);

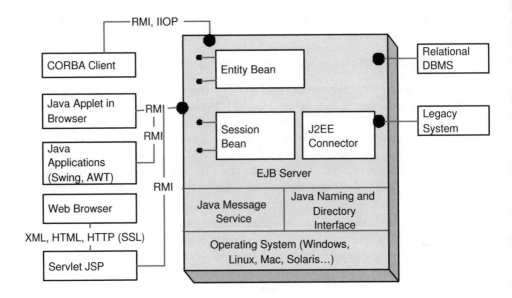

Figure 4.1: The J2EE architecture

- Naming and Directory (JNDI);

- Remote Method Invocation (RMI);

- Security (JNDI Security); Java Authentication and Authorization Service (JAAS);

- Java Messaging Service (JMS);

- J2EE Connector Architecture (JCA);

- Security (realms, access control lists);

- Web Services;

- XML (JAXP);

- Caching;

- Web based management and monitoring.

4.3   .NET

.NET refers to Microsoft's overall framework for Web services software [Meyer, 2001]. Its architecture is shown in Figure 4.2. The framework includes an interpreter and compiler, the Common Language Runtime (CLR) engine, which receives application code expressed in the Microsoft Intermediate Language (MSIL). The engine converts the MSIL code into native code, using just-in-time compilation techniques. There are interpreters for many programming languages, such as C++, Cobol, and Visual Basic, that generate MSIL. The result is that applications written in a variety of languages can be compiled into MSIL and then made part of a Web service. For example, Listing 4.1 shows a simple COBOL program repackaged as a Web service—this is accomplished via the webservice directive in the first line. The program, which returns the result of multiplying two numbers, is merely a Cobol class definition with a method called MULTIPLY.

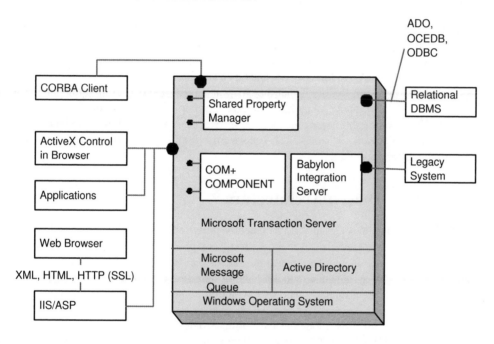

Figure 4.2: The Microsoft .NET architecture

Listing 4.1: A Cobol program for multiplication as a service

```
<\%@ webservice language="COBOL" \%>
CLASS-ID.
MULTIPLICATIONSERVICE.
FACTORY.
PROCEDURE DIVISION.
METHOD-ID.
MULTIPLY.
DATA DIVISION.
LINKAGE SECTION.
```

```
01 VAL−1 PIC S9(9) COMP−5.
02 VAL−2 PIC S9(9) COMP−5.
01 PRODUCT PIC S9(9) COMP−5.
PROCEDURE DIVISION
USING BY VALUE VAL−1 VAL−2 RETURNING PRODUCT.
COMPUTE PRODUCT = VAL−1 * VAL−2.
END METHOD MULTIPLY.
END FACTORY.
END CLASS MULTIPLICATIONSERVICE.
```

The Microsoft Transaction Server (MTS) corresponds to the J2EE server. This hosts the ActiveX and COM+ objects, which correspond to EJBs. Further correspondences are quite direct. The .NET Active Directory Service (ADS) corresponds to J2EE's JNDI, Microsoft Message Queue (MSMQ) to J2EE's JMS, and Open Database Connectivity (ODBC) to JDBC.

The interesting point to note is that the .NET and J2EE approaches are quite similar to each other. As shown above, they have architectural components that correspond to each other mostly in a straightforward manner. A difference is that .NET is a product from a single vendor (Microsoft), whereas J2EE is a specification for which products can be fielded (and are) by different vendors.

4.4   Model-Driven Architecture

The Object Management Group (OMG) recently proposed the Model-Driven Architecture (MDA) as a way to generalize over the existing component architectures to consider the entire system lifecycle in a unified, platform-independent framework. The lifecycle includes capturing business requirements, modeling and design, implementation and testing, configuration and deployment, management, and evolution induced by changing requirements or changing technology.

MDA is based on three modeling capabilities. The first is the famous *Unified Modeling Language* (UML), which includes sublanguages for a variety of software modeling needs. The key sublanguages of UML are those for capturing class diagrams, activities, and statecharts. The second is the *Meta Object Facility* (MOF), which is built as a subset of the constructs of UML that are expressive enough to capture models of interest. The third is the *Common Warehouse Metamodel* (CWM), which standardizes the data warehouse application lifecycle, i.e., design, build, and manage.

Having a language such as MOF enables the exchange of models among development tools and middleware. Model exchange is standardized through *XML Metadata Interchange* (XMI), which provides the DTDs for UML, MOF, and CWM.

MDA distinguishes between a *platform-independent model* (PIM) and a *platform-specific model* (PSM). A PIM formally captures the essence of the structure and function of the modeled system (or artifact or activity), whereas a PSM captures details specific to an implementation or class of implementations. For example, in platform-independent terms, a payment protocol involves the transfer of funds from one party to another. Payment via a credit card, which involves a credit card company and the concomitant messages, is thus a special case of the above but remains platform independent. That is, the PIM is the conceptual model of the given system. However, in platform-specific terms, payment could be mapped into an exchange of requests and responses implemented in SOAP or via remote method invocations. Similar examples could be constructed for intraenterprise settings such as internal billing processing.

The idea behind MDA is to enable conceptual models that are independent of platforms and thus to make for easier interoperation of heterogeneous systems and easier portability and evolution of such systems as the underlying platforms evolve. MDA would include the so-called profiles for the various platforms of interest, e.g., Common Object Request Broker Architecture (CORBA), .NET, and J2EE. When new standards are proposed, the MDA process requires them to be specified in terms of a PIM and one or more PSMs.

In practical terms, MDA can be used to model services formally. The models would be constructed using UML, possibly augmented with stereotypes specific to services. Following the terminology of Section 2.3.3, the MDA PIM would correspond to the service interface and the MDA PSM would correspond to the service implementation.

4.5   Legacy Systems

Legacy systems is a phrase, often pejorative, for describing computing systems that are obsolete in some manner and typically not easy to modify or modernize. The term has undergone some modification over the years. In the early days of client-server computing, the older mainframe systems with their rigid hardware constraints and vertically integrated applications were often the culprits. Now those constraints have mostly been relaxed and the problems that remain are the more insidious ones based on the semantics or meaning of the data, the applications, and the user interfaces.

Typically, legacy systems would be those that run on obsolete hardware architectures and operating systems and over nonstandard communication networks. Traditionally, systems based on mainframe computers were like that. The operating systems were proprietary to the hardware vendor as were the communication networks. But the industry has evolved in some respects. The communication networks are now all able to support IP. Legacy systems also run poorly documented, unmaintainable software created by *ad hoc* patches to handle bugs, changing regulations, and updated business needs. Mainframes do not have a monopoly on unmaintainable software and the criticism may easily be levied against other systems developed in a patchwork manner and without adequate modeling.

Legacy systems often involve a substantial database component, which is poorly modeled if at all. Early on, the databases were hosted on hierarchical or network database management systems and required custom programming. Nowadays, adapters are available to access data with a relational interface from hierarchical or network databases, so the problem is no longer due to the underlying data representation. However, even relational databases are often not modeled well, so the data that can be accessed is only imperfectly understood.

Legacy systems tend to support rigid user interfaces. Traditionally, these were proprietary "screen-based" interfaces that were hardwired into applications and relied on particular terminal models manufactured by particular vendors, e.g., IBM's 3270 or Digital's vt100. Again, now almost all of these screens can be mapped into browser-based interfaces, and potentially run over any browser and any monitor. However, the semantics of the interfaces remains arcane and that is by far the bigger problem.

On the positive side, legacy systems fulfill crucial business functions, such as the majority of banking systems, business data processing, and airline reservation systems. They represent huge investments that cannot easily be discarded. We advocate that legacy systems be modeled as Web services, so that modern applications can interoperate with them to share data and preserve integrity.

4.6   Notes

Because J2EE is an interface, it is implemented by several vendors and open-source projects. The following are some well-known implementations.

- BEA WebLogic Server 8.1 from BEA Systems. BEA provides good integration between tools and server, easy setup, Web services standards support, and a well-rounded feature set for enterprise deployments. WebLogic Workshop includes an integrated development environment.

- IBM WebSphere Application Server 5.0 from IBM. WebSphere has an extensive set of features, especially in regard to enterprise-level deployments. WebSphere allows granular control of the application server and the Web services running on it from either a GUI or a command-line interface.

- JBoss 3.2.1 and Apache Tomcat 5.0 from JBoss and the Apache Project, respectively, are a combination of open source systems provides a powerful feature set and excellent support for importing and exporting existing Web services, as well as support for Web service standards. On the downside, the JBoss and Apache tools require more expertise and initial configuration than their commercial counterparts. They also lack many of the simpler, enterprise-class configuration and management tools found in BEA WebLogic Server and IBM WebSphere Application Server.

- Novell exteNd Application Server 5.0 and Workbench. Novell's exteNd is a plain application server with an accompanying Workbench IDE. Novell has announced plans for associated Director and Composer tools, which will make the Novell system more capable and easier to use.

- Sybase EAServer 4.2. This J2EE platform features good integration between application server and tools, a serviceable interface, and a reasonable set of features that is comparable to the set in JBoss.

4.7   Exercises

4.1. Install a development framework, based on either .NET or J2EE (using IBM Web-sphere, Sybase EAServer, Novell exteNd, BEA WebLogic Server, or JBoss and Apache Tomcat). Using your framework, construct a simple method, such as the GetTemp method from Chapter 2, and expose it as a Web service. To test it you will have to write a client application that connects to your server and requests the service.

4.2. Construct a Web service for unit conversion between units describing length, mass, and time. For example, the Web service should be able to convert from centimeters to inches. The WS would accept three inputs—a value, its unit, and the desired output unit—and it would return one output—the value in terms of the output unit. Write a WSDL description for this Web service, and then deploy the service on a Web server by using one of the frameworks described in Exercise 4.1.

4.3. Construct a Web service for unit conversion into MKS units. This Web service would be able to convert any length measure into meters, any mass measure into kilograms, and any time measure into seconds. The WS input would be a value and its unit, and the output would be the appropriate MKS unit and its value. Write a WSDL description for this Web service, and then deploy the service on a Web server by using one of the frameworks described in Exercise 4.1.

Chapter 5

Principles of Service-Oriented Computing

The preceding chapters have taken us from a historical perspective of Web services to the basic standards for realizing them, as well as current enterprise computing architectures and programming approaches for implementing Web services. If we were content with using and fielding simple Web services, the above topics would be more than adequate. However, if we would like to develop and manage systems of real-life complexity, the above concepts are merely a prologue to the underpinnings of the technology that we will need to develop.

A question often arises about the benefits of service-oriented computing and specific approaches for it. There are two major answers to this question. One, service-oriented computing enables new kinds of flexible business applications of open systems that simply would not be possible otherwise. When new techniques improve the reaction times of organizations and people from weeks to seconds, they change the very structure of business. This is not a mere quantitative change, but a major qualitative change. It has ramifications on how business is conducted. Two, service-oriented computing improves the productivity of programming and administering applications in open systems. Such applications are notoriously complex. By offering productivity gains, new techniques make the above kinds of applications practical, thus helping bring them to fruition.

5.1 Use Cases

In order to motivate the full expanse of service-oriented computing that this book presents, it would help to take a closer look at its main use cases—that is, its major application settings. The interesting thing about these use cases is they are simple and straightforward. This supports the view that service-oriented computing is not a set of futuristic technologies for applications that may or may not matter, but a set of existing and emerging technologies that

be processed via robust commercial tools, rather than through custom software. In recent years, industry has converged on XML as the data format of choice. This is clearly a success. However, it opens up questions of how the content of the data being communicated can be understood and processed.

Service-oriented computing provides the same benefits as for intraenterprise interoperation above. In addition, it provides the ability for the interacting parties to choreograph their behaviors so that each may apply its local policies autonomously and yet achieve effective and coherent cross-enterprise processes.

5.1.3 Application Configuration

Imagine that the hospital purchases an anesthesia information management system (AIMS) to complement its existing systems. An AIMS would enable anesthesiologists to manage anesthetic procedures on patients in surgery better and to monitor, record, and report key actions better, such as the turning on or turning off of various gases and drips. Such information can help establish compliance with government regulations, ensure that certain clinical guidelines are met, and support studies of patient outcomes. It is easy to purchase an AIMS, but not as easy to install and use it (or any such system: this point is not just about AIMS). To introduce such a system requires that the right interface be exposed by the new system and by the existing systems. Since these systems would have been developed on different platforms and may, in fact, run on different operating systems, adherence to standards for interconnections is required for each.

We can assume that low-level connectivity is taken care of. However, it still leaves the following challenges as in intraenterprise interoperation. One is *messaging* so that the systems can be operationally connected. Another is semantics so that the components understand each other.

However, there is another challenge with introducing a new application. This is to configure and customize its behavior. In the case of an AIMS, it must be populated with a hospital-specific data model or terminology so that it displays the right user interface screens to the hospital staff and logs the right observations. This model is governed by hospital procedures as well as the requirements imposed by the insurance companies and government agencies with whom they interact. If the application is designed with the considerations of service-oriented computing in mind, then it can be quickly configured and introduced into existing business processes.

Service-oriented computing enables the customization of new applications by providing a Web service interface that eliminates messaging problems and by providing a semantic basis to customize the functioning of the application.

5.1.4 Dynamic Selection

Imagine that a hospital wishes to purchase supplies such as catheters. To carry out such purchases efficiently requires that the hospital be able to interoperate with the catheter vendor—a case of interenterprise interoperation. Now suppose that the hospital would like to purchase catheters from whichever vendor offers it the best terms. In other words, the business partner—the other enterprise with which to interoperate—would be chosen on the fly. Such dynamic selection is becoming increasingly common as the possible gains of such flexibility are recognized. If business partners can be selected flexibly, then they can be selected to optimize any kind of quality-of-service criteria, such as performance, availability, reliability, and trustworthiness.

Service-oriented computing enables dynamic selection of business partners based on quality-of-service criteria that each party can customize for itself.

5.1.5 Software Fault Tolerance

Suppose a hospital is carrying out a business transaction with a partner and encounters an error. It would be great if the interaction could be rewired to an alternative business partner dynamically in a manner that is transparent to the overall process. To the extent that the state of the interaction is lost, some means of recovery would be needed to restore a consistent state and resume the computation with new business partners.

Service-oriented computing provides support for dynamic selection of partners as well as abstractions through which the state of a business transaction can be captured and flexibly manipulated; in this way, dynamic selection is exploited to yield application-level fault tolerance.

5.1.6 Grid

Grid computing refers to distributed computing where several resources are made available over a network, and are combined into large applications on demand. Grid computing is a form of *metacomputing* and arose as the successor of previous approaches for large-scale scientific computing. Building complex applications over Grid architectures has been difficult, which has led to an interest in the more modular kinds of interfaces based on services. Accordingly, Grid services have been proposed in analogy with Web services.

Service-oriented computing enables the efficient usage of Grid resources.

5.1.7 Utility Computing

Following up on Grid-like environments, there has been a recent expansion of *utility computing*, where computing resources are modeled as a utility analogous to electric power or telecommunications. The idea is that enterprises would concentrate on their core business and out-source their computing infrastructure to a specialist company. Leading companies such as IBM and HP have made utility computing offerings; the one from IBM is called *autonomic computing*, although that term is starting to be used generically as a technical area for fault-tolerant computing. Utility computing presupposes that diverse computational resources can be brought together on demand and that computations can be realized on physical resources based on demand and service load. In other words, service instances would be created on the fly and automatically bound to configure applications dynamically.

Service-oriented computing facilitates utility computing, especially where redundant services can be used to achieve fault tolerance.

5.1.8 Software Development

Software development remains a challenging intellectual endeavor. Improvements are realized through the use of superior abstractions. Services offer programming abstractions where different software modules can be developed through cleaner interfaces than before. When the full complement of semantic representations are employed, the resulting modules are not only more easily customizable than otherwise, but the following holds:

Service-oriented computing provides a semantically rich and flexible computational model that simplifies software development.

5.2 Service-Oriented Architectures

The above use cases provide a challenging set of requirements for any approach to computing. While there are no free lunches in computer science, the requirements can be satisfied more easily through an architecture that matches the essential properties of the above use cases. Let us term such an architecture a service-oriented architecture (SOA).

The emphasis falls on the architecture because many of the key techniques are already well understood in isolation. Practical success would depend on how well these techniques can be placed in a cohesive framework—an architecture—and translated into methodologies and infrastructure so they can be applied in production software development. Recent progress on standards and tools is extremely encouraging in this regard. There can be several SOAs provided they satisfy the key elements of service-oriented computing, which are introduced below.

The current incarnation of Web services emphasizes a single provider offering a single service to a single requester. This is in keeping with a client-server architectural view of the Web.

5.2.1 Elements of Service-Oriented Architectures

To realize the above advantages, SOAs impose the following requirements:

Loose coupling. No tight transactional properties would generally apply among the components. In general, it would not be appropriate to specify the consistency of data across the information resources that are parts of the various components. However, it would be reasonable to think of the high-level contractual relationships through which the interactions among the components are specified.

Implementation neutrality. The interface is what matters. We cannot depend on the details of the implementations of the interacting components. In particular, the approach cannot be specific to a set of programming languages.

Flexible configurability. The system is configured late and flexibly. In other words, the different components are bound to each other late in the process. The configuration can change dynamically.

Long lifetime. We do not necessarily advocate a long lifetime for our components. However, since we are dealing with computations among autonomous heterogeneous parties in dynamic environments, we must always be able to handle exceptions. This means that the components must exist long enough to be able to detect any relevant exceptions, to take corrective action, and to respond to the corrective actions taken by others. Components must exist long enough to be discovered, to be relied upon, and to engender trust in their behavior.

Granularity. The participants in an SOA should be understood at a coarse granularity. That is, instead of modeling actions and interactions at a detailed level, it would be better to capture the essential high-level qualities that are (or should be) visible for the purposes of business contracts among the participants. Coarse granularity reduces dependencies among the participants and reduces communications to a few messages of greater significance.

Teams. Instead of framing computations centrally, it would be better to think in terms of how computations are realized by autonomous parties. In other words, instead of a participant commanding its partners, computation in open systems is more a matter of business partners working as a team. That is, instead of an individual, a team of cooperating participants is a better modeling unit. A team-oriented view is a consequence of taking a peer-to-peer architecture seriously.

Researchers in multiagent systems (MAS) confronted the challenges of open systems early on when they attempted to develop autonomous agents that would solve problems cooperatively, or compete intelligently. Thus, ideas similar to service-oriented architectures were developed in the MAS literature. Although SOAs might not be brand new, they address the fundamental challenges of open systems. Clearly the time is right for such architectures to become more prevalent. What service-oriented computing adds to MAS ideas is the ability to build on conventional information technology and do so in a standardized manner so that tools can facilitate the practical development of large-scale systems.

5.2.2 RPC versus Document Orientation

There are two main views of Web services. Services can be understood in terms of the *RPC-centric view* or the *document-centric view*. The former treats services as offering a set of methods to be invoked remotely, i.e., through remote procedure calls. The latter treats services as exchanging documents with one another. In both views, what is transmitted are XML documents and what is computed with are objects based on or corresponding to the XML documents. However, there is a significant conceptual difference.

The RPC view sees the XML documents as incidental to the overall distributed computation. The documents are merely serializations of the business objects on which the main computation takes place. The document-centric view considers the documents as the main representations and purpose of the distributed computation. Each component reads, produces, stores, and transmits documents. The documents are temporarily materialized into business objects to enable computing, but the documents are the be all and end all of the computation.

The RPC view thus corresponds to a thin veneer of Web services over an existing application. The application determines what functionality the services will support. The document view more naturally considers Web services as a means of implementing business relationships. The documents to be processed (and their relationships) determine the functionality of the services. The business objects, such as there are, on either side of a relationship are local, and should not be exposed to the other side.

For this reason, the document-centric view coheres better with our primary use case of applying services in open environments. The RPC view is more natural for the use case of making independently developed applications interoperate. What happens is that application developers expose their application interface in the form of Web services, which can then be bound to in the usual manner. If the applications are designed for method integration, then the RPC view of services is natural for such interoperation. However, if the applications are designed—as they should be—to function as independent components, then the document-centric view would be natural even for application interoperation.

5.3   Major Benefits of Service-Oriented Computing

It is worth considering the major benefits of using standardized services here. Clearly anything that can be done with services can be done without. So what are some reasons for using services, especially in standardized form? The following are the main reasons that stand out.

- Services provide higher-level abstractions for organizing applications in large-scale, open environments. Even if these were not associated with standards, they would be helpful as we implemented and configured software applications in a manner that improved our productivity and improved the quality of the applications that we developed.

- Moreover, these abstractions are standardized. Standards enable the interoperation of software produced by different programmers. Standards thus improve our productivity for the service use cases described above.

- Standards make it possible to develop general-purpose tools to manage the entire system lifecycle, including design, development, debugging, monitoring, and so on. This proves to be a major practical advantage, because without significant tool support, it would be nearly impossible to create and field robust systems in a feasible manner. Such tools ensure that the components developed are indeed interoperable, because tool vendors can validate their tools and thus shift part of the burden of validation from the application programmer.

- The standards feed other standards. For example the above basic standards enable further standards, e.g., dealing with processes and transactions.

5.4   Composing Services

Although there can be some value in accessing a single service through a semantically well-founded interface, the greater value is clearly derived through enabling a flexible *composition* of services. Composition leads to the creation of new services from old ones and can potentially add much value beyond merely a nicer interface to a single preexisting service. The new services can be thought of as *composite services*.

From a business perspective too, intermediaries that primarily offer access to a single service would have a tough time thriving or even surviving. Airline travel agents are a case in point. Traditional travel agents provide a nice user interface: friendly and with a human touch, but little more. However, airlines do not like to pay commissions for services that merely repackage their offerings. As a result, the airlines compete with the travel agents and reduce their commissions, slowly squeezing them out of business. This is as one would expect where the offerings are conceptually simple, especially for frequent customers. By contrast, package tour operators, who combine offers from airlines and other vendors, can prosper. In other words, the increased complexity due to subtle compositions is essential for intermediaries to flourish, because it provides an opportunity for offering greater value to customers.

Service composition concepts involve enough intricacy so as to attract considerable interest and to demand a careful analysis of the underlying principles. The need for principles is greater as the basic infrastructure for Web services becomes more common. We address these principles herein. Sometimes, the term *composition* is taken to mean a particular approach to achieving composition, for example, by invoking a series of services. In the present usage, however, composition refers to any form of putting services together to achieve some desired functionality.

Composed Web services find application in a number of practical settings. For example, portals aggregate information from a number of sources and possibly offer programmatic facilities for their intended audience. The challenge to making an effective portal is to be able to personalize the information presented to each user. Electronic commerce is another major scenario where users would like to aggregate product bundles to meet their specific needs. Virtual enterprises and supply-chain management reflect generalizations of the consumer-oriented e-commerce scenarios, because they include more subtle constraints among a larger number of participants.

using HTTP, or directly with server objects via CORBA's Internet Inter-ORB Protocol (IIOP) (where an ORB is CORBA's object request broker) or Remote Method Invocation (RMI).

- If there are too many orders to process synchronously, they could be put in a message queue, managed by a Message Oriented Middleware (MOM) server (which guarantees message delivery or failure notification), and customers would be notified by email when the transaction is complete.

Email is typically used for people to communicate with each other, so in using email, the server is behaving like an intelligent agent. We will have more to say about the emerging agent-like aspects of the Web and its services in later chapters.

Notice that although the above example considers a user dealing with a particular enterprise, the problem arises in more acute form in business-to-business settings. If our little camera store were considered as merely a component in a large supply network, it would have no hope of forcing the other parties to conduct their local transactions in any particular manner or to reliably converge to a state that would be consistent across the system. Deeper models of transactions and of business processes are needed to ensure that the correct behavior is realized in such cases.

The current specifications for Web services do not address transactions or specify a transaction model. The Organization for the Advancement of Structured Information Standards (OASIS) is developing one, but the view of most implementors is that SOAP will manage transactions—somehow. Without guidance from a standard or an agreed-upon methodology by the major vendors, transactions will be implemented in an *ad hoc* fashion, thus defeating the hopes for interoperability and extensibility.

Some of the other problems for composed services are

- Security will be more difficult, because more participants will be involved and the nature of their interactions and their needs might be unanticipated by the designers of the services.

- There will be incompatibilities in vocabularies, semantics, and pragmatics among the service providers, service brokers, and service requesters.

- As services are composed dynamically, performance problems might arise that were not anticipated.

- Dynamic service composition will make it difficult to guarantee the quality of service (QoS) that applications require.

Two fundamental styles for delivering Web services are emerging, characterized as RPC-style (favored by Sun) and document-style (favored by Microsoft, and supported by Sun). In the latter style, the body of a SOAP message would not have the call-response semantics of most programming languages, but rather would consist of arbitrary XML documents that use WSDL to describe how a service works. In the long term the document-style is likely to prevail, because it is more declarative (rather than procedural), more asynchronous, and more consistent with the document-exchange underpinnings of the Web.

5.5 Spirit of the Approach

Figure 2.1 shows the generic architecture for Web services. Although this is a simple picture, it radically alters many of the problems that must be solved in order for the architecture to become viable on a large scale.

- To publish effectively, we must be able to specify services with precision and with greater structure. This is because the service would eventually be invoked by parties that are not from the same administrative space as the provider of the service and differences in assumptions about the semantics of the service could be devastating.

- From the perspective of the registry, it must be able to certify the given providers so that it can endorse the providers to the users of the registry.

- Requestors of services should be able to find a registry that they can trust. This opens up challenges dealing with considerations of trust, reputation, incentives for registries and, most importantly, for the registry to understand the needs of a requestor.

- Once a service has been selected, the requestor and the provider must develop a finer-grained sharing of representations. They must be able to participate in conversations to conduct long-lived, flexible transactions. Related questions are those of how a service level agreement (SLA) can be established and monitored. Success or failure with SLAs feeds into how a service is published and found, and how the reputation of a provider is developed and maintained.

The keys to the next-generation Web are *cooperative services, systemic trust, and under-standing based on semantics, coupled with a declarative agent-based infrastructure.*

The size and dynamism of the Web presents problems, but it fortuitously provides a means for solving its own problems. For example, for a given topic there might be an overload of information, with much of it redundant and some of it inaccurate, but a system can use voting techniques to reduce the information to that which is consistent and agreed upon. For another example, there might be many potential service providers competing for many potential clients, and some of the providers might not be trustworthy, but a system can use a Web-based reputation network to assess credibility. Finally, different sites might use different ontologies, but a multiplicity of ontologies can yield a global, dynamically formed, consensus ontology. The subsequent chapters of this book address each of these concepts.

5.6 Exercises

5.1. Consider the basic Web service architecture and its main components introduced above: WSDL, SOAP, and UDDI. List and briefly explain a total of six shortcomings of these three components. Two sentences each would be adequate to convey the essential points, although you could certainly present more in-depth analyses.

Part II

Description

Chapter 6

Modeling and Representation

Services have several important static and dynamic aspects, roughly dealing with their underlying information and process models, respectively. When services are to be understood, implemented, modified, discovered, selected, engaged, or composed, it is essential that these key aspects be captured perspicuously. This is the primary motivation for *modeling* services.

The main motivation for modeling a resource in an enterprise is that the model captures the requirements for the resources and the rationale behind their design. Thus models facilitate reuse of the resource within the enterprise. Further, they enable integrity constraints on the given resource to be stated so that the resource (or instances of it) can be validated. Even more importantly, when a model exists, consistency analysis to discover errors can be performed on the model itself, perhaps in conjunction with the models of other resources, instead of on the actual resources involved. Models can be used as a basis for tracking changes in requirements and to analyze the impact of changes. Lastly, the models can be taken as inputs by various software engineering frameworks and tools to help in the generation of the necessary data models, application stubs, and interfaces.

But one important respect in which services go beyond traditional settings is that they are meant to be used in multiple contexts. That is, models for services apply both in terms of how the services are implemented and how they are used by others. For complex service implementations, we would need to model the databases and knowledge bases, applications, workflows, and the organizational roles involved. For complex service usages, formal markup is needed not only for the services themselves, but also for the needs and preferences of the users for whom the services are composed and executed.

As explained in Section 6.2.1 below, the most perspicuous models are those that are declarative, capturing the content of what is being represented without over-constraining how the representations are used.

6.1   Modeling to Enable Interoperation

When we build a conceptual model for a service, we rely upon our knowledge of the domain in which the service will be embedded, i.e., of the universe of discourse. A model of a service is useful for all the reasons discussed at the beginning of this chapter. Most importantly, the model helps us design the proper interactions among services, generally through a composition consisting of or otherwise involving the modeled services.

To achieve the desired interactions among services, we must back off from the service implementations and consider their conceptual schemas. Relating conceptual schemas essentially involves finding common ground between the universes of discourse of the services. When the conceptual schemas are sufficiently rich as to capture the dimensions of abstraction introduced in Section 6.3, the interrelationships among the involved services can support their effective interaction.

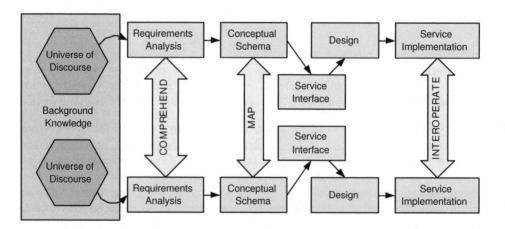

Figure 6.1: Modeling the composition of services at a conceptual level

Figure 6.1 shows how two services may be enabled for interoperation. Going left to right along the top branch shows how a traditional application that maps to a single service may be modeled and constructed. It helps to review this figure going from right to left. If we would like the two services to interoperate, we must make sure that their models are interrelated. The point of modeling is to extract dynamic behavior into a static representation. How the models relate to one another statically determines how the services interoperate dynamically. But the models cannot be interrelated unless we can establish that they share the same universe of discourse. Ultimately, it is the universe of discourse that determines the overlaps between the models and thus makes the interoperation meaningful. Below we discuss how the universe of discourse can be computationally represented in an ontology, and how such shared universes can be built and used.

6.2   Integration versus Interoperation

Often we hear discussion of integrating schemas, databases, workflows, or services. Integration refers to the idea of putting diverse concepts together to create an integrated whole. Having an integrated model would facilitate the services working well together. This contrasts with interoperation, which refers to making services work together by sharing the appropriate messages and using narrow, agreed-upon, interfaces, but without any single conceptual integration. In general, interoperation is what we desire and integration is often the wrong way to go about trying to obtain it. This is because integration can be expensive to achieve. Also, integration is fragile, meaning that when one of the integrated services changes, it may affect the integrated whole. Therefore, it is wiser to motivate service composition from the perspective of interoperation.

Interoperability can be discussed at the levels of syntax and semantics. Standard information interchange formats, e.g., those based on XML, solve the syntax problem, but the semantics problem has several aspects.

6.2.1   Declarative versus Procedural Representations

As before, there is a question of procedural versus declarative representations. Declarative representations have the advantages of enabling standardization, optimization through sophisticated tools, and improved productivity in terms of capturing and reusing knowledge. However, declarative approaches sometimes have a greater startup cost, because they require deeper and cleaner conceptualization of the domain of interest. Further, declarative conceptualizations rely upon a reasoning engine for processing and such engines can have poor performance.

Throughout the history of computing, declarative approaches have been developed many times, usually in the face of opposition from those who consider performance as paramount. Each time, the productivity gains they offer have far outweighed the loss in performance. Tools with higher performance are soon developed and improvements in hardware and systems reduce the effect of any loss of performance as well. For example, higher-level programming languages and compilers were initially slower than expert assembly language programmers, but today no one would seriously consider developing a large business application in assembly language. Likewise, database query languages such as the Structured Query Language (SQL) faced much opposition initially, but soon went on to deliver sufficient performance with greater productivity than procedural techniques.

Web applications, because of the autonomy and heterogeneity of the constituents, demand flexibility and open representations. Consequently, they greatly shift the trade-off in favor of declarative modeling.

6.2.2   Interoperation

Figure 6.2 describes how interoperation approaches have evolved over the years, bringing us closer to a service-oriented architecture (SOA) with each step. The earliest approach to achieving interoperation involved forcing the interoperating parties to fit into each other's requirements. Typically, one of the parties would dominate because of its business or political clout and the others would comply with whatever arbitrary encoding and networking standards it adopted. With this approach, labeled direct integration, the master computation must invoke methods in the slave computation. To pass parameters to invoke such methods and to process their results presupposes a tight integration of the data structures between the interacting parties.

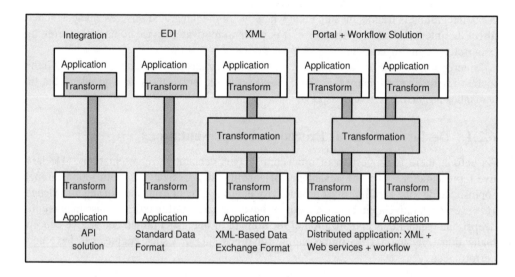

Figure 6.2: Interoperation scenarios

Traditional enterprise application integration (EAI) is a step toward increased flexibility wherein there is a common data exchange format. The format would be something proprietary, but having it explicit enables interacting parties to be added more easily than in the previous case. A more recent approach bases the common data exchange format on XML, and is viewed as part of the middleware. It can be produced, parsed, validated, and transformed through commercial, general-purpose tools.

The state-of-the art approach for EAI would involve distributed workflows in which a number of application components are orchestrated over the Web. The components can be thought of as exposing functionality in the style of Web services and of composing functionality as in a portal. Here the middleware does more than merely translate data formats. It also handles process dependencies among the components. This is clearly an improvement, because it separates not only the data flow, but also the top-level control flow, from the application logic.

6.2.3   Layered View

Interoperation between several applications or processes can conveniently be viewed as a set of layers, where each layer provides a higher-level model of the processes.

For distributed applications to interoperate, they must be able to exchange information successfully. For the information to be communicated, it must be represented in low-level data, but the desired semantics needs to be associated with the information. Putting aside *ad hoc* approaches, the semantics should capture the dimensions of abstraction discussed above. As discussed in Chapter 2, for service-oriented computing, XML has become the low-level representation for information across components. Consequently, information from the application layer should be conveyed down to the data systematically when it is serialized into an XML document; conversely, the data should be conveyed up to the application layer when the XML document is materialized.

The generation and consumption of the XML documents can be understood in terms of a series of transforms from the semantic to the syntactic layers. Consider an application executing at one of the business partners in an instantiation of an SOA. The application, which incorporates the business logic and specifies internal object representations, can be thought of as the top layer. In principle, the application itself may expose further representations to its users and a rich organizational structure could cover the various business partners, but those would take us from computational representations into the human organizational aspects and cannot readily be modeled, at least for the present purposes.

We can imagine that the application at a given business partner creates and reasons with an object graph representing the entities and their relationships. The application is supported by an object representation layer. The application invokes that layer by constructing meta-data for its object graph. The metadata is sufficiently expressive as to convey which ontology languages and ontologies are used, how they are implemented, and how cardinality and aggregation are expressed. The metadata describes, e.g., how ordered relationships or n-ary relationships are implemented, or how typing of nodes is represented in the object graph. The application passes both the object graph and its metadata to the object layer. This information is forwarded to the syntax layer, which generates an XML document containing the object graph, and references an XML schema document needed to extract the object graph from the main XML document. On the other end of the communication link, the software modules of the interacting business partner begin executing. They reverse the above process, culminating in the delivery of a high-level object graph to the application in that business partner.

In this scheme, the idea of the business partners interacting through documents that are described via separate schema documents comes from the SOA. The idea of separating the conceptual, logical (or representational), and syntactic layers is a simple variant of the three-level models in information systems, where a conceptual schema is separated from a logical schema, which is itself separated from the physical layer. In a typical database application, the object graph resides in the application code and is modeled via UML or entity-relationship diagrams; the representational or object layer corresponds to the relational schemas, and the syntactic representation to the physical rendition of the database on disk.

6.2.4   Interoperation Trends

Ever since enterprise integration came into its own, interoperation has gone through a series of steps in its evolution. The earliest generation involved point-to-point communication; the latest, which is only now beginning to emerge, allows flows of computation that respect the autonomy and heterogeneity of the various components.

Table 6.1: A historical view of interoperation levels

Generation	First	Second	Third	Fourth
Communication	TCP/IP	CORBA	HTTP	Messaging
Information	SQL	XML	RDF	OWL
Application	RPC	EDI	SOAP	Protocols
Configuration	Hard-coded	Directories	UDDI	Selection

Table 6.1 illustrates the historical development of the main levels of interoperation. The term *protocols* as used in Table 6.1 refers to *business protocols*, such as for payment and negotiation, and not to low-level protocols such as for request-response interactions. Of course, business protocols would be realized using such lower-level protocols. Likewise, *selection* refers to the trusted selection. Also, messaging refers to reliable messaging, which would build on underlying protocols such as HTTP.

6.3   Common Ontologies

A shared representation is essential for the mutual understanding of communications. For humans, the physical, biological, and social world that they inhabit provides a basis for mutual understanding. For computations, a common ontology provides the basis.

An *ontology* is a kind of a knowledge representation describing a conceptualization of some domain. An ontology specifies a vocabulary including the key terms, their semantic interconnections, and some rules of inference.

In general terms, as a representation for a universe of discourse, an ontology need apply not merely to information that is stored and manipulated, but also to any area of intellectual endeavor. For example, we can have an ontology for security concepts, which formalizes terms such as roles, credentials, authentication, privileges, granted privileges, and revoked privileges. Such an ontology could be used to compare security approaches and perhaps make them function in a cohesive manner. As another example, we might build an ontology of student life, which encodes home, work, commuting, courses, prerequisites, theses, graduation, internships, financial aid, and so on. Such an ontology provides a basis for students to talk intelligibly with each other about their lives even if they have not been acquainted previously. An ontology might delve into its universe of discourse to the desired level of detail. For example, it might or might not be worthwhile to represent teaching assistants in the above ontology of student life. However, if we are going to talk about teaching assistants, we would have to represent them in the ontology. The same consideration applies in computational settings: an ontology facilitates conversations and must be (or become) complete enough to sustain the desired conversations.

If there were a central authority with a global ontology to which all Web components adhered, and if the components of the Web were static, and if the identity of the components were fixed, and if there were a small fixed number of component types, then the challenges of interoperation and understanding would disappear, but the Web would no longer be the vibrant useful place upon which the global economy and modern society increasingly rely. However, two architectural approaches have emerged for achieving mutual understanding that do not place such strong restrictions on the Web. The first, a client-server approach to information management, has produced a plethora of search and query tools that are mostly based on keywords. Keywords are better for text than for the structured data found in most databases, but are completely unsuitable for information sources that do not adhere to a uniform semantics, especially the autonomously maintained databases and services being deployed widely.

A second, more compelling, approach, which achieves interoperation among information sources, applications, and users, introduces software agents. An *agent* can serve as a mediator, translator, or information broker. Here the term agent may be understood as an active, autonomous software component that serves as a critical part of the middleware. The deeper technical consequences of this definition are revisited in Chapter 15. In simple terms, the major task for the agents is to reconcile the varied semantics of the mostly autonomous resources in a manner that is scalable across large numbers of sources. This is the essence of cooperation that must be incorporated in a viable SOA.

For either approach, ontology-based interoperation provides the best solution, especially in environments with heterogeneous semantics. Ontologies can capture both the structure and semantics of information environments. An ontology-based search engine can handle both simple keyword-based queries as well as complex queries on structured data. Reconciliation can be accomplished through the use of a global ontology to which the semantics of the individual resources can be related.

Ontologies are described at greater length below. The subsequent chapters describe some emerging approaches that seek to satisfy the above considerations in a decentralized, flexible manner that is compatible with service-oriented computing.

6.3.1   Ontologies: A Definition

An *ontology* is a computational model of some portion of the world. It is often captured in some form of a *semantic network*—a graph whose nodes are concepts or individual objects and whose arcs represent relationships or associations among the concepts. The network is augmented by properties and attributes, constraints, functions, and rules, which govern the behavior of the concepts.

Formally, an ontology is an agreement about a shared conceptualization, which includes frameworks for modeling domain knowledge and agreements about the representation of particular domain theories. Definitions associate the names of entities in a universe of discourse (for example, classes, relations, functions, or other objects) with human-readable text describing what the names mean, and formal axioms that constrain the interpretation and well-formed use of these names.

For individual information systems, or for the Internet at large, ontologies can be used to organize keywords and database concepts by capturing the semantic relationships among the keywords or among the tables and fields in a database. The semantic relationships give users an abstract view of an information space for their domain of interest.

6.3.2   A Shared Virtual World

How can such an ontology help our software agents? It can provide a shared virtual world in which each agent can ground its beliefs and actions. When we talk with a human travel planner, we rely on the fact that we all live in the same physical world containing planes, trains, and automobiles. We know, for example, that a Boeing 777 is a type of airliner that can carry us to our destination.

When our agents talk, the only world they share is one consisting of bits and bytes— which does not allow for a very interesting discussion. An ontology (see Figure 6.3) gives the agents a richer and more useful domain of discourse. A communication protocol specifies the syntax but not the semantics of a message. However, a more flexible protocol would allow the agents to state which ontology they are presuming as the basis for their messages.

Suppose our agent interacts with our travel planner's agent. Suppose both agents have access to a common ontology for travel (Figure 6.3), and suppose their agent tells our agent about a flight on a Boeing 777. Suppose further that the concept "Boeing 777" is not part of the travel ontology. How could our agent understand? The travel planner's agent could explain that a Boeing 777 is a kind of "Commercial Transportation Device," which is a concept in the travel ontology. Our agent would then know the general characteristics of a Boeing 777. The relationships from the common ontology to the local representation of a service consumer or provider are termed *articulation axioms* or *mappings*. These axioms help us infer how a term, e.g., Airliner, used in the schema of a service corresponds to another term, e.g., Airplane, used in the schema of an application.

Other categories and examples of mappings from our wire example are given below. Here $O1$ and $O2$ are the ontologies whose concepts are being compared and related. We can imagine that each of the ontologies is for an application or information resource.

- Term-to-term (one-to-one), e.g.,
  hookupWire$_{O1}$ ≡ wire$_{O2}$

- Many-to-one, e.g.,
  solidWire$_{O1}(x, size, color)$ ∧ strandedWire$_{O1}(x, size, color)$
  ≡ wire$_{O2}(x, size, color, (Stranded|Solid))$

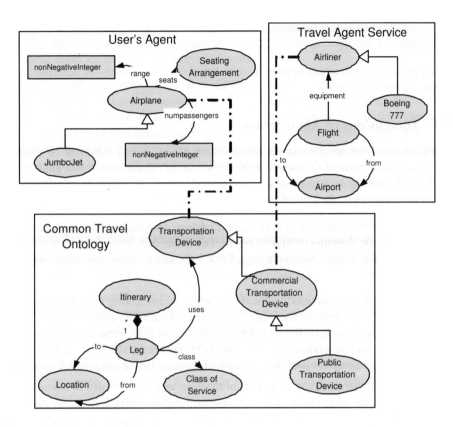

Figure 6.3: Ontologies and articulation axioms. Here a service user's application and a service provider's information source are related to a common ontology (dotted lines represent articulation axioms). Consequently, their schemas become comprehensible to each other

- Many-to-many, e.g.,
  $\mathsf{solidBlueWire}_{O1}(x, size) \wedge \mathsf{solidRedWire}_{O1}(x, size) \wedge$
  $\mathsf{strandedBlueWire}_{O1}(x, size) \wedge \mathsf{strandedRedWire}_{O1}(x, size)$
  $\equiv \mathsf{solidWire}_{O2}(x, size, (Red|Blue))$
  $\wedge \mathsf{strandedWire}_{O2}(x, size, (Red|Blue))$

Finding such mappings is difficult, and there have been many research attempts to provide automated help. All of the attempts extend one of the following two basic approaches:

- *Extensional approach.* If there are a number of instances of predicate assertions from different ontologies, as might be obtained from a database, then clustering can be used to identify and match enumerated predicate terms.

- *Intensional approach.* Substring matching can be used on predicate names, where

Table 6.2: Grading schemes used by five leading insurance company rating agencies in the USA [Wiegold, 1997, page 134]

A. M. Best	Duff & Phelps	Moody's	Standard & Poor's()	Weiss
A++	AAA	Aaa	AAA	A+
A+	AA+	Aa1	AA+	A
A	AA	Aa2	AA	A−
A−	AA−	Aa3	AA−	B+
B++	A+	A1	A+	B
B+	A	A2	A	B−
B	A−	A3	A−	C+
B−	BBB+	Baa1	BBB+	C
C++	BBB	Baa2	BBB	C−
C+	BBB−	Baa3	BBB−	D+
C	BB+	Ba1	BB+	D
C−	BB	Ba2	BB	D−
D	BB−	Ba3	BB−	E+
E	B+	B1	B+	E
F	B	B2	B	E−
	B−	B3	B−	F
	CCC	Caa	CCC	
	DD	Ca	R	
		C		

business reasons to relate the different ratings, e.g., to build consolidated lists of insurance companies that have been rated by one or more of the rating agencies. How can we relate the ratings of the different agencies?

Notice that Duff & Phelps and Standard & Poor's (S&P) use sets of grades that are almost identical (the only difference is in the lowest grade), but that does not entail that the meanings of the grades (i.e., the reasoning behind the grades) are in agreement. However, it might be reasonable to assume that the agencies are referring to the same inherent concept of reliability. Thus we can try to construct a mapping among the different rating schemes. In the case of Duff & Phelps and S&P, a straightforward mapping may appear reasonable and would be technically obvious.

However, when the agencies use different numbers of grades, a trivial mapping between their grades is not possible. Interpreting the ratings as recommendations, we can infer at best that any acceptable mapping will be order-preserving.

This discussion can be cast more generally in the form of value sets and value mappings. A grading scheme in the above sense is a kind of value set. To formalize this concept, let each value set be defined as a pair $\langle V, \prec \rangle$, where V is the set of values allowed in the given set and $\prec$ is the strict partial order among the values. Then a mapping m_{ab} from a value set

$\langle V_a, \prec_a \rangle$ to another value set $\langle V_b, \prec_b \rangle$ is given by a (partial) function from V_a to V_b. Since a mapping is a function, it must be unambiguous, meaning that a value from one set may be mapped to no more than one value in the other set.

Typically, for a mapping m_{ab}, there will also be an inverse mapping from V_b to V_a, which we can notate m_{ba}. A mapping and its inverse would generally need to be designed together.

A mapping m_{ab} and its inverse m_{ba} may be subject to the following properties:

Totality. $(\forall v \in V_a : m_{ab}(v) \in V_b)$. We would usually require totality to ensure that we could map any value that came about from the given value sets. Unless the value sets in question had the same cardinality, both a mapping and its inverse would not be total unless at least one of the mappings was not an injective (i.e., not a one-to-one) function.

Order preservation. $(\forall v_1, v_2 \in V_a : v_1 \prec_a v_2 \Rightarrow m_{ab}(v_1) \preceq_b m_{ab}(v_2))$. Notice that the consequent uses $\preceq$ instead of $\prec$, because the mapped values may need to coincide— especially when the target value set has a smaller cardinality, but even otherwise.

Consistent inversion. $(\forall v \in V_a : m_{ab}(m_{ba}(m_{ab}(v))) = m_{ab}(v))$. That is, a mapping should be such that the inverse of the inverse of the result of mapping a value is the same as the result of mapping a value. In other words, the inverses cancel out when applied to the result of mapping a value. Notice that, in general, we *cannot* require $(\forall v \in V_a : m_{ba}(m_{ab}(v)) = v)$, because often we may wish to map two or more values to the same target value; thus the inverse of a mapped value may not be the original value. That is, a mapping and its inverse do not necessarily cancel out.

The above definitions are stated for one direction of the mapping, from value set $\langle V_a, \prec_a \rangle$ to value set $\langle V_b, \prec_b \rangle$. In general, we would need to ensure explicitly that the necessary properties hold in both directions. When each of a pair of mappings satisfies consistent inversion, then we say that the mappings are *consistent inversions of each other*. Consistent inversions do not entail order preservation.

Figure 6.4 shows possible value maps between the ratings of A. M. Best and Moody's. Although the two value maps are similar, they have a subtle difference. If you compare the mappings from Moody's to A. M. Best, you will notice that, in Figure 6.4(a), Aaa and Aa1 are mapped to A++, Aa2 to A+, and Aa3 to A. By contrast, in Figure 6.4(b), Aaa is mapped to A++, Aa1 to A+, Aa2 to A, and Aa3 to A−. This is only a slight shift in the mappings. Why is it potentially important? This shift causes the map of Figure 6.4(b) to violate the property of consistent inversion in each direction. When that property is violated, it means we cannot partition the values into those that map well to each other. Exercise 6.8 asks you to study these value maps more closely.

Notice that just because two value sets have the same cardinality does not mean that we must map their values sequentially. Figure 6.5 shows a less obvious value map between two of the grading schemes introduced above. This particular value map is not parsimonious. However, it can be justified as follows. Imagine that the S&P grading scheme corresponds to the natural numbers 1, 2, and so on, whereas the Duff & Phelps grading scheme corresponds (in the same order) to the following real numbers: 1, 1.4, 3, 3.4, and so on. Now, going from

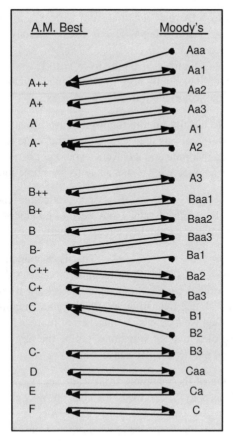

(a) A consistent value map

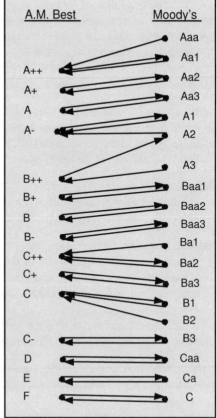

(b) A value map that violates
consistent inversion

Figure 6.4: Consistent and inconsistent value maps between A. M. Best and Moody's ratings (note the changed mapping for A. M. Best's B++ rating

Duff & Phelps to S&P, the best match for AAA (i.e., 1) and AA+ (i.e., 1.4) would be AAA (i.e., 1), and nothing would map to AA+. However, in the reverse direction, we would need to map AA+ (i.e., 2) to something, and the the best match for it would be AA+ (i.e., 1.4).

The value sets of Table 6.2 are all finite. In general, this need not be the case. Value sets may be infinite, in which case they may be modeled as discrete and unbounded on one end (like the natural numbers), discrete and unbounded on both ends (like the integers), dense (like the rational numbers), or continuous (like the real numbers). The above assumes that the value sets use total ordering relationships. In general, if a value set represents data pairs, then the ordering relationship defined for it would not be total. For instance, if a service rates

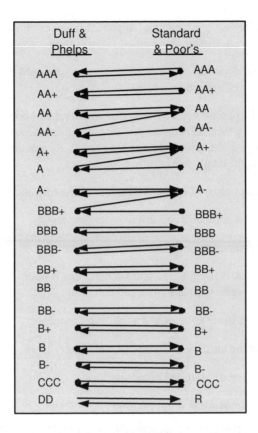

Figure 6.5: An example of an unintuitive value map between the grading schemes of Duff & Phelps (D & P) and S&P. Notice the interesting pattern

insurance companies with separate grades for their claims-paying ability and the friendliness of their customer response, the value set for that service would not be totally ordered in a natural manner.

6.4 Knowledge Representations

Many efforts are underway to devise classification schemes and to use the schemes to build and populate classification structures. The following list includes four main types of classification schemes of varying power that provide semantics for messages among services or agents. Each scheme has particular strengths and weaknesses, and provides a foundation upon which particular capabilities can be built.

Keywords. Keywords are a quick way for agents to locate potentially useful information.

Thesauri. Thesauri offer a more structured approach than keywords, arranging descriptive terms into *broader*, *narrower*, and *related* classification categories.

Taxonomies. Taxonomies provide classification structures that add the power of inheritance of meaning from generalized *taxa* to specialized *taxa*.

Ontologies. Ontologies permit a richer variety of structural and nonstructural relationships than taxonomies, which are limited just to generalization. Ontologies provide more complete and precise domain models as are needed, for example, by software applications that implement intelligent information services.

There are additional purposes for developing classifications for concepts. Among them are:

- helping users find a particular concept from among many;

- facilitating the administration of information systems;

- through inheritance, conveying semantic content that is often only incompletely specified by other attributes, such as names and definitions;

- deriving and formulating abstract and application concepts;

- ensuring appropriate attribute and attribute-value inheritance;

- deriving names from a controlled vocabulary;

- disambiguating communicated information;

- recognizing superordinate, coordinate, and subordinate concepts;

- recognizing relationships among concepts;

- assisting in the development of modularly designed names and definitions.

Chapter 8 discusses some widely used representation languages for ontologies.

6.4.1   Relationships Represented

Most ontologies represent and support relationships among classes. Among the most important of these relationships are the following:

Generalization and inheritance are powerful abstractions for sharing similarities among classes while preserving their differences. Generalization is the relationship between a class and one or more refined versions of it. Each subclass inherits the features of its superclass, adding other features of its own. Generalization and inheritance are transitive across an arbitrary number of levels. They are also antisymmetric.

Aggregation is the part–whole or part–of relationship, in which classes representing the components of something are associated with the class representing the entire assembly. Aggregation is also transitive, as well as antisymmetric. Some of the properties of the assembly class propagate to its component classes.

Instantiation is the relationship between a class and each of the individuals that constitute that class.

Some of the other relationships that occur frequently in ontologies are owns, causes, and contains. Causes and contains are transitive and antisymmetric; owns propagates over aggregation, because when you own something, you also own all of its parts.

6.4.2 Frames versus Descriptions

We now discuss two main approaches behind ontologies. The approaches are appropriate for an intensional view, whereby a class or concept is defined by a set of either membership conditions or properties.

Frames directly express knowledge in terms of graphs. They involve defining frames (a structured object corresponding to classes or instances) and relating them explicitly to other frames through labeled edges. The definition of a frame is expressed in terms of a set of properties and their allowed values.

Description logic is a family of languages that formally express certain constraints on knowledge representation. Since they have a precise semantics and axiomatization, they are amenable to automatic processing in a manner that is unambiguous across implementations. Description logic begins with primitive concepts and defines further concepts in terms of formal descriptions. Concepts are computed from these descriptions. Subsumption, that is, the specialization-generalization hierarchy among concepts, can be determined from the descriptions. The potential advantage of description logic is that it can help determine if a concept is redundant or how it relates to other concepts.

Frame representations are intuitive for people to build and do not rely on any special mathematical training. However, such representations provide no principled basis for inferring relationships between frames. Frame representations provide no intrinsic account of meaning, but rely on names of classes and properties to indicate meaning. This may be natural for humans, but can introduce problems when used computationally.

Description logic, by contrast, is difficult for people. Often, descriptions are created only to capture a frame hierarchy that a person has in mind. Managing the trade-off between ease of use and rigor is a major challenge for knowledge representation.

The above are both what might be termed discrete and vivid approaches to knowledge representation. Yet another alternative for knowledge representation is a connectionist approach, which stores knowledge in terms of (internal) weights on the edges of a graph of concepts. Such approaches keep their knowledge implicit. In general, such representations are difficult to engineer and particularly difficult to share and reconcile with other representations.

has been the most successful to date. In demonstrations, PSL has enabled descriptions of processes that are operating on different CAD/CAM tools to be related, even if the individual descriptions are encoded in different languages. The resultant processes can then be made to work in concert. Chapter 13 addresses process modeling in general.

6.5   Elementary Algebra: Relations

There are two main examples of hierarchies in knowledge representation. One, class or inheritance hierarchies correspond to the *isA* relation, where a class extends another class. Two, part–whole hierarchies correspond to the *isPartOf* relation, where a class aggregates other classes, i.e., instance of a class are said to be parts of the instances of another class.

Relationships in ontologies are naturally modeled as binary relations. For example, a class being a subclass of another is a binary relationship. Other domain-specific relationships or properties can be captured as binary relations. For example, a student taking a course expresses a relationship (taking) between student and course.

A binary relation R between a set S_d and a set S_r relates zero or more members of S_d with zero or more members of S_r. That is, R can itself be modeled as a set of pairs, each of which consists of a member of S_d and a member of S_r. Recall that $S_d \times S_r$ refers to the Cartesian product of S_d and S_r, meaning the set of all possible pairs whose first component is drawn from S_d and whose second component is drawn from S_r. Formally, we can write $R \subseteq S_d \times S_r$. A binary relation naturally corresponds to a graph as shown in Figure 6.6.

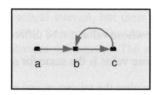

Figure 6.6: A binary relation on a set corresponds to a graph whose vertices are the underlying set and whose edges represent the relation instances. Here the set is $\{a, b, c\}$, and the relation is $\{(a, b), (b, c), (c, b)\}$

Often properties that indicate binary relationships occur in pairs, indicating the polarity of the relationship. For example, if c_1 is a subclass of c_2, then c_2 is a superclass of c_1. Likewise, if students take courses, courses are taken by students: i.e., "take" and "taken by" exhibit reverse polarities. Such pairs of relationships are called *inverses* of each other. A relationship may be explicitly documented as being the inverse of another. A relationship is much like its inverse, except that their domains and ranges are interchanged. Thus any instance of one corresponds to exactly one instance of the other. Formally, a binary relation $R^{-1} \subseteq S_r \times S_d$ is defined as an *inverse* of R if and only if the following holds: $(\forall d \in S_d, r \in S_r : (d, r) \in R \equiv (r, d) \in R^{-1})$.

A *partial order* is a binary relation for which the following properties hold. Partial orders are frequently notated via an infix operator such as $\prec$ and are termed *precedence* relations. That is, let us write $x \prec y$ to mean that x relates to y via the given relation.

- *Antisymmetry.* If $x \prec y$ and $y \prec x$, then $x = y$. In other words, we cannot have two distinct objects such that each precedes the other. However, it is acceptable (but not required) for an object to precede itself.

- *Transitivity.* If $x \prec y$ and $y \prec z$, then $x \prec z$. If x precedes y and y precedes z, then x precedes z.

In addition, some partial orders can also satisfy asymmetry, which also forces them to satisfy irreflexivity:

- *Asymmetry.* If $x \prec y$ then $y \nprec x$. Asymmetry is a stronger form for antisymmetry, because it forbids two objects (whether the same or distinct) from preceding each other. That is, an object cannot precede itself.

- *Irreflexivity.* $x \nprec x$. This simply states that an object cannot precede itself. Irreflexivity is entailed by asymmetry.

Lastly, a *total order* or a *linear order* is a partial order that also satisfies the linearity (also known as totality) property:

- *Linearity.* $x \prec y$ or $y \prec x$ or $x = y$. This states that for any two distinct objects, one must precede the other. That is, for any two objects, the ordering relation must hold one way or the other.

6.6 Hierarchies

A hierarchy is naturally modeled as a binary relation that is a *partial order*.

6.6.1 Taxonomy

The *isA* relation is interpreted to indicate subclassing and it defines a *taxonomy*. For example, a human *isA* mammal. The *isA* relation is antisymmetric and transitive. The combination of antisymmetry and transitivity means that if you ever have a cycle of hierarchic relationships, then all the classes that occur on the cycle are equal.

6.6.2 Meronomy

The *isPartOf* relation means that the given individual is part of another. For example, an engine *isPartOf* an automobile. The *isPartOf* relation applies to instances and forms a part-whole hierarchy or *meronomy*. Meronomic hierarchies are also partial orders in the above sense. However, we would not encounter any cycles in a correct part-whole hierarchy, i.e., an object would not be a part of itself. For this reason, meronomic hierarchies are *asymmetric* and *irreflexive*.

Meronomic hierarchies are not considered a native part of most existing ontology languages. In these languages, if needed, the part-whole hierarchy can be captured via an explicit property. But, for such added properties, no special mathematical constraints can be automatically assumed and there would be no tool support available. In other words, the modelers would be on their own in ensuring correctness for such constructs. The lack of support for part-whole representation and reasoning is a significant limitation of current ontology languages.

6.7   Modeling Fundamentals

A conceptualization consists of the following components:

- a universe of discourse, which is the set of entities under consideration;

- concepts that identify sets of entities;

- relationships among these entities;

- functions that map from entities to entities.

Conceptual modeling languages are languages that support the above components. This holds whether a given language is designed to model information systems or real-world processes. That is, a conceptual modeling language would include symbols that may be associated in different ways with entities, concepts, relationships, and functions.

Associating a conceptualization with the "real world" gives semantics to statements in a conceptual modeling language. Such an association of symbols to the world is termed an *interpretation*, and an interpretation characterizes the meaning of a statement in the given conceptual modeling language. Typically, we are interested in the semantics of the language in general from which the semantics of its statements can be inferred. To capture the semantics of a language, we would nail down some, but not all, details of its intended interpretations. For example, a conceptual modeling language for document structure might fix the meanings of symbols such as "chapter," "section," "title," and "author," but let the other symbols in a statement be given meanings according to the particular needs of an application.

Modeling languages must provide explicit representations for *concepts*, which can be real or imaginary, tangible or intangible, and concrete (actual) or abstract (prototypical). For a few simple examples of concepts, your car is real, tangible, and concrete; a Porsche 911S is real, tangible, and abstract; your last vacation is real, intangible, and concrete; and a unicorn is imaginary, tangible, and abstract.

Second, modeling languages must provide means for representing *relationships* among the concepts. The representations might involve such relationships as ownership, causation, generality, instantiation, and meronomy (part-whole). For example, you *own* your car, a steering wheel *is part of* a car, and a Porsche *is a kind of* sports car.

Third, models can include *constraints* on the relationships that can occur among the concepts. The constraints might involve cardinality, temporal existence, business rules, or higher-order functions. Some examples are: your car has four wheels; in most states in the USA, a person should be older than 16 years to obtain a driver's license; and the perimeter of a square is four times the length of its side.

Finally, the planned uses and applications of the model determine the concepts that must be represented. Consider the following different conceptualizations of a piece of electrical hook-up wire that is solid (versus stranded), is uniquely labeled ID5, has size 22 in the American Wire Gauge (AWG), and has blue insulation:

- awg22SolidBlueWire(ID5)

- blueWire(ID5, AWG22, Solid)

- solidWire(ID5, AWG22, Blue)

- wire(ID5, AWG22, Solid, Blue)

Here ID5 is a symbol whose interpretation is an entity in the universe of discourse. Predicates such as awg22SolidBlueWire are interpreted as the appropriate concepts. Let us hold off discussing the other symbols until Section 6.7.3. The first conceptualization, a unary predicate, simply tests if the entity denoted by ID5 belongs to the concept denoted by awg22SolidBlue. In the second conceptualization, the wire core type is extracted from the predicate name and added to the symbols representing objects in the universe of discourse; in the third the wire color is extracted. The fourth conceptualization is the most flexible, since it allows the properties of the wire to be explicitly expressed as separate arguments. That is, the concepts of wire size, wire type, and color are elevated from being "bound" in predicated names to representing entities in the universe of discourse. The entities can then be reasoned about, whereas concepts embedded within predicate names cannot.

Attributes that can be enumerated (e.g., color, type of wire core, and the small range of integer wire sizes) can possibly be included in predicate names. Nonenumerated attributes are never good candidates for inclusion in predicate names. For example, we would not expect to use a predicate such as blue13.245ftWire for what happens to be a 13.245 foot length of wire. We would need uncountably many predicates just to capture the possible real number lengths of wires. Section 6.7.3 discusses other variations and aspects of the above example.

6.7.1   Perspectives for Conceptualization

Even if you settle upon the most flexible of the styles of conceptualization, there are a number of choices for one. How do you choose the best conceptualization? It would be dependent on your application's view. For example, the following views might be reasonable when talking about electrical wires.

- Design view: the color and core type of a wire are not important. Wire size and breakdown voltage of the insulation, however, are important considerations.

- Manufacturing view: the color and core type of a wire are important. Breakdown voltage is not important, if the proper wire has already been selected.

- Sales view: the number of feet on a spool and shipping weight are important. The above discussion did not mention weight, so it would have to be added to the conceptualization to accommodate the sales view.

In other words, not every conceivable property need be represented explicitly, and thus some properties would need to be added as the uses of the conceptualization develop.

6.7.2   Guidelines for Conceptualization

The best overall conceptualization is one that obeys the following guidelines:

- A concept must have instances, directly or through some subconcepts.

- A concept must contain all properties common to the instances in its extension.

- Classification should obey *cognitive economy*—instances of a concept must share some, but not all of their properties. Otherwise, if two instances were at all distinguishable based on their properties, then they would have to belong to different concepts, leading to a proliferation of concepts.

- Classification should enable inference of properties based on the membership of an instance in a concept.

- Restrictions on concept relationships:

  - *Completeness*: every property must be used in the definition of at least one concept.
  - *Nonredundancy*: a subconcept must be defined by at least one property not in any of its superconcepts. The result is that a subconcept is always a specialization of any of its superconcepts, i.e., it has more properties or restrictions, and has fewer instances.

However, even if a rich conceptualization such as

$$\mathsf{wire(id, size, solidity, color)}$$

is a part of a model, interoperation with systems modeled by other conceptualizations might be desired, which would require extensive reasoning to map between them. A challenge is how do we perform the mappings if the conceptualizations are different? That is, what must we do to go beyond simple term matching schemes, such as hookUpWire in model 1 matches wire in model 2? For example, suppose there are two conceptualizations for a two-dimensional point:

- *Point(x, y)*, based on a rectangular coordinate system.

- *Point(radius, theta)*, based on a polar coordinate system.

Section 6.7.3 considers interoperation and mappings in more detail.

6.7.3 Modularity and Extensibility

We discussed how to map the symbols ID5 and wire. A problem lurking behind the above approach is how to map the remaining symbols. For instance, to what should Blue be mapped? It is tempting to map it to some set of entities in our universe of discourse, namely, the set of entities that are blue. In other words, it would then be a concept, i.e., have the same kind of denotation as the denotation of a unary predicate. In that case, placing it as an argument to another predicate seems suspect. An alternative is that Blue be mapped to an entity, which means that we have different sorts of entities: concrete objects such as wire as well as abstract objects such as clear. Either of these approaches can work depending on other assumptions in the conceptualization.

The conceptualization described above suffers from one major shortcoming. As we invent new attributes, we would not be able to accommodate them in our conceptualization without continually revising our predicates. For example, if we wish to add the fact that the given wire is heavy, we would have to modify the predicate wire to take an additional argument as in the expression below.

- wire(ID5, AWG22, Solid, Blue, Heavy)

For this reason, a more parsimonious predicate, as in the expression below, proves desirable.

- wire(ID5) ∧ AWG22(ID5) ∧ solid(ID5) ∧ blue(ID5) ∧ heavy(ID5)

Here we conceptualize the main entity as a predicate, but also have separate predicates for each of the attributes that we are interested in. Indeed, this is how natural languages handle complex noun phrases. They use nouns for entities and adjectives for properties of nouns. New attributes can be inserted with as much ease as adding a new adjective to a phrase. Moreover, additional domain-specific constraints can be captured and reasoned about perspicuously. For example, we might state that wires above a certain product of length and thickness are always heavy.

Although it is more modular than the naive approach, the above approach can be improved further. Whenever we declare a predicate such as blue or heavy, we are implying that nothing more can be said about the feature that is reflected in that predicate. A predicate is the end of the road for modeling, as it were. Now if we wanted to capture that the blue was a particular deep shade and was permanent and would not stain the clothes of people handling the blue wires, we would be at a loss. We may generate additional predicates such as permanentBlue and use those predicates instead of blue. But then it would no longer be obvious that a permanent blue wire was a blue wire. The above discussion leads to yet another formulation where the predicates are chosen to represent a given feature concept or attribute, not any of the values that may be assigned to the given attribute. The values are modeled as objects and further assertions can then be made about them. The following formulation illustrates this for our example, where we have added that the Blue in question is permanent.

- wire(ID5) ∧ thickness(ID5, AWG22) ∧ type(ID5, Solid) ∧ color(ID5, Blue) ∧ weight(ID5, Heavy) ∧ permanent(Blue)

An analogous situation occurs when modeling activities or processes. For example, we may wish to model a transaction, such as a payment, being performed. As above, we might encode the amount of the payment, its currency, its mode (cash or check), and so on, into the predicate name. For the same reasons as above, such an approach is inappropriate because it hides semantically relevant properties into a text label. As before, an approach that exhibits *reification* by upgrading the important attributes into entities improves reasoning about them. Further, an approach that captures separate predicates that associate different attributes with the payment provides the same kind of extensibility as above. For example, we may have

- payment(P1) ∧ by(P1, Alice) ∧ to(P1, Bob) ∧ value(P1, 5) ∧ currency(P1, USD) ∧ mode(P1, cash) ∧ on_time(P1)

Now we can capture additional constraints such as that payments below a certain amount must be cash. In our natural language analogy, activities are verbs (e.g., P1, but read on), the objects they involve are nouns (e.g., Alice, Bob, 5, and USD), the properties of the objects are adjectives (none here), and the properties of the activities themselves are adverbs (e.g., on_time and mode). In conceptual terms, there is often a thin line between different parts of speech such as verbs and nouns, and adjectives and adverbs. That is, we can easily write equivalent natural language sentences that present the same idea using differing parts of speech. Indeed, in the above example P1 is a way of thinking of a payment as a noun, although in the real world, making a payment is an activity. This phenomenon is called *nominalization* and is common in natural languages. It corresponds to the use of *gerunds*—simply, verbs that act as nouns—in English. In our example, it is as if to capture that "Alice paid Bob USD five in cash on time" we are saying "there was a paying (a payment event), whose actor was Alice, whose beneficiary was Bob, which involved the USD currency and an amount of five; moreover, this paying was in cash and was on time." Here, "paying" is a gerund. The long-winded formulation has the advantage that it is modular and extensible. That is, we can readily question or deny specific components of it and we can stick more facts into it should we need to.

The moral of the above discussion is that the more explicit we can make our formulation, and doing so in a manner that facilitates inferencing about the representations, the better it is in terms of extensibility (adding more facts) and flexibility (adding features that further finesse the properties of interest).

6.8   UML as an Ontology Language

The Unified Modeling Language is a language for building conceptual models. Although targeted originally towards software development, and especially *visual* modeling, it has the constructs needed to specify ontologies. Table 6.3 summarizes the basic features provided by UML for modeling objects in ontologies. As can be seen there, although it does not provide a description logic, it is richer than OWL in some ways. Moreover, it enables a form of aggregation to be captured directly and has a means for modeling activities. Overall, UML would be a credible ontology language for many practical circumstances, as we briefly describe next.

UML provides five different architectural views of a system (use-case, design, process, implementation, and deployment), but the most relevant for ontology development is the design view. This view enables both the static and dynamic properties of a system to be described. We address the static aspects here, and the dynamic aspects in Section 13.2.

Many of UML's structures for supporting the static aspects of a domain are exemplified in the small ontology shown in Figure 6.7. In particular, showing aggregation, the parts of a purchase order are item details, shipping details, and billing details. Buyers can be either organizations or persons, showing specialization and generalization, and organizations can be both buyers and sellers, showing multiple inheritance. There is an association between a buyer and an account, and they are both parts of billing details. Although the textual serialization of a UML diagram captures the graphical layout well, it is cumbersome and generally unsuitable for the automated understanding of the diagram's concepts.

6.9 Alternative Terminology

Variations of the above terminology also exist in the literature. An entity is a class or an individual. *Classes* are also called *concepts* or *categories*.

Members or *instances* of typical classes are *individuals*.

Classes can be members of classes, especially of predefined classes such as owl:Class.

Properties are also called *slots* (in frame languages) and *roles* (in description logic).

Restrictions on properties are called *facets* or *role restrictions*.

6.10 Notes

An idea similar to object graphs mapping to objects, which map to XML documents is developed in Melnik and Decker [2000] in the context of XML-based communication. Our discussion of language features is based on Melnik and Decker [2000] and Gomez-Perez and Corcho [2002].

Protégé is an ontology tool available for free download from http://protege.stanford.edu.

6.11 Exercises

6.1. The Semantic Web vision calls for ontologies to (choose one):

- be created incrementally;

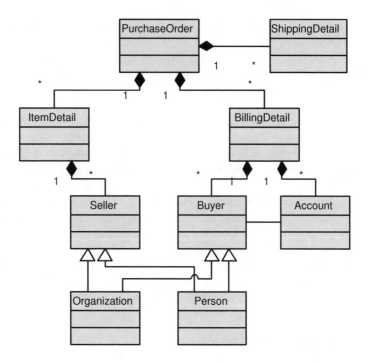

Figure 6.7: The Unified Modeling Language, with its support for aggregation, generalization, association, and inheritance, can be used to model a domain and depict an ontology

- reside in a central location;
- go through a standardization process;
- be defined with any encoding language available to the user;
- obey a top-level ontology.

Discuss and provide a justification for your answer.

6.2. Which one of the following statements about Semantic Web documents is not true?

- They can reference terms without an ontology.
- They are written in XML.
- They are written in RDF.
- They can reference many ontologies.
- They can define new ontologies.

Justify your answer.

6.3. Consider the case of software development, involving software systems, project super-visors, and project teams. The following knowledge is given:

- Each supervisor can be a member of several project teams.
- Each supervisor can supervise several project teams.
- Each software system is developed by exactly one project team.
- Each project team develops exactly one software system.
- Each supervisor can manage several software systems.
- Each project team has exactly one supervisor.

(a) Using a tool such as Rational Rose or Visio, construct a UML representation corre-sponding to the above description.

(b) In the above description, one of the relationships involving software systems is redundant and can be removed. Identify it, and justify your choice.

6.4. Consider the problem of managing the waiting lists for courses in an academic depart-ment. Construct a UML model for a system that could help an administrator in this task. Your system should enable a student to be added to a waiting list for a course. Students should be able to check their place in a list. The administrator should be able to find the name of the student who is first in a list and add that student to a course. Professors should be able to find out how many students are waiting to be added to their course. Construct a use-case diagram for this system. Then construct an interaction diagram, state-machine diagram, and class diagram for it. Use a tool such as Rational Rose or Visio to generate these diagrams.

6.5. Consider the problem of managing the computer systems in your department. Con-struct a UML model for a system that could help your computer systems administrator in this task. Computers are either OK, broken, or being repaired. Your system should enable a student to report a broken computer. Students and staff should be able to check which computers are not broken. The computer administrator should be able to change the status of a computer to OK after it gets repaired. The maintenance staff should be able to find which computers are broken so that they can begin repairing them. Construct a use-case diagram for this system. Then construct an interaction diagram, state-machine diagram, and class diagram for it. Use a tool such as Rational Rose or Visio to generate these diagrams, and turn in the results.

6.6. A rectangular two-dimensional object can be represented by a set of line segments or a set of boxes, as indicated in Figure 6.8. Fill in the following table by indicating which representation is easier or harder for calculating perimeter and area, and then define a mapping that could be used to interoperate between conceptual models based on each.

Chapter 7

Resource Description Framework

The previous chapter argued that we need deeper representations of meaning so as to enable interoperation among services. These representations require us to employ more sophisticated languages than XML. Why do we need another language? It is simply because we need to express knowledge structures that have a mathematically well-formed semantics, and which our computational tools can process in a well-defined predictable manner.

However, this is not to suggest that we need a syntax other than that of XML. Any language that we come up with will be expressible in XML. This is because computational (or, for that matter, logical) languages are given context-free syntactical specifications, typically in a notation such as the Backus Naur Form (BNF). XML is adequate to express any such syntax: it is just not adequate to capture the semantics. In other words, XML is expressively complete in one sense (syntax) and expressively limited in another sense (semantics).

It is the latter limitation that drives the development of additional languages. The reason we need other languages is that XML offers no special constraints on the meanings associated with the specific constructs of the given languages. What other languages offer are higher-level abstractions. These abstractions streamline certain aspects of representations, in essence standardizing the range of meanings that can be expressed. Thus, instead of another *ad hoc* encoding of knowledge via XML, we would use a high-level language that encodes the desired knowledge in a manner that can be understood by all. Thus each language enables two major productivity enhancements:

- Service descriptions created by one party become readily comprehensible to another party, because the standard semantics for the given language substitutes for the out-of-band communication between the interacting parties that would otherwise be necessary.

- Tools can be built that accommodate the given language at its level of abstraction, thus enabling more efficient creation, validation, and processing of documents in the given language.

The first language we consider is the Resource Description Framework (RDF). RDF provides an ability to define graph-like structures; thus, any discrete knowledge structures can be encoded in RDF. However, the encodings would need to be invented for each case. The RDF Schema solves this problem by providing the abstractions with which new vocabularies can be readily described. Next, the Web Ontology Language (OWL) proposes a specific vocabulary that provides selected frame and description logic primitives to capture ontologies.

7.1 Motivation for RDF

Consider a simple XML document that purports to describe a purchase order for two wires, both of gauge AWG22, blue, and stranded. Wire lengths are specified as subelements.

Listing 7.1: An example purchase order in XML

```xml
<purchase buyer='Alice'>
   <price>50</price>
   <date>31 December 2004</date>
   <wire gauge='AWG22' color='blue' type='stranded'>
     <length>12.5</length>
     <length>6.25</length>
   </wire>
</purchase>
```

Knowing or guessing the meanings of the English words used to name the various elements and attributes, we can infer the meaning of the document. However, the inherent structure of the document is no more than the parse tree to which it corresponds. In other words, the structure is not expressive of the content of the document. Indeed, there might be another representation, which is syntactically quite different from the above and yet somehow closely related in terms of meaning. For example, consider the following encoding for the above purchase order.

Listing 7.2: An alternative XML representation of a purchase order

```xml
<purchase buyer='Alice'>
   <price>50</price>
   <date>31 December 2004</date>
   <wire gauge='AWG22' color='blue' type='stranded'>
     <length>12.5</length>
   </wire>
   <wire gauge='AWG22' color='blue' type='stranded'>
     <length>6.25</length>
   </wire>
</purchase>
```

Listings 7.1 and 7.2 yield different parse trees and yet may have the same semantic contents. Clearly, a large number of other variations are possible. Typically, the lower layers of document processing would accommodate some designated set of syntactic variations. We cannot expect all variations to be processed automatically without some effort by a designer. However, we can shift the emphasis from the syntax of the document to the structure of the content that it represents. This is the role of the RDF, which provides us with a standardized means to capture the structure of information, enabling other parties to understand the document's contents.

In simple terms, RDF provides a simple language in which to capture knowledge. RDF incorporates a number of well-known ideas from knowledge representation, but standardizes them and applies them on the Web. (RDF can be applied in settings other than the Web, of course, and often is.) RDF is built on top of the Web notion of a URI, which is described in greater detail in Appendix B. URIs uniquely identify resources. Notice that, because the underlying services are context sensitive, a resource may not be a unique entity and may change its interpretation over time. For example, the page created by a news service or a book catalog service may be a different entity on each invocation, but could still be conceptualized as a single resource. Importantly, URIs are employed in RDF to remove ambiguity about terms and to enable multiple perspectives to coexist. The use of URIs gives RDF additional expressive power.

7.2   RDF Basics

An RDF document is a collection of *statements*, each of which is expressed as a triple involving a *subject*, a *predicate*, and an *object*. The subject and predicate are each a resource, and the object can be a *resource* or a *literal*. Sometimes the predicate is called the property of the given triple. The RDF term for a predicate is rdf:Property. In effect, a statement makes an assertion about its subject, namely that the subject is related to the object through its predicate.

7.2.1   Resources

A resource is something with identity. Resources are identified via URIs or, rather, to be more precise, by *URI References*. Importantly, URIs need not be absolute, in that they need not correspond to the name of any actual object to be accessed via any specific protocol. URLs are a kind of absolute URIs. Moreover, even the absolute URIs need not correspond to a physical object. That is, resources in RDF can be abstract and might not be mapped to a network address. rdf:Resource is the set of resources.

Some useful examples of resources are entities that exist on the Web, such as documents, images, video clips, and, most importantly, services. Resources can also aggregate other resources. Lastly, some resources can reside decidedly outside the realm of the Web. Good examples of such resources are people, social entities, and physical objects.

Resources map conceptually to entities or sets thereof. A resource would be considered unchanging as long as its identity holds, even if the underlying contents change. Intuitively, this is a powerful aspect of having names. For example, General Motors will remain the same conceptual entity even as it changes physically, e.g., as all of its staff retire and are replaced by new people. To resolve a resource involves finding the entity to which it binds currently.

7.2.2   Literals

Besides resources, the other kind of object that RDF deals with are literals. Literals can be expressed explicitly. Plain literals are expressed as a string with an optional language identifier (the language identifier determines which human language is being considered). Typed literals, for which a type is provided, are specified in terms of a lexical representation (a string) and a URI Reference to the XML Schema describing the type.

7.2.3   Properties

RDF properties are two-place predicates. That is, they are resources that are interpreted as predicates. Predicates with two arguments form statements. The first argument of the predicate becomes the subject and the second argument the object of the given statement.

7.2.4   Statements

An RDF statement corresponds to a logical assertion based on a two-place predicate whose first place is taken by the subject and second place by the object. For example, the notation in Listing 7.3 expresses several statements, namely, that "SOC is a resource, whose title is Service-Oriented Computing, whose creators are Munindar and Michael, and whose publisher is Wiley."

Listing 7.3: An example RDF snippet

```
<?xml version='1.0' encoding='UTF-8'?>
<rdf:RDF
    xmlns:rdf="http://www.w3.org/1999/02/22-rdf-syntax-ns#"
    xmlns:dc="http://purl.org/dc/elements/1.1/">
    <rdf:Description rdf:about="http://www.wiley.com/SOC">
      <dc:title>Service-Oriented Computing</dc:title>
      <dc:creator>Munindar</dc:creator>
      <dc:creator>Michael</dc:creator>
      <dc:publisher>Wiley</dc:publisher>
    </rdf:Description>
</rdf:RDF>
```

This listing illustrates the standard XML serialization or rendering for RDF. Section 7.4 presents an additional discussion of rendering RDF documents in XML.

7.3   Key Primitives

RDF is intended for capturing information about resources. To this end, it defines primitives for identifying and referring to resources in various ways. The main concept is that of a resource identifier, expressed as rdf:ID. The rdf:ID attribute comes with a constraint, namely, that two or more elements in an RDF document cannot have the same value for it (unless they use different base URIs: see below).

The rdf:ID values are interpreted as fragment identifiers relative to the base URI of the current scope. This means that an rdf:ID value, when prepended with a #, corresponds to a relative URI, which when appended to the base URI yields an absolute URI.

The rdf:about attribute functions quite like the rdf:ID except that it takes a URI as a value. The URI can be relative (if it begins with #) or absolute. However, the intent behind rdf:about is different from that behind rdf:ID. Whereas rdf:ID introduces a resource being defined, rdf:about refers to a resource defined elsewhere.

The main primitive predicate that RDF includes is rdf:type. This is used to assert that its object is the type of its subject. In effect, the object is a class and the subject is an instance of that class.

Still other primitives deal with containers and the reification of statements and are discussed next.

7.3.1   Containers and Collections

RDF statements involve exactly three components. Therefore, an additional mechanism is needed to make assertions that involve aggregations of various kinds. For this purpose, RDF provides a means to define structured *containers* of resources. Containers enable the representation of a relationship where more than one resource may be required to participate.

Bag. An rdf:Bag is an unordered collection. As usual, a bag is like a set, except that it may have duplicate members. An example of a bag follows:

```
<rdf:Bag ID="group1">
  <rdf:li >One</rdf:li >
  <rdf:li >Two</rdf:li >
  <rdf:li >Three </rdf:li >
</rdf:Bag>
```

Seq. An rdf:Seq is an ordered collection, which, being a sequence, obviously may have duplicates.

```
<rdf:Seq ID="sequence1">
  <rdf:_1 >First </rdf:li >
  <rdf:_2 >Second </rdf:li >
  <rdf:_3 >Third </rdf:li >
</rdf:Seq>
```

terminate on the same vertex.

Thus it might seem that a graph in general cannot be represented as a tree. However, it turns out that representing a graph as a tree is quite straightforward, provided the vertices of the tree can be referred to from other vertices. Such references must be symbolic, because a tree will not allow multiple edges that terminate in a given vertex. The solution is to label the multiply referenced vertices uniquely within the graph and to refer to them as needed. RDF provides the necessary primitives to label and refer to vertices.

- rdf:Description is the main element used to capture statements about resources. To serialize an RDF document, start with an rdf:Description element corresponding to a resource vertex in the graph. If possible, this should be a root vertex (which means that the graph in this case would have no edges coming into this vertex). The description element would have an rdf:about attribute giving the URI for this vertex.

- The vertex element would contain a subelement for every property of which the given vertex is the subject.

- If the object of the property is a literal, the text of the literal would be placed within the property element.

- If the object of the property is a resource, the property element would contain a subelement for the object, and the rdf:about of this subelement would be the URI of the object resource.

- A useful abbreviation of the above syntax applies when the object of a property element has no further properties applying to it (i.e., the object element would be empty): the object element can be eliminated altogether. Its URI (otherwise to be set as the value for rdf:about of the object) can be placed as the rdf:resource of the property element itself. Since XML syntax requires that there can be no more than one copy of an attribute within a given element, this means that separate property elements are needed for each copy of the given property that applies to the subject resource.

- The attribute rdf:nodeid, when applied on rdf:Description elements instead of rdf:ID or rdf:about, is used to identify blank nodes internally within a document. The same attribute is used on property elements, instead of rdf:resource, to refer to blank nodes.

- Literals can be typed through the rdf:datatype mechanism. These are written like any other literals, except that the property element is given an attribute rdf:datatype with a value as the URI for the desired datatype. Then the text enclosed within the scope of the property element is interpreted according to the named datatype. RDF mostly leaves the datatypes to be defined externally, although it offers a built-in datatype rdf:XMLLiteral to represent XML content in RDF literals. Such literals can use the standard XML Schema datatypes. An advantage of RDF's hands-off treatment of datatypes is that, because RDF has no native datatypes, it does not require that the various application-specific datatypes be translated into a particular set of datatypes.

Conceptually, RDF distinguishes three components of a datatype: its *value space* or set of legal values, its *lexical space* or legal string representations, and a *lexical to value mapping* or how a value can be materialized from or serialized into a lexical string.

- An rdf:Description element with an rdf:type property can be replaced by an element whose tag is the object of the rdf:type property (which is then not needed). For example,

```
<!-- The Wire type is defined in the EX namespace -->
<!-- ex is the declared abbreviation for EX -->
<rdf:Description rdf:about="wire-9">
  <rdf:type rdf:resource="&EX;#Wire"/>
  <ex:color>blue</ex:color>
</rdf:Description>
```

can be replaced by

```
<ex:Wire rdf:about="wire-9">
  <ex:color>blue</ex:color>
</ex:Wire>
```

In other words, the defined type acts as a custom vocabulary.

- RDF uses the xml:base attribute to specify a base URI for a given document. When it is used, the values of the RDF URI references, i.e., rdf:ID, rdf:about, rdf:resource, and rdf:datatype, are interpreted as relative references within the base URI.

- Reification can be accomplished quite simply by placing an rdf:ID attribute on a property element. This attribute in essence names the statement being asserted by the given property element. The value of the rdf:ID attribute can be used as the value of an rdf:about attribute so as to capture additional properties about the statement.

- The XML serialization of RDF includes an attribute rdf:parseType, which enables further compaction of the syntax. For example, when we have an attribute value rdf:parseType='Collection' on an element, its subelements are interpreted as forming a collection list.

7.5 The N-Triples Notation

RDF is not tied exclusively to XML, and can be given different syntaxes. A mapping to XML is, of course, defined and is popularly used. The N-Triples notation simply expresses each RDF statement as a single triple, in text. Each triple, as the name suggests, consists of three components: the subject, predicate, and object of the corresponding RDF statement. Each component is placed on a separate line, with the third line terminating in a period. N-Triples are easier to read and often more succinct than XML. Listing 7.4 shows the N-Triples corresponding to the XML document in Listing 7.3.

- rdfs:Resource is an instance of rdfs:Class and is the top member of the class hierarchy induced by rdfs:subClassOf. All other classes are subclasses of it. That is, all RDF resources are instances of rdfs:Resource.

- rdfs:Literal is the class of literals, i.e., strings and integers. Thus, rdfs:Literal is an instance of rdfs:Class and a subclass of rdf:Literal.

- rdfs:Datatype is the class of datatypes. Each of its instances is a subclass of rdfs:Literal. A special instance of rdfs:Datatype is rdf:XMLLiteral, the class of XML literals. Interestingly, rdfs:Datatype is both an instance and a subclass of rdfs:Class.

- rdf:Property, the class of properties, is an instance of rdfs:Class.

- rdfs:subPropertyOf forms statements that mean that their subject property is a subset of their object property. Thus rdfs:subPropertyOf forms a hierarchy of properties. rdfs:subPropertyOf is also defined to be reflexive.

- The properties rdfs:range and rdfs:domain have properties as subjects and classes as objects. When multiple domains are defined for a property, then the resources used as a subject for the property must be instances of all the domains—i.e., the domain assertions are interpreted conjunctively. Similarly, multiple ranges are interpreted conjunctively.

  Interestingly, rdfs:range and rdfs:domain apply to themselves. The rdfs:range of both is rdfs:Class and the rdfs:domain of both is rdf:Property.

- rdf:type is a property that states the class of a given resource.

- rdfs:label, rdfs:comment, rdfs:seeAlso are used to provide a user-friendly label, text comments, and cross-references, respectively.

- rdfs:Container is a superclass of the RDF container classes, namely, rdf:Bag, rdf:Seq, and rdf:Alt. The property rdfs:member captures membership in any of the containers.

7.8   Vocabularies in RDF Schema

Some useful vocabularies built on RDF and RDF Schema are the following.

- Dublin Core defines an annotation system for documents. It is an important namespace that standardizes several key metatags, including author and title. It is conventionally abbreviated as dc. Also by convention, capitalized terms refer to vertices and lowercase terms refer to edges.

- RDF Graph.

- RDF Site Summary.

- RDF Directory Description Language.

- XML Topic Maps.

7.9 Notes

Jena is an open-source RDF and RDFS toolkit [McBride, 2002]. Jena includes XML readers, writers, storage, querying, and inference mechanisms.

7.10 Exercises

7.1. Which of the following best characterizes the difference between the rdf:ID and rdf:about attributes?

- They are equivalent.
- rdf:ID is used when defining an object, rdf:about is used when referring to an object.
- rdf:ID is used for resources, rdf:about is used for predicates.
- rdf:ID can have any value, rdf:about's value must be a URN.
- rdf:ID does not support the use of namespaces, but rdf:about does.

7.2. Of the following RDF snippets which, if any, are equivalent to the English phrase "Anything that is a student is a person?"

- ```
  <!-- rdfs is the URI for the RDF Schema namespace -->
  <rdf:Description rdf:ID="student">
    <rdf:type
       resource="&rdfs;#Class"/>
    <rdfs:subClassOf rdf:resource="#person"/>
  </rdf:Description>
  ```

- ```
  <rdf:Description>
    <student>
      <rdfs:subClassOf rdf:resource="#person"/>
    </student>
  </rdf:Description>
  ```

- ```
  <rdf:Description rdf:about="student">
    <rdfs:isA rdf:resource="#person"/>
  </rdf:Description>
  ```
  ```

- ```
  <rdf:Class rdf:ID="student">
    <rdfs:subClassOf rdf:resource="#person"/>
  </rdf:Description>
  ```

7.3. Choose an HTML-encoded page from the Web—the simpler the better—and construct a representation of its content using RDF. For concreteness, consider the Web pages maintained by the authors, specifically, http://www.csc.ncsu.edu/faculty/mpsingh/books/ or http://www.cse.sc.edu/~huhns/ (with either page, ignore the links in the top bars and consider only a few of the entries).

7.4. Using a tool such as Protégé, construct RDFS and RDF representations for the following made-up report from an intelligence agency:

> CIA Report Date 1 April, 2004 (from MI5): The British Special Branch arrested suspect XYZ at his residence at 11 St. Mary's Place, London. Found in XYZ's bedroom was a small container holding 8 ounces of pentaerythritol (PETN) and triacetone triperoxide (TATP). This is the same explosive that a terrorist tried to use on American Airlines flight #63 from Paris to Miami on 22 December 2001.

That is, produce two files: one in RDFS that describes the general concepts and classes involved in the domain of the report, and a second in RDF that describes the specific instances of the concepts.

7.5. The objective for this homework is to construct a model (ontology) in RDF for a domain of your choice. Specifically,

- Choose a site on the Web that has links to other pages as your starting domain. Some general possibilities are a university department, a company portal, a government organization, a city or country, and a zoo.
- Create a node-and-link diagram containing at least 15 resources, with associated properties and statements.
- Convert the diagram into RDF.
- Turn in the page(s) from the Web site you chose, the RDF diagram, and the RDF statements.

7.6. Programming Project: An XML Database

- Create a database in MS Access, MySQL, Oracle, or any other DBMS that you have available. Your database should have at least three tables, with at least three fields (columns) per table.
- Write an XML Schema that describes the tables.
- Add at least two tuples to each table and write the XML document that describes the data. Verify the document against its XML Schema.
  ```

- Write an interface program in Java or C# that will allow a user to send a query to the database and then display the resultant data from the tables. The query and results will be in XML, i.e., there will be XML tags surrounding an SQL command and the results that are returned. An example command user command is

  <query>Select * from Product</query>

7.7. Repeat Exercise 7.6, except describe the database using RDFS, describe the data in the tables using RDF, and write the query in RDQL or RQL.

7.8. For the database you constructed in Exercise 7.6, write an XML mediator that will receive a query, remove the XML tags, send the query to your database using JDBC, receive the results from the database, wrap the appropriate XML tags around the data (to match the XML Schema), and then send the tagged data to the interface.

7.9. Compare XML Schema's type extension mechanism with RDF Schema's subclassOf. Are members of an XML Schema subtype members of the corresponding supertype?

7.10. Express the following English assertions using RDF in its XML serialization.

- Amitoj is funny.
- Amitoj thinks that Amitoj is funny.
- Amitoj is funny and Amitoj thinks that Amitoj is funny.
- Amitoj is not funny and Amitoj thinks that Amitoj is funny.

7.11. Repeat Exercise 7.10, but using the N-triples notation.

7.12. Using the N-triples notation and some of the following semantic primitives for objects and properties, describe the brick-and-board bookshelf shown in the picture below.

standing	lying	aTypicalMemberIs	Board
Pyramid	Object	hasProperty	onePartIs
subclass	group	supportedBy	Brick

7.13. Give the rdfs:range and rdfs:domain statements whose subjects are rdf:type, rdfs:range, and rdfs:domain, respectively.

7.14. Define a vocabulary to capture the assertions of Exercise 7.10. Use your vocabulary to express those assertions.

7.15. A *relation schema* (in the sense of the relational model for databases) is an association of attributes or columns, each defined with a name and a datatype. The names of the different attributes must be distinct. Each relation schema has a key formed from the aggregation of one or more of its attributes. A tuple in a relation conforming to a given

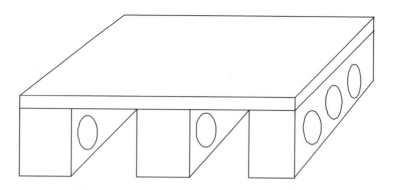

Figure 7.2: A brick-and-board bookcase to be described by a set of RDF N-triples

relation schema associates data values, each value being a value of the datatype of the corresponding column. The tuples in a relation cannot be duplicated. Describe a generic approach to represent a single tuple using RDF. *Hint:* represent the key for the schema and the entire set of attributes as objects and assert the main statement between them. The problem comes down to representing the key and the tuple accurately.

7.16. Which one of the following best represents the basic modeling primitives of RDF Schema?

- Class, Property, and ConstraintProperty.
- Elements and attributes.
- Subject, verb, and object.
- Things and ideas.
- Namespaces, elements, and doctypes.

7.17. Develop an RDF Schema for *assertions* as shown below. Listing 7.5 shows an example RDF document in which the last part would use your schema.

- The ⟨isTrue⟩ of a statement is an assertion. For example, assuming that a statement of rdf:ID="S1" is given, ⟨isTrue rdf:resource="#S1"⟩ is an assertion.
- The ⟨isFalse⟩ of a statement is an assertion. For example, assuming that a statement of rdf:ID="S1" is given, ⟨isFalse rdf:resource="#S1"⟩ is an assertion.
- The ⟨and⟩ of two assertions is an assertion.

The solution involves defining classes Assertion, and, isTrue, and isFalse, and suitable properties.

Listing 7.5: Examples of assertions

```
<rdf:RDF
```

```
   xmlns:rdf  = "as␣usual"
   xmlns:rdfs = "as␣usual"
   xmlns:bool = "your␣solution">

<rdfs:Class  rdf:ID="Person"/>
<rdfs:Class  rdf:ID="Univ"/>
<Univ  rdf:ID="NCSU"/>
<Univ  rdf:ID="USC"/>

<rdf:Property  rdf:ID="gradOf">
  <rdfs:domain  rdf:resource="#Person"/>
  <rdfs:range  rdf:resource="#Univ"/>
</rdf:Property>

<Person  rdf:ID="Jie">
  <gradOf  rdf:ID="G1"  rdf:resource="#NCSU"/>
  <gradOf  rdf:ID="G2"  rdf:resource="#USC"/>
</Person>

<bool:and  rdf:ID="Main">
  <bool:left  rdf:ID="yesNCSU">
    <bool:isTrue>
      <bool:ofStatement  rdf:resource="#G1"/>
    </bool:isTrue>
  </bool:left>
  <bool:right  rdf:ID="noUSC">
    <bool:isFalse>
      <bool:ofStatement  rdf:resource="#G2"/>
    </bool:isFalse>
  </bool:right>
</bool:and>

</rdf:RDF>
```

7.18. Add the following Boolean operators to your solution to Exercise 7.17.

- The ⟨or⟩ of two assertions is an assertion.
- The ⟨not⟩ of an assertion is an assertion.

7.19. Recall that an AND-OR graph is a directed graph whose nodes are interpreted as tasks. The root node indicates the main overall task. Each leaf node is an atomic task. Non-leaf nodes can be AND-nodes or OR-nodes. The out-edges of a node indicate its subtasks. For each AND-node, all subtasks must be performed; for each OR-node, at least one of the subtasks must be performed. There is no ordering among the tasks.

- Give an RDF Schema description for AND-OR graphs.

analogous to AND, NOT, and OR, as well as the constraints on various properties. A special form of constraint involves the existence of properties. For example, a parent may be defined as an adult who has at least one child, i.e., participates as a subject in one or more instances of the hasChild property. Further, the languages define constructs such as inverse relations, transitivity, and disjointness.

The following is a detailed description of an OWL model for a domain of animals. An OWL specification document—in loose terms, this would correspond to a file—has rdf:RDF as its top-level element. This element includes attributes declaring the key namespaces. These would almost always include the OWL, RDF, and RDF Schema namespaces, and often XML Schema, plus others that are specific to the given domain. The following listing shows the top element for our example:

```
<rdf:RDF
   xmlns:owl="http://www.w3.org/2002/07/owl#"
   xmlns:rdfs="http://www.w3.org/2000/01/rdf-schema#"
   xmlns:rdf="http://www.w3.org/1999/02/22-rdf-syntax-ns#"
   xmlns:xsd="http://www.w3.org/2001/XMLSchema#">
<!— more here —>
</rdf:RDF>
```

It then contains an assertion, made via the owl:Ontology element, that the given document is an ontology. Here it is customary to specify the version and a comment. OWL also supports a set of primitives to encode versions and relationships among versions. It also supports importing ontologies into other ontologies. Section 9.5 reviews these primitives.

```
<owl:Ontology rdf:about="Life">
   <owl:versionInfo>$Id: Life.owl,v1.0 2003/12/01
                  12:35:31 huhns Exp$
   </owl:versionInfo>
   <rdfs:comment>An Ontology for Life</rdfs:comment>
</owl:Ontology>
```

An ontology defines classes, and relationships among those classes, as in the following definitions. The first definition asserts that there is a class named "Animal," and the second that there is another class named "Mammal," which is a subclass of "Animal" and, further, that it is disjoint with a class named "Reptile."

```
<owl:Class rdf:ID="Animal">
   <rdfs:label>Animal</rdfs:label>
   <rdfs:comment>
     This class represents the animal kingdom.
   </rdfs:comment>
</owl:Class>

<owl:Class rdf:ID="Mammal">
```

```
   <rdfs:subClassOf rdf:resource="#Animal"/>
   <owl:disjointWith rdf:resource="#Reptile"/>
</owl:Class>
```

Multiple superclasses are allowed in OWL, so we can state that a "Reptile" is both an "Animal" and an "OxygenUser" (not defined within this ontology snippet) as

```
<owl:Class rdf:ID="Reptile">
  <rdfs:subClassOf rdf:resource="#Animal"/>
  <rdfs:subClassOf rdf:resource="#OxygenUser"/>
</owl:Class>
```

An OWL class can be further defined by its properties. Like properties in RDF, OWL properties are binary relations between two classes. Like in RDF Schema, these properties are defined via their domain and range. For example, we may have a parent property that states that the parent of an "Animal" is also an "Animal."

```
<owl:ObjectProperty rdf:ID="hasParent">
  <rdfs:domain rdf:resource="#Animal"/>
  <rdfs:range rdf:resource="#Animal"/>
</owl:ObjectProperty>
```

Further, as in RDF Schema, a property can have subproperties. For example, "hasFather" is a subproperty of "hasParent" that is restricted to have only male animals as its objects. However, OWL enables us to state additional restrictions on the allowed property values. For example, we can state that an animal must have at most one father by declaring it to be a member of (i.e., having an rdf:type of) the following class:

```
   <owl:Restriction>
    <owl:onProperty rdf:resource="#hasFather"/>
    <owl:maxCardinality rdf:datatype="xsd:nonNegativeInteger">
     1
    </owl:maxCardinality>
   </owl:Restriction>
```

Finally, an individual can be defined as an object in a class. For example, the fact that "Rover" is a "Mammal" whose father is the "MaleAnimal" instance "Spot" would be represented as

```
<Mammal rdf:ID="Rover">
  <hasFather><MaleAnimal rdf:ID="#Spot"/></hasFather>
</Mammal>
```

The above discussion gives us a start on an OWL ontology. However, OWL has a number of other constructs that make it an interesting and powerful language. These are introduced below.

8.2   OWL Dialects

OWL is a set of three languages or *dialects* of different expressiveness. Playing on the OWL pun, these dialects are sometimes referred to as *species*.

- OWL Lite provides a classification hierarchy and limited constraints. For example, it limits cardinality restrictions, and does not allow complementation (negation) or union (disjunction) operators.

- OWL DL provides maximum expressiveness while ensuring computational completeness (all valid conclusions can be inferred) and decidability (the inferences take finite time). For example, classes cannot be used as instances.

- OWL Full provides maximum syntactic freedom, limited only by RDF. For example, classes can be used as instances. However, OWL Full makes no guarantees of completeness or decidability.

An ontology expressed in OWL Lite is automatically expressed in OWL DL, and an ontology expressed in OWL DL is automatically expressed in OWL Full. All conclusions drawn in OWL Lite are valid in OWL DL, and all conclusions drawn in OWL DL are valid in OWL Full.

In terms of what can be expressed, OWL Full is an extension of RDF, whereas OWL Lite and OWL DL build on a restricted view of RDF. Thus every OWL document (of any dialect) is an RDF document and every RDF document is an OWL Full document.

Section 8.6 returns to a discussion of the dialects and compares their expressiveness. For now, the discussion introduces the main constructs without specific attention to the distinctions among the dialects.

8.3   OWL Constructors

OWL distinguishes between constructors and axioms, which are discussed at length in Section 8.4. In simple terms, the OWL constructors are the primitives that help us specify new classes and axioms are the primitives that help us make additional assertions about classes and properties.

The OWL dialects provide class constructors that are based on description logic. These constructors build on the datatypes defined in XML Schema. These include the primitive datatypes such as integer, string, and floating point number, which are defined in the XML Schema namespace that is conventionally abbreviated to xsd. These are sometimes referred to as the datatype domain and the object domain, respectively.

8.3.1   Classes

As before, classes correspond to sets of objects (their instances). Object types are distinct from datatypes. That is, the instance of an object class cannot be the instance of a datatype.

This is a useful practical requirement, because most times you would need to specify constraints on the values of the data types, but the values of the data types cannot constrain the objects. You would rarely (perhaps never) need to form new datatypes using the object types. On the other hand, allowing such variation would complicate the representation and reasoning significantly. It makes more sense to set up the predicates defined on the datatypes to be responsible for the datatypes. The reasoners built for OWL dialects would typically involve a component for reasoning about the XML datatypes, which would be kept separate from the component for reasoning about the object properties.

OWL includes a class called owl:Class, which is defined as a subclass of rdfs:Class. All OWL object classes are members of owl:Class. OWL supports a variety of datatypes, including those based on RDF datatype specifications, RDF Schema literals, and enumerations based on objects. An example declaration, repeated from above, is

```
<owl:Class rdf:ID="Reptile">
  <rdfs:subClassOf rdf:resource="#Animal"/>
  <rdfs:subClassOf rdf:resource="#OxygenUser"/>
</owl:Class>
```

A built-in class owl:Thing is at the top of the class hierarchy. Individuals are instances of classes. OWL includes a built-in class called owl:Nothing that cannot have any instances and which is, therefore, a subclass of every class. Formally, owl:Thing and owl:Nothing are complements of each other.

We can declare individuals to be members of specified classes by using the class name as the element name: this is a special case of the RDF XML syntax convention where the rdf:type of an individual can be used as the element name:

```
<Reptile rdf:ID="BobsLizard"/>
<owl:Thing rdf:ID="BobsLizardCage"/>
```

Here no special class is being asserted for BobsLizardCage. It is simply an owl:Thing, which all individuals in OWL are.

8.3.2 Properties

Properties relate pairs of individuals. Object properties, instances of owl:ObjectProperty, relate instances of two classes (possibly the same class). Datatype properties, instances of owl:DatatypeProperty, relate an instance of a class with an instance of a datatype. Whereas in RDF we can declare an rdf:Property, in OWL we must explicitly choose the kind of property we declare: owl:ObjectProperty or owl:DatatypeProperty.

Domain and range are global restrictions since they apply to a property by itself independently of the class to whose instances the property is being applied. That is, the domain of a property must be an instance of owl:Class, and is owl:Thing unless otherwise specified. The range of an object property is also an instance of owl:Class, and is owl:Thing unless otherwise specified. However, the range of a datatype property is an owl:DataRange, which is a subclass of rdfs:Datatype.

OWL allows multiple assertions about the domains and ranges of properties. These assertions are interpreted conjunctively, i.e., all must be true. In other words, something is in the domain of a property if it is in the intersection of all the specified domains. The range of a property is treated similarly.

An rdf:Property element refers to a property name as a URI. As explained above, the instances of a property are pairs of individuals, the first of which must be an object and the second may be an object or a data value. An rdf:Property element may contain the following:

- Zero or more rdfs:subPropertyOf elements, each containing a property name. Any pair of individuals that is an instance of the property named in the main rdf:Property element must be an instance of the property named in each rdfs:subPropertyOf element. An owl:ObjectProperty cannot be the subproperty of an owl:DatatypeProperty and vice versa.

- Zero or more rdfs:domain elements. The first component of each instance for which the given property applies must be in the stated domain. Multiple domain elements are thus interpreted conjunctively.

- Zero or more rdfs:range elements. The second component of each instance for which the given property applies must be in the stated range. Multiple range elements are thus interpreted conjunctively.

- Zero or more owl:equivalentProperty elements, which assert the equivalence of the two properties. This is a subproperty of rdfs:subPropertyOf. Therefore, it must satisfy the same constraint (mentioned above) about not relating an owl:ObjectProperty with an owl:DatatypeProperty.

- Zero or more owl:inverseOf elements, each naming a property of which the given property is taken as an inverse. Because all properties must have owl:Class as their domain, the owl:inverseOf element applies only to an owl:ObjectProperty. Section 6.5 defines inverses.

The following listing brings the above primitives together in a single example.

```
<owl:ObjectProperty rdf:ID="livesIn">
  <rdfs:domain rdf:resource='#Animal'/>
  <rdfs:range rdf:resource='#Locale'/>
  <rdfs:subPropertyOf rdf:resource='#hasHabitat'/>
  <owl:inverseOf rdf:resource='#hasDenizen'/>
  <owl:equivalentProperty rdf:resource='#hasHome'/>
</owl:ObjectProperty>
```

8.3.3 Class Expressions

The above constructors are not much different from the primitives of RDF Schema. OWL comes into its own through a sophisticated set of class expression constructors. Listed in order of increasing sophistication, a class expression may be any of the following things:

- Simply a class name, as given via a URI. The above examples are illustrations of this variant.

- A Boolean combination of class expressions, enclosed in the owl:Class element.

- An enumeration enclosed in the owl:Class element.

- A property restriction, which provides an interesting technique (unique to description logic) for deriving class definitions from properties.

As one would expect, except for the explicitly named class, the other expressions yield *anonymous classes*. The owl:equivalentClass axiom introduced in Section 8.4 provides a means to assign names to anonymous classes.

8.3.3.1 Boolean Expressions

OWL includes the Boolean operators captured as three elements:

- The owl:intersectionOf element includes a list of class expressions and defines a class equal to the intersection of the given expressions.

- The owl:unionOf element includes a list of class expressions and defines a class equal to the union of the given expressions.

- The owl:complementOf element includes a single class expression and defines a class equal to the complement of the given expression. One can think of this as producing owl:Thing minus the given class expression.

The Boolean expressions have the following syntax:

```
<owl:Class rdf:ID='SugaryBread'>
  <owl:intersectionOf rdf:parseType='Collection'>
    <owl:Class rdf:about='#Bread'/>
    <owl:Class rdf:about='#SweetFood'/>
  </owl:intersectionOf>
</owl:Class>
```

Notice that the above syntax yields a definition for the SugaryBread class. This is to be contrasted with merely asserting subclass relations about it. For example, if we define RaisinBread as follows

```
<owl:Class rdf:ID='RaisinBread'>
  <rdfs:subClassOf rdf:resource='#Bread'/>
  <rdfs:subClassOf rdf:resource='#SweetFood'/>
</owl:Class>
```

then we would be able to infer that raisin bread is a subclass of sugary bread, but not that they are equivalent.

8.3.3.2 Enumerations

An enumeration is given by the owl:oneOf element, which literally enumerates the objects that are the instances of the anonymous class being described by the given expression. A key feature of this constructor is that it provides an exhaustive list of its members.

```
<owl:Class rdf:ID='OvenType'>
  <owl:oneOf rdf:parseType='Collection'>
    <OvenType rdf:about='#Conventional'/>
    <OvenType rdf:about='#Convectional'/>
    <OvenType rdf:about='#Microwave'/>
    <OvenType rdf:about='#Tandoor'/>
  </owl:oneOf>
</owl:Class>
```

The owl:oneOf element can be combined with axioms such as rdfs:subClassOf. For example, we can write

```
<owl:Class rdf:ID='OvenType'>
  <rdfs:subClassOf rdf:resource='#EquipmentType'/>
  <owl:oneOf rdf:parseType='Collection'>
    <!-- as above -->
  </owl:oneOf>
</owl:Class>
```

which not only specifies the members of oven type, but also states that it is a subclass of equipment type.

8.3.3.3 Restrictions

OWL enables classes to be constructed out of properties through the mechanism of the *property restriction*. In essence, the objects involved in a property that satisfy the stated restriction form an *anonymous class*. A property restriction is modeled as an owl:Restriction, which is a subclass of owl:Class. A property restriction can be of two varieties: owl:ObjectRestriction (applies on object properties) or owl:DatatypeRestriction (applies on datatype properties). Object and datatype restrictions are created with the same syntax: the former uses a class element, the latter a datatype reference.

OWL enables stating restrictions on the type of values for a property that are related to a class on which the property is defined. A property restriction acts as the constructor for a class, namely, the class of individuals that satisfy the given restriction. The class is anonymous, but may be declared to the same class as a named class, and hence be named. Property restrictions provide local restrictions on a property in the context of a given class. Each restriction applies to a single property; the property in question is specified through owl:onProperty.

An owl:Restriction element takes an owl:onProperty subelement, which specifies the property being restricted. This is usually written as the first subelement of owl:Restriction. Thus the following is a typical syntax fragment:

```
<owl:Restriction>
  <owl:onProperty rdf:resource='#bakes'/>
<!-- actual restriction here -->
</owl:Restriction>
```

As remarked above, the owl:Restriction functions as a class. Consequently, this element can be dropped into an OWL expression where a class would be expected. Restrictions are a *local* way of stating constraints in that they would apply only to some of the range individuals or values for a property. Based on these range individuals or values, we can induce the domain individuals for which the constraint holds. These domain individuals, in essence, form an anonymous class that is identified by the restriction expression.

The following are the major property restriction types that can be plugged into the above syntax:

- owl:someValuesFrom forms the class of individuals such that at least one of the objects or values associated to any of these individuals via the given property is a member of the specified class or datatype. Notice that if no objects or data values are associated with a given individual, then it automatically fails to qualify as a member of this class. This is simply the traditional trivial case for the existential quantifier in logic. For example, we may define a bread oven as a subclass of oven that can be used to bake a kind of bread.

```
<owl:Class rdf:ID='BreadOven'>
  <rdfs:subClassOf rdf:resource='#OvenType'/>
  <rdfs:subClassOf>
    <owl:Restriction>
      <owl:onProperty rdf:resource='#bakes'/>
      <owl:someValuesFrom rdf:resource='#Bread'/>
    </owl:Restriction>
  </rdfs:subClassOf>
<owl:Class>
```

- owl:allValuesFrom forms the class of individuals such that all objects or data values associated to any of these individuals via the given property are members of the specified class or datatype. Notice that if no objects or data values are associated with a given individual, then it automatically qualifies as a member of this class. This is simply the traditional trivial case for the universal quantifier in logic. For example, we may define a European company as a company all of whose offices are located in countries in Europe.

```
<owl:Class rdf:ID='EuropeanCompany'>
  <rdfs:subClassOf rdf:resource='#Company'/>
  <rdfs:subClassOf>
    <owl:Restriction>
      <owl:onProperty rdf:resource='#locatedIn'/>
      <owl:allValuesFrom rdf:resource='#EuropeanCountry'/>
    </owl:Restriction>
  </rdfs:subClassOf>
<owl:Class>
```

- owl:hasValue forms the class of individuals such that at least one of the objects or values associated to any of these individuals via the given property is the specified individual or data value. Note that owl:hasValue refers to a particular instance, whereas owl:someValuesFrom refers to a class expression. For example, we may define a USA company as a company that has an office located in the USA.

```
<owl:Class rdf:ID='USACompany'>
  <rdfs:subClassOf rdf:resource='#Company'/>
  <rdfs:subClassOf>
    <owl:Restriction>
      <owl:onProperty rdf:resource='#locatedIn'/>
      <owl:hasValue rdf:resource='#USA'/>
    </owl:Restriction>
  </rdfs:subClassOf>
<owl:Class>
```

Another example is for specifying countries that belong to Europe. Here we identify an individual continent called Europe. Notice that if a country falls partly in Europe and partly in another continent (as do Russia and Turkey), and if the property isInContinent is interpreted as being multivalued, then such countries would be identified as European countries.

```
<owl:Class rdf:ID='EuropeanCountry'>
 <owl:equivalentClass>
  <owl:Restriction>
   <owl:onProperty rdf:resource='#isInContinent'/>
   <owl:hasValue rdf:resource='#Europe'/>
  </owl:Restriction>
 </owl:equivalentClass>
<owl:Class>
```

- owl:minCardinality: Consider the number of instances to which a given instance of a class must be related via the given property. This number must be equal or above the minimum cardinality stated for that property. If nothing is stated about minimum cardinality, then a minimum cardinality of 0 is implied. A minimum cardinality of 0 indicates that participation in the given property is optional, whereas a minimum cardinality of 1 indicates that participation in the given property is mandatory.

  As a restriction, this selects the domain individuals for a property for which the given minimum cardinality constraint holds. For example, we can define a distributed company as one that has at least two locations. All the cardinality constraints are based on the XML Schema Datatype of nonNegativeInteger.

  ```
  <owl:Class rdf:ID='DistributedCompany'>
    <rdfs:subClassOf rdf:resource='#Company'/>
    <rdfs:subClassOf>
      <owl:Restriction>
        <owl:onProperty rdf:resource='#locatedIn'/>
        <owl:minCardinality
          rdf:datatype='&xsd;#nonNegativeInteger'>
          2
        </owl:minCardinality>
      </owl:Restriction>
    </rdfs:subClassOf>
  <owl:Class>
  ```

- owl:maxCardinality: Consider the number of instances to which a given instance of a class must be related via a given property. This number must be equal or below the maximum cardinality stated for the property. If nothing is stated about maximum cardinality, then an unbounded maximum cardinality is implied. A maximum cardinality of 1 indicates a functional property (defined in Section 8.4). A maximum cardinality of 0 indicates that the property cannot apply on the given class—typically, this would be a subclass of a class where the property does apply. For example, an unemployed person may be modeled as one with no employer.

  ```
  <owl:Class rdf:ID='UnemployedPerson'>
    <rdfs:subClassOf rdf:resource='#Person'/>
    <rdfs:subClassOf>
      <owl:Restriction>
        <owl:onProperty rdf:resource='#hasEmployer'/>
        <owl:maxCardinality
          rdf:datatype='&xsd;#nonNegativeInteger'>
          0
        </owl:maxCardinality>
      </owl:Restriction>
    </rdfs:subClassOf>
  <owl:Class>
  ```

- owl:cardinality specifies the minimum and maximum cardinalities to be equal to the stated value. Thus it abbreviates owl:minCardinality and owl:maxCardinality.

8.3.4   Collections

OWL has support for collections of entities. For a given collectionID, we can declare it to be of rdf:type owl:Collection. Further we can assert for an entity that it is the owl:memberOf of a given collectionID.

8.4   OWL Axioms

The above section covered the main OWL class constructors. Other key OWL primitives behave as axioms or assertions about classes and properties. These are introduced next.

8.4.1   Individuals

Instances of both classes (objects) and properties (pairs of objects) are written in the XML syntax for RDF.

- Individuals can be stated to be of the rdf:type of a class. Properties can be asserted for an individual by explicitly specifying an individual or data value with which the given individual is being associated.

- The owl:sameAs primitive applies to individuals. In OWL Full, classes are treated as individuals, so owl:sameAs applies to them as well. When applied to classes, it means that they denote the *same* concept, not that they are distinct concepts that happen to have the same instances. For example, the singleton class of morning stars could be asserted to be equivalent to the singleton class of evening stars, even though they are not asserted to be the same class.

- The OWL semantics makes no assumptions about unique names. In other words, distinct URIs are obviously distinct, but the resources they point to might in fact be the same or distinct objects. But it is often essential to assert that two resources are in fact the same individual. The identity of such pairs of resources can possibly be inferred, but it is often essential and always good practice to record known such identities explicitly. (Section 8.5 discusses OWL inference.) This is achieved through owl:sameIndividualAs. For example, the following asserts that the country Persia is the same as Iran.

```
<ex:Country rdf:ID='Iran'/>
<ex:Country rdf:ID='Persia'>
  <owl:sameIndividualAs rdf:resource='#Iran'/>
</ex:Country>
```

Such an assertion of identity would be most useful if Iran and Persia were defined as terms in different ontologies.

- owl:differentFrom is used to assert that two individuals are different from each other. This is valuable because OWL makes no assumption about the uniqueness of the names of individuals. For example, we may state that

```
<ex:Country rdf:ID='Russia'/>
<ex:Country rdf:ID='India'>
  <owl:differentFrom rdf:resource='#Russia'/>
</ex:Country>
```

- The above works for more individuals, but requires a separate assertion for each pair. For n individuals, we would need on the order of n^2 entries. Clearly, a more compact notation is called for. The owl:AllDifferent primitive along with owl:distinctMembers is used to assert that a number of individuals is pairwise distinct.

```
<owl:AllDifferent>
  <owl:distinctMembers rdf:parseType='Collection'>
   <ex:Country rdf:ID='Russia'/>
   <ex:Country rdf:ID='India'/>
   <ex:Country rdf:ID=USA'/>
  <owl:distinctMembers/>
</owl:AllDifferent>
```

8.4.2 Data Values

Data values are written in RDF syntax with a string representation of the desired value along with a URI for the XML Schema datatype and a mechanism to parse the string and produce the desired value. The XML Schema datatype is the rdf:type of the data value and the string representation is the rdf:value of the value. Thus the decimal 7.9 would be expressed as an RDF literal as ⟨xsd:decimal rdf:value="7.9"⟩. For data values, XML Schema Datatype identity is used to determine whether they are the same or different.

8.4.3 Classes

A pair of classes or a pair of properties may be stated to be identical or equivalent: in this case they behave as synonyms and one may be substituted for the other. OWL class properties are

- rdfs:subClassOf, already introduced, asserts that its subject is a subclass of its object.

- owl:equivalentClass is a subProperty of rdfs:subClassOf. When applied to classes, it has the same meaning as owl:sameAs. However, because owl:equivalentClass is a subproperty of rdfs:subClassOf, this indicates that a part of its meaning is interpretable from RDF Schema without necessarily involving any knowledge of OWL. In other words, an RDF Schema processor could partially understand owl:equivalentClass but not owl:sameAs. Therefore, the owl:equivalentClass formulation is preferred over owl:sameAs when classes are involved.

- owl:disjointWith asserts that its subject class has no members in common with its object class. For example, the following asserts that no mammal may be a reptile, bird, or amphibian.

```
<owl:Class rdf:ID="Mammal">
  <rdfs:subClassOf rdf:resource="#Vertebrate"/>
  <owl:disjointWith rdf:resource="#Reptile"/>
  <owl:disjointWith rdf:resource="#Bird"/>
  <owl:disjointWith rdf:resource="#Amphibian"/>
</owl:Class>
```

8.4.4   Properties

Properties can be introduced as inverses of other properties. This only works for object properties, because only a class may be the domain of a property. Further, properties may be declared to be transitive or symmetric. Obviously, these specifications too make sense only for object properties. In the above and other such cases, a reasoner is authorized to make the corresponding inferences.

Functional or unique properties are those that have at most one value for a given instance to which they apply. This specification holds for all properties. These correspond to the so-called partial functions of elementary algebra, because they are not defined for every member of their domain. In other words, the minimum cardinality of such properties is 0 and their maximum cardinality is 1. Properties can be inverse functional, meaning that their inverses must be functional, that is, the inverse of such a property would have at most one value for each instance to which it applies. In other words, it would be unambiguous.

Inverse functional properties are those for which each range value relates to no more than one domain value. This specification holds only for object properties.

- owl:equivalentProperty is a subProperty of rdfs:subPropertyOf. When applied to properties, it has the same meaning as owl:sameAs. However, for reasons similar to those discussed for owl:equivalentClass, when properties are involved it is preferable to use owl:equivalentProperty instead of owl:sameAs.

- owl:inverseOf is a property of object properties. It asserts that its subject and object properties are inverses of each other. The formal definition for owl:inverseOf follows the definition of inverse binary relations in Section 6.5.

- The owl:TransitiveProperty element, is a subclass of owl:ObjectProperty, and asserts that the given property is transitive. In other words, if the pairs (x, y) and (y, z) are instances of the given property, then so is the pair (x, z). As for transitive binary relations in general, owl:TransitiveProperty requires that its domain and range be equal so that the transitivity is well-defined.

- The owl:SymmetricProperty element, is a subclass of owl:ObjectProperty, and asserts that the given property is symmetric. In other words, if the pair (x, y) is an instance of the given property, then so is the pair (y, x). As above, owl:SymmetricProperty requires that its domain and range be equal so that symmetry is well-defined.

- The owl:FunctionalProperty element asserts that the given property can have at most one pair instance with a given first component. This behaves as a global cardinality restriction.

- The owl:InverseFunctionalProperty element asserts that the given property can have at most one pair instance with a given second component. This behaves as a global cardinality restriction. As is perhaps obvious, owl:InverseFunctionalProperty applies only on object properties, that is, it is an rdfs:subClassOf owl:ObjectProperty.

8.4.5 Elementary Algebra: Functions

It is helpful to relate ontologies to binary relations as in elementary, high-school algebra. Let us review some basic definitions, which support some interesting results.

Recall from elementary algebra that given sets S_d (domain) and S_r (range, also known as *codomain*), a binary relation R between these sets is defined as any subset of $S_d \times S_r$. Informally, R associates each member of S_d with zero or more members of S_r. Consider the following definitions.

- A *partial function* f from S_d to S_r (written $f : S_d \mapsto S_r$) is a relation that associates each member of S_d with at most one member of S_r. For $d \in S_d$, $f(d)$ denotes the member of S_r (if any) associated with S_d.

- A *total function* $f : S_d \mapsto S_r$ is a relation that associates each member of S_d with at most one member of S_r. A total function is a partial function that is defined for all members of its domain.

- We say that a total function $f : S_d \mapsto S_r$ is *injective* or *one-to-one* if $(\forall d_1, d_2 \in S_d : f(d_1) = f(d_2) \Rightarrow d_1 = d_2)$. That is, an injective function is one whose targets are different if its sources are different.

- We say that a total function $f : S_d \mapsto S_r$ is *surjective* or *onto* if $(\forall r \in S_r : (\exists d \in S_d : f(d) = r))$. That is, a surjective function is one that "uses up" all of its range.

- We say that a total function $f : S_d \mapsto S_r$ is bijective if it is both injective and surjective. Bijective functions are also known as bijections.

Exercises 8.6 and 8.8 ask you to establish some simple results and give natural examples of the above.

8.5   OWL Inference

The essential theme behind knowledge representation is that it is about computing with the specifications of conceptual models that are given. In particular, a form of computing—*inference*—can be specified as part of the semantics of the language. The dialects of OWL, RDF Schema, and RDF are all syntactically expressible in XML. However, the inferences they support go beyond those of XML to capture increasingly subtle shades of meaning. Each well-formed OWL document legitimizes a number of inferences.

A consequence of the inference view is that we cannot consider OWL documents merely in terms of their syntax, but should also consider their semantics. In other words, two superficially different OWL documents may have the same semantics because they legitimize the same inferences. For example, we can infer the meaning of the following listing:

```
<owl:Class rdf:ID="Mammal">
  <rdfs:subClassOf rdf:resource="#Vertebrate"/>
</owl:Class>
```

from the following listing:

```
<owl:Class rdf:ID='Vertebrate'>
  <owl:unionOf rdf:parseType='Collection'>
    <owl:Class rdf:about="#Amphibian"/>
    <owl:Class rdf:about="#Reptile"/>
    <owl:Class rdf:about="#Bird"/>
    <owl:Class rdf:about="#Mammal"/>
  </owl:unionOf>
</owl:Class>
```

An advantage of thinking in terms of inferences is that we can put together knowledge from different sources. In other words, differently developed conceptual models can be combined or mapped to each other. The inferencing mechanism can detect potential inconsistencies, thus helping resolve errors early.

A special case of the above is that classes can be defined piecemeal. Thus an owl:Class element may contain only part of the definition of the given class. Additional parts of the definition can be added via other elements. The following snippets illustrate this.

```
<owl:Class rdf:ID="Mammal">
  <rdfs:subClassOf rdf:resource="#Vertebrate"/>
</owl:Class>
```

Notice that whereas the snippet above uses the rdf:ID attribute to declare the identifier for a class, the snippets below use the rdf:about attribute to refer to the same class.

```
<owl:Class rdf:about="#Mammal">
  <owl:disjointWith rdf:resource="#Reptile"/>
```

```
<owl:disjointWith rdf:resource="#Bird"/>
<owl:disjointWith rdf:resource="#Amphibian"/>
</owl:Class>
```

```
<owl:Class rdf:about="#Mammal">
  <rdfs:subClassOf>
    <owl:Restriction>
      <owl:onProperty rdf:resource='#hasSkinCovering'/>
      <owl:hasValue rdf:resource='#Hair'/>
    </owl:Restriction>
  </rdfs:subClassOf>
  <rdfs:subClassOf>
    <owl:Restriction>
      <owl:onProperty rdf:resource='#temperatureControl'/>
      <owl:hasValue rdf:resource='#Warm'/>
    </owl:Restriction>
  </rdfs:subClassOf>
<owl:Class>
```

This is not to say that we should purposely scatter the definition of a class in different places, but that we can begin with a definition and possibly augment it. The above case might occur if one ontology simply introduces mammals as a kind of vertebrate; another one remarks they are disjoint from reptiles, birds, and amphibians; and a third source states that they are covered with hair and are warm-blooded. If the sources came about independently, then there would be a step where it is recognized that the different documents are about the same concept. Chapter 9 revisits this point. If the extensions were inconsistent, then the class would end up being subsumed by owl:Nothing.

There are can be two main ways to specify a class. One involves necessary, but in general not sufficient, constraints. An owl:Class element may include zero or more subclass elements, each of which contains a resource, which is a class expression. It may also have zero or more disjointWith elements. Each disjointWith element specifies a class expression and indicates, as described above, that the given class does not overlap with the specified expression.

The other way to specify a class involves necessary and sufficient conditions for class membership. The simplest means to do so is by using the owl:equivalentClass element. An owl:Class element may use as many of these as necessary. The class is then taken to be equivalent to *each* of the class expressions. An owl:Class element may contain Boolean class expressions, indicating intersection, union, and complementation. Lastly, an owl:Class element may include some individuals that are treated as being its exact set of instances: no more, no less.

The combination of the axiom primitives with the constructors yields enhanced expressive power. For example, OWL DL enables us to state axioms, such as rdfs:subClassOf, between arbitrary class expressions. An advantage of this expressiveness is that many of the primitives can be understood theoretically as abbreviations of other primitives. We can still use the primitives for their convenience, but they do not add to the overall expressive power.

8.6 OWL Dialects Compared

We can now summarize the main entities in the OWL metamodel. Figure 8.1 presents an RDF Schema that includes the main entities and relationships of OWL.

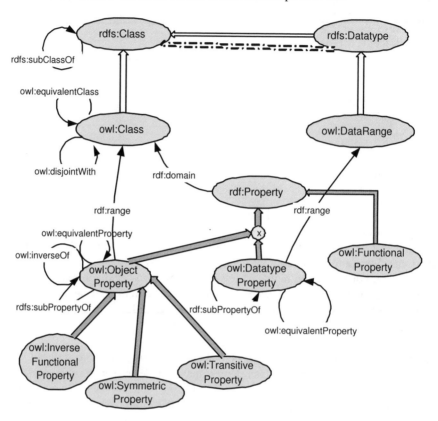

Figure 8.1: The main OWL entities and relationships. Here a double arrow indicates rdfs:subClassOf, a dashed arrow indicates rdf:type, and all other arrows indicate the labeled properties. The ⊗ connector symbol indicates mutual exclusion of the subclasses

The OWL dialects vary in their treatment of cardinality. OWL Lite limits all cardinality values to be 0 or 1. For minimum cardinality, specifying nothing is equivalent to specifying 0, so it is only necessary to specify mandatory participation. For maximum cardinality, specifying nothing is equivalent to leaving it unbounded. If the maximum cardinality is 1, then that means there can be at most one associated value. In general, limiting the maximum cardinality to 0 would be unlikely, since it suggests that participation is forbidden. In OWL DL and in OWL Lite, only an owl:ObjectProperty can be an owl:InverseFunctionalProperty; in OWL Full, it may be an owl:ObjectProperty or an owl:DatatypeProperty.

Interestingly, OWL Lite and OWL DL impose the additional requirements that transitive properties and their superproperties may not have a maximum cardinality of 1. Without such a requirement, these languages would become undecidable [Horrocks et al., 1999].

As explained in Section 8.3.3.3, the above restrictions can be thought of as generating anonymous classes. OWL Lite supports intersection of named classes and restrictions. It also limits rdfs:subClassOf and owl:equivalentClass statements to use individual class names as their subjects, as opposed to arbitrary expressions denoting classes; the objects of these statements can be named classes or restrictions.

OWL Lite respects the limitations imposed by OWL DL. Further, OWL Lite eliminates some useful primitives from the following categories, namely, enumeration (owl:oneOf), set operations (owl:unionOf and owl:complementOf), data ranges (owl:DataRange), class axioms (owl:disjointWith), and property restrictions (owl:hasValue and cardinalities other than 0 and 1). Moreover, in OWL Lite, intersections must involve more than one class (each named or an expression). Further, the objects of the properties rdfs:range, owl:allValuesFrom, and owl:someValuesFrom must be named classes or datatypes; the objects of rdfs:domain must be named classes; and the objects of rdf:type must be named classes or restrictions.

OWL DL includes some further valuable constructs. Enumerated classes can be defined in terms of a set of enumerated individuals that form their set of instances—this is the exact set of instances.

OWL Full can state that two classes are mutually disjoint, that is, they have no common instances. It allows cardinalities to be arbitrary nonnegative integers. Importantly, it allows complex class descriptions to be used for the constructs where a class name may be used. Further, OWL Full allows classes to be used as instances.

8.7 An OWL Example

We now discuss a short example ontology that brings together some of the concepts introduced above. Consider an academic setting where students take courses and courses are offered by departments. Further, assume that each course is offered by exactly one department, CS is a department, a student must take at least one course, and a full-time student must take between three and five courses. How might we capture this scenario in OWL? The following listing illustrates the basic entities and relationships.

```
<owl:Class rdf:ID="Student"/>
<owl:Class rdf:ID="Course"/>
<owl:Class rdf:ID="Department"/>

<Department rdf:ID='CS'/>

<owl:ObjectProperty rdf:ID='takes'>
  <rdfs:domain rdf:resource='#Student'/>
  <rdfs:range rdf:resource='#Course'/>
</owl:ObjectProperty>
```

8.8   Expressiveness

When designers of Web services are faced with choosing a representation for their services, there are a number of trade-offs that they must consider. The trade-offs must be made among the expressive power, the rigor, the ease of use, and the computational tractability of a representation.

8.8.1   Tree Model Definitions

A limitation of OWL is that, in essence, it allows the formulation of restrictions that consider no more than one individual member of a class. Formally, description logic (as remarked above, the intellectual basis of languages such as OWL) satisfy the so-called *tree model property*. In other words, we cannot define classes or properties whose members are related via an anonymous variable.

Consider this scenario about products stored in warehouses and shipped on trucks:

```
<owl:Class rdf:ID="Product"/>
<owl:Class rdf:ID="Warehouse"/>
<owl:Class rdf:ID="Truck"/>
<owl:Class rdf:ID="Location"/>

<owl:ObjectProperty rdf:ID='inWarehouse'>
  <rdfs:domain rdf:resource='#Product'/>
  <rdfs:range rdf:resource='#Warehouse'/>
</owl:ObjectProperty>

<owl:ObjectProperty rdf:ID='fitsOnTruck'>
  <rdfs:domain rdf:resource='#Product'/>
  <rdfs:range rdf:resource='#Truck'/>
</owl:ObjectProperty>

<owl:ObjectProperty rdf:ID='wLoc'>
  <rdfs:domain rdf:resource='#Warehouse'/>
  <rdfs:range rdf:resource='#Location'/>
</owl:ObjectProperty>

<owl:ObjectProperty rdf:ID='tLoc'>
  <rdfs:domain rdf:resource='#Truck'/>
  <rdfs:range rdf:resource='#Location'/>
</owl:ObjectProperty>
```

The challenge is to define a class Shippable of products that are in a warehouse where a suitable truck is available. In predicate logic, this could be represented as

$$\text{shippable}(x) \equiv (\exists y, z, w : \text{inWarehouse}(x, y) \wedge \text{fitsOnTruck}(x, z) \wedge \text{wLoc}(y, w) \wedge \text{tLoc}(z, w))$$

Such a simple formula cannot be expressed in OWL, because it involves a nontree graphical

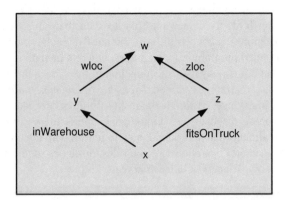

Figure 8.2: A schematic representation for the warehouse and shipping example

model. Figure 8.2, where a cycle is apparent, illustrates the desired formula schematically.

8.8.2   Constraints among Individuals

A representation such as OWL that is based on description logic allows ontological reasoning but not constraint reasoning. For example, an AntiqueBook can be described as a subclass of Book, enabling a Web service to provide information either about all books or just about antique books. Unfortunately, this would require the designer of a bookstore Web service to classify each book as an antique or not. However, because each book has a publication date, it would be easier to define an antique book as one that is published (printed) before, say, 1850. Then, each book would not have to be individually classified. The Web service would only have to reason about whether or not a publication date is prior to 1850, a simple kind of constraint reasoning, but one that OWL does not allow.

8.8.3   Specialized Properties

OWL allows properties to be subproperties of others. It also captures restrictions. Now consider the following setting. Assume we have modeled a class Animal and various subclasses of it, such as Mammal and Reptile. Assume we have defined a property hasChild from Animal to Animal. Going further, we would like to state that the child of a Mammal must be a Mammal and the child of a Reptile must be a Reptile. It turns out this is not possible in OWL. We can, however, define different subproperties of hasChild, such as hasMammalChild and hasReptileChild, but doing so would proliferate the properties in the model.

8.8.4   Defeasible Concepts

In the same vein, a lot of commonsense knowledge has the flavor of defaults. A classical example is whether birds fly. We would like to state that birds fly and yet be able to accommodate that penguins, a subclass of birds, do not fly. Such reasoning is operationally reflected in object-oriented programming languages, where a method in a subclass (imagine a method canFly on penguins) may override a method of the same name in a superclass. The logical treatment of such reasoning—termed *defeasible* or *nonmonotonic*—is nontrivial. The ontology framework, in essence, partitions the responsibilities between ontologies and other general-purpose reasoning mechanisms. Ontologies capture the basic definitions whereas other application-specific mechanisms handle how to reason with the ontologies when given specific data and usage context. Section 15.7, on rules, discusses such reasoning in practical settings and includes some aspects of defeasible reasoning.

8.9   Notes

The OWL proposed standard was created by a standards committee called the Web ontology (WebOnto) group, whose main input was the representation language called DAML+OIL. DAML+OIL itself arose from the synthesis of two preceding efforts, namely, the DARPA Agent Modeling Language (DAML) and the Ontology Interchange Language (OIL). OIL sometimes is said to stand for the Ontology Inference Layer. DAML is discussed in [DAML] and RDF is discussed in Decker *et al.* [2000a].

A number of tools for OWL have been implemented. Jena, described in Section 7.9, has libraries that support using OWL on top of RDF. Protégé, introduced in Section 6.10, supports editing ontologies and generates OWL documents automatically.

A number of effective validators for OWL now exist. These include the BBN OWL Validator, which is available from http://owl.bbn.com/validator/. It is also known as the *vOWLidator*. vOWLidator has both a downloadable tool and a Web form interface. Another validator is RACER: Renamed ABox and Concept Expression Reasoner, available from http://www.cs.concordia.ca/~haarslev/racer/. RACER integrates with Protégé.

The theoretical study of ontology modeling languages is a major research area. Interested readers can pursue the references given above to learn more about specific algorithms, the complexity of the algorithms and of the decision problems, and the trade-offs involved in designing the representation languages.

8.10   Exercises

8.1. Which one of the following best characterizes what OWL can be used for?

- Creating ontologies.
- Describing Web services.

- Describing applications.
- Defining XML Schema types.
- Creating multicast SOAP messages.

8.2. Which one of the following best represents the main entities we define in OWL?

- Classes and properties.
- Elements and attributes.
- Services and bindings.
- Ports and services.
- Strings and integers.

8.3. Define a property parentOf relating animals and a class Parent as a restriction on parentOf of minimum cardinality 1. Further, write an OWL specification for grandparent.

8.4. Beginning from Exercise 8.3 and adding a property sex, write an OWL specification for grandmother and for maternal grandmother.

8.5. Produce an OWL description of the following scenario. Express the solution as a graph with suitable annotations of labels or constraints. Use the following abbreviations: student (S); faculty member (F); regular faculty member (R); department (D); thesis committee (T).

- An S belongs to exactly one D.
- An R is an F.
- An R advises zero or more Ss.
- An F is affiliated with one or more Ds.
- An S is advised by exactly one R.
- An S is evaluated by exactly one T.
- A T evaluates exactly one S.
- A T has three or more Fs as its members.
- Exactly one of the members of a T is its chair, who is an R.

8.6. Prove or disprove that an OWL property is transitive if and only if its inverse is transitive.

8.7. Section 6.6.2 described the *isPartOf* relation. Formalize this relation as an OWL property, capturing the formal properties discussed for it in that section.

8.8. Give distinct examples of OWL specifications for describing properties that behave as a partial function, a total function, an injective function, a surjective function, and a bijection, respectively. Make sure that the examples are natural or realistic, i.e., they are about some real-world (or even computational) phenomenon that you did not just conjure up for this exercise.

8.9. Suppose you were asked to augment your solution to Exercise 8.5 to capture the following additional facts and constraints. How would you go about capturing this knowledge?

- An S is advised by exactly one R, who is affiliated with the D to which the S belongs.
- A T has three or more Fs as its members, each of whom is affiliated with the D to which the evaluated S belongs.
- Exactly one of the members of a T is its chair, who is an R and belongs to the same department as the student.
- A T consists of zero or one nondepartmental members, each of whom is affiliated with a D different from the one to which the evaluated S belongs.

8.10. Construct an OWL-DL ontology for a small domain of your choice.

8.11. Produce an OWL document covering the following university scenario. Assume the classes student, course, and department, as well as the properties takes and offeredBy. You can define auxiliary classes and properties if you like. Use the RDF-XML syntax.

- Define a cross-listed course as one that is offered by two or more departments.
- A student is a member of exactly one department.
- Define an unlucky student as one who takes a course that is taken by a hundred or more students.

8.12. How many instances of rdfs:Class are defined by the current standard for OWL?

8.13. Which one of the following is an example of a statement that can be defined in OWL but not in RDF or RDFS?

- Things that have the isTall property cannot have the isShort property.
- Anything that has the isTall property is a human.
- All things that are *Men* are *Human*.
- The property hasIncome maps a *Worker* to a *Real*.
- There is a class called *Firefly*.

8.14. Develop an OWL model for basic aspects of Web services: include consideration of service provider, requester, registry, and the publishing, discovery, and invocation of services. State what you can about the parameters of services.

Chapter 9

Ontology Management

In early work on ontologies, there was often an implicit assumption that the process would proceed in the following manner. You would build or find an ontology for your domain. Possibly, this would be a standard ontology. Then you would relate the models for your information sources with this ontology by defining appropriate formal mappings. The mappings between the ontology and each information source would be composed to yield direct mappings between the information sources themselves. Thus, the ontology would have assisted in relating the given information sources.

This approach is intuitive and it yields one significant advantage over its precursor, namely attempting to create direct mappings between information sources from scratch. If you have n sources, you would need a total of n^2 mappings (one for each direction of each pairing). However, if you have a shared domain ontology, you only need $2n$ mappings. Since the creation of mappings is a human-intensive task, this reduction in complexity is enormous, going from what is practically intractable to what seems quite feasible.

However, the above approach has one significant shortcoming. It calls for the existence of an ontology that is expressive enough to yield mappings from the n sources. In practice, such an ontology would be difficult to come by. However, such an ontology can be built provided you are willing to augment it as you go along. In other words, instead of starting with a sufficiently complete ontology, you start with an ontology that is the best one you can locate. In the worst case, this could be an empty ontology. Next you attempt to map your information sources one by one to this ontology. As you proceed, you enhance the ontology itself. Of course, you do not need to create all the mappings in one shot. You might create a mapping and then post the enhanced ontology. Another person may create another mapping and post their enhanced ontology, and so on.

The above is clearly attractive as an approach for building ontologies. However, it opens up the following challenges. It is conceivable, and indeed likely, that the enhancements to an ontology will modify its structure sufficiently to call the previous mappings into question. Likewise, different people modifying an ontology could make conflicting modifications. Therefore, a resulting challenge is to reconcile the modifications to an ontology. The above

of new knowledge-based applications, which otherwise would place onerous demands for knowledge acquisition. By amortizing the costs of acquiring knowledge, Cyc can potentially assist in the development of knowledge-based systems. Cyc has been used to support inter-operation among information resources [Huhns et al., 1998]. An open-source version of the upper ontology of Cyc, containing approximately 6 000 concepts and 60 000 assertions about the concepts, is available at http://www.opencyc.org/.

9.2.3   IEEE Standard Upper Ontology

The IEEE Standard Upper Ontology (SUO) is the proposed outcome of a working group whose purpose is to specify a standardized set of concepts, axioms, and relationships that describe a domain of interest. The resultant common ontology will be suitable for auto-mated reasoning, will form the basis for interoperability among software and database appli-cations, and will support applications from e-commerce to natural language understanding. The working group has produced a metaontology, encoded in the Knowledge Interchange Format (KIF). (KIF is now included in the proposed ISO standard called Common Logic.) Using this metaontology, other ontologies can be related. In particular, it enables upper onto-logies and domain ontologies to be aligned and unified. An example upper ontology is shown in Figure 9.2.

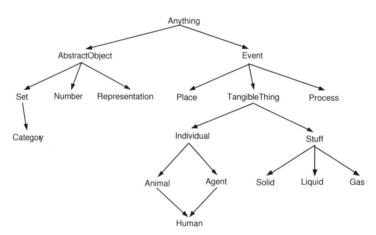

Figure 9.2: An example upper ontology. The links represent specialization

9.3   Standardization versus Semantic Reconciliation

Let us consider the following basic problem: a search will typically uncover a large number of independently developed information sources—some relevant and some irrelevant. The sources might be ranked, but they are otherwise unorganized, and there are too many for a user to investigate manually. The problem is familiar and many solutions have been proposed, ranging from requiring the user to be more precise in specifying search criteria, to constructing more intelligent search engines, or to requiring sources to be more precise in describing their contents. A common theme for all of the approaches is the use of ontologies for describing both requirements and sources. Unfortunately, ontologies are not a panacea, and are not effective unless everyone adheres to the same one, and no one has yet constructed an ontology that is comprehensive enough (in spite of determined attempts to create one, such as the Cyc Project, described in the previous section). Moreover, even if one did exist, it probably would not be adhered to, considering the dynamic and eclectic nature of the Web and other information sources.

Different communities of practice often tend to have standardized vocabularies and concepts. So it is natural to expect that the ontologies of different domains be created and ratified by the appropriate standards bodies or professional societies. Good standards are clearly desirable because they can reduce the wasted effort and unnecessary discrepancies among models of services. For instance, if our interest is in modeling the databases of a chemical factory, we would use modern knowledge of chemistry and would not need to build models that, say, discussed the phlogiston theory (discredited a long time ago with the discovery of oxygen).

However, there are some major challenges indicating that we cannot rely exclusively upon standards.

- Universally agreed-upon standards address only a small body of the knowledge for dealing with different applications. For example, in chemistry, chemical bonds might be well understood, but the specifics of many processes, especially those that are still of competitive value, are not likely to be standardized yet.

- Even if a standard emerges in a given domain, we still must accommodate resources that exist prior to the creation of the standard.

- Further, even if a standard is established for some domain, new work will have to go beyond the standard and thus will venture into uncharted territory.

In other words, we would claim that heterogeneity is the normal state of affairs, with small bursts of homogeneity when specific standards are adopted. Consequently, we must always be able to handle heterogeneous models. For this reason, techniques and tools that support reconciliation of ontologies will remain important.

9.4   Consensus Ontologies

Ontologies are critical for interoperation but are difficult to build. Further, even if an ontology is available, it may not necessarily be accepted by all the parties involved. An approach that suggests itself is to attempt to develop a unified ontology from a set of ontologies. These ontologies would be heterogeneous and might be inconsistent, but if they apply to the same universe of discourse, they might have sufficient overlap to enable a consensus ontology to be effectively induced.

Organizational knowledge typically comes from many independent sources, each with its own semantics. Corporate information searches can involve data and documents both internal and external to the organization. We describe below a methodology by which information from large numbers of such sources can be associated, organized, and merged.

9.4.1   Analysis

The approach hypothesizes that a multiplicity of ontology fragments, representing the semantics of the independent sources, can be related to each other automatically without the use of an existing global ontology. That is, any pair of ontologies can be related indirectly through a semantic bridge consisting of many other previously unrelated ontologies, even when there is no way to determine a direct relationship between them. The relationships among the ontology fragments indicate the relationships among the sources, enabling the source information to be categorized and organized.

The methodology relies on sources that have been annotated with ontologies [Pierre, 2000]; such annotation is consistent with the semantic Web, e.g., Heflin and Hendler [2000]. The domains of the sources must be similar—else there would be no interesting relationships among them—but they will undoubtedly have dissimilar ontologies, because they will have been annotated independently.

Other researchers have attempted to merge a pair of ontologies in isolation, or merge a domain-specific ontology into a global, more general ontology [Wiederhold, 1994]. To our knowledge, no one has previously tried to reconcile a large number of domain-specific ontologies. A preliminary evaluation of the methodology has been conducted by relating 53 small, independently developed ontologies for a single domain. One of these small ontologies is shown in Figure 9.3. A nice feature of the methodology is that common parts of the ontologies reinforce each other, while unique parts are deemphasized. The result is a consensus ontology.

9.4.2   Reconciling Ontologies

In agent-assisted information retrieval, a user will describe a need to his agent, which will translate the description into a set of requests, using terms from the user's local ontology. The agent will contact on-line brokers and request their help in locating sources that can satisfy the requests. The agents must reconcile their semantics in order to communicate about the request. This will be seemingly impossible if their ontologies share no concepts. However, if their ontologies share concepts with a third ontology, then the third ontology might provide a *semantic bridge* to relate all three. Note that the agents do not have to relate their entire ontologies, only the portions needed to respond to the request.

The difficulty in establishing a bridge will depend on the semantic distance between the concepts involved, and on the number of ontologies that comprise the bridge. Our methodology is appropriate when there are large numbers of small ontologies—the situation we

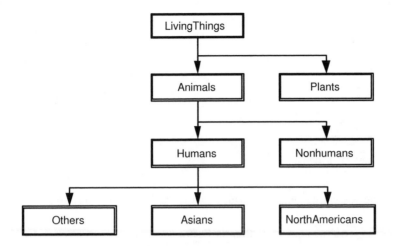

Figure 9.3: A typical small ontology used to characterize an information source about people (all links denote subclasses)

expect to occur in large and complex information environments. Our metaphor is that a small ontology is like a piece of a jigsaw puzzle, as depicted in Figure 9.4. It is difficult to relate two random pieces of a jigsaw puzzle until they are constrained by other puzzle pieces. We expect the same to be true for ontologies.

In Figure 9.4, the ontology fragment on the left would be represented as partOf(Wheel, Truck), while the one on the right would be represented as partOf(Tire, APC). There are no obvious equivalences between these two fragments. The concept Truck in the first ontology could be related to APC in the second by equivalence, partOf, hasPart, subclass, superclass, or other. There is no way to decide which is correct. When the middle ontology fragment partOf(Wheel, APC) is added, there is evidence that the concepts Truck and APC, and Wheel and Tire could be equivalent.

- A concept in one ontology can have one of seven mutually exclusive relationships with a concept in another:

  - subclass
  - superclass
  - partOf
  - hasPart
  - sibling
  - equivalence
  - other

- Each ontology adds constraints that can help to determine the most likely relationship

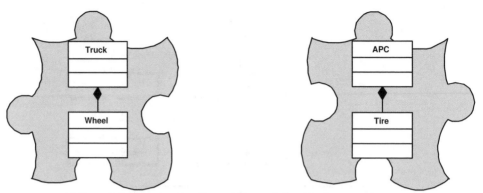

(a) Two ontology fragments with no obvious relationships between them

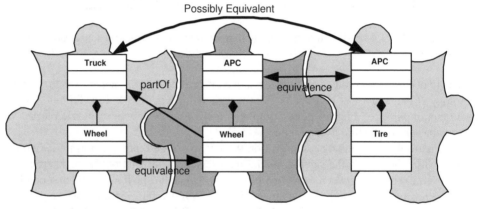

(b) The introduction of a third ontology reveals equivalences between components of the
original two ontology fragments

Figure 9.4: Ontologies can be made to relate to each other like pieces of a jigsaw puzzle

In attempting to relate two ontologies, a system might be unable to find correspondences
between concepts because of insufficient constraints and similarity among their terms. However, trying to find correspondences with other ontologies might yield enough constraints to
relate the original two ontologies. As more ontologies are related, there will be more constraints among the terms of any pair, which is an advantage. It is also a disadvantage in that
some of the constraints might be in conflict. However, we can make use of the preponderance
of evidence to resolve these conflicts statistically.

The following process is described in the context of a merger for a Human (alternatively,
People or Person) ontology.

1. Using string-matching and other such heuristics, merge the ontologies by identifying
   various classes.

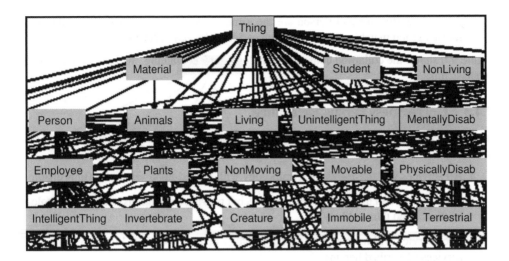

Figure 9.5: A portion of the ontology formed by merging 53 independently constructed onto-logies for the domain Humans-People-Persons. The entire ontology has 281 concepts related by 554 subclass links

2. Construct a consensus ontology by counting the number of times classes and subclass links appeared in the component ontologies when performing the merging operation. The component ontologies described 864 classes, while the merged ontology contained 281 classes in a single graph with a root node of the OWL concept #Thing. All of the concepts were related, i.e., there was some relationship (path) between any pair of the 281 concepts (see Figure 9.5).

3. The *reinforcement* of a concept refers to how often a class or subclass relationship appears in the given ontologies (after they have been matched). For example, the class Person and its matching classes appeared 14 times. The subclass link from Mammals (and its matches) to Humans (and its matches) appeared 9 times.

4. Redundant subclass links are removed and the corresponding transitive closure links were reinforced. That is, if C has subclass A with reinforcement 2, C has subclass B reinforced m times, and B has subclass A reinforced n times, then the link from C directly to A is removed and the remaining link reinforcements are increased by 2. Next, any classes or links that are not reinforced are then removed from the merged ontology.

5. An equivalence heuristic can be applied for collapsing classes that have common rein-forced superclasses and subclasses. For example, the equivalence heuristic found that all reinforced subclasses of Person are also reinforced subclasses of Humans, and all reinforced superclasses of Person are also reinforced superclasses of Humans. It thus

refer to other OWL ontologies, whose definitions are then imported into the current ontology. This is in contrast with RDF Schema documents where a schema document does not need to be explicitly imported.

An ontology automatically imports the ontologies imported by the ontologies that it directly imports—this is only natural, because otherwise the imported ontologies would not be interpretable. Importing drops the definitions of the imported ontology into the imported ontology. If two ontologies are defined as importing each other, then the sets of assertions in each is the same. Thus, they are considered equivalent.

Listing 9.1: An example of ontology imports in OWL

```
<?xml version="1.0"?>
<rdf:RDF xmlns:rdfs="http://www.w3.org/2000/01/rdf-schema#"
  xmlns:owl="http://www.w3.org/2002/07/owl#"
  xmlns:rdf="http://www.w3.org/1999/02/22-rdf-syntax-ns#">

<owl:Ontology rdf:about="Life">
  <owl:versionInfo>$Id: Life.owl,v1.0 2003/12/01
                   12:35:31 huhns Exp$
  </owl:versionInfo>
  <rdfs:comment>An Ontology for Life</rdfs:comment>
  <owl:imports
    rdf:resource="http://www.w3.org/2002/07/owl"/>
</owl:Ontology>

<!-- Definitions go here -->

</rdf:RDF>
```

Recall that an XML namespace is simply a namespace. It provides a means for uniquely naming terms, such as elements. The definitions of the terms are not included as they are with OWL imports. In fact, the URI given in a namespace declaration may not even correspond to a real resource. Conversely, an OWL import is simply an import. It does not set up a namespace. Thus, it is possible to import ontologies that define terms but do not introduce any new names. Typically, however, we would see matching pairs of namespace declarations and import statements.

OWL provides a property owl:priorVersion applied to owl:Ontology. OWL also provides properties to declare backward compatibility or incompatibility. The owl:versionInfo applies not only to the owl:Ontology element, but also to other elements, e.g., those describing classes and properties.

```
<owl:Ontology rdf:about="Life">
  <owl:versionInfo>$Id: Life.owl,v1.0 2003/12/01
                   12:35:31 huhns Exp$
```

```
  </owl:versionInfo>
  <owl:priorVersion rdf:resource='URI_to_prior_version'/>
  <owl:backwardCompatibleWith rdf:resource='URI_to_prior_version'/>
  <owl:incompatibleWith rdf:resource='URI_to_prior_version'/>

</owl:Ontology>
```

Further, there is an ability to declare classes and properties as deprecated.

9.6 Notes

Ontologies are typically constructed with the aid of editors. Common ones for the modern languages are Protégé, WebOnto, and OilEd. Of these, Protégé has a new version for OWL.

The Cyc project http://www.cyc.com has developed a large-scale ontology to support commonsense reasoning.

9.7 Exercises

9.1. Compare XML Schema's type extension mechanism with RDF Schema's subclassOf. Are members of an XML Schema subtype members of a supertype?

9.2. Extend the consensus ontology approach in the following ways.

- Improve the algorithm for relating ontologies, based on methods for partial and inexact matching, making extensive use of common ontological primitives, such as subclass and partOf. The algorithm should take as input ontology fragments and produce mappings among the concepts represented in the fragments. Constraints among known ontological primitives should be used to control computational complexity.

- Develop metrics for successful relations among ontologies, based on the number of concepts correctly related, as well as the number incorrectly matched. The quality of a match will be based on semantic distance, as measured by the number of intervening semantic bridges.

9.3. Search the Web for publicly available DTDs or XML Schemas. Find at least two that have at least some degree of overlap in the concepts they include. For example, one might be about widgets and one might be about buying and selling things, so the overlap would be the buying and selling of widgets. Turn in a listing of the DTDs or schemas, a written description of what each covers, and a written description of the overlap.

9.4. Relate the concepts (each element and attribute) in each DTD or schema from the previous exercise to the concepts in the Cyc ontology that they best match. A relationship

Chapter 10

Execution Models

Interoperation can occur at multiple levels of processing and abstraction. *Effective* interoperation must occur at all of the levels listed below; this is made easier by the availability and widespread adoption of standards (also listed, where appropriate), but it also requires surmounting a series of challenges at the process level and above, as later sections of this text address.

Transport. HTTP, SMTP, and SIP.

Formatting. XML.

Messaging. SOAP.

Data and structure. WSDL.

Finding and binding. UDDI.

Semantics. Ontologies, as expressed in RDF, RDFS, and OWL.

Transaction. WS-AtomicTransaction, WS-BusinessActivity, WS-Coordination, and BTP.

Process. OWL-S, BPEL4WS, PSL, and WSCI.

Policy. XACML.

Dynamism. Agents.

Cooperation. Multiagent systems.

It is worth considering the architectural elements over which services are engaged. At the lowest level, which is included in commercial application servers, we find support for directories and messaging. Above this, in the data and process interoperation layer, we find metadata and transformations, message routing, and possibly rule engines to handle routing and "transactional" qualities, ensuring that all or none of a certain set of messages is sent or received.

Message-oriented middleware represents a major class of interaction, which conceptually differs from invocation. In practice, it is usually implemented through an invocation-based interface wherein callbacks are registered by the parties interested in receiving notifications. There are two main varieties:

Queues. Queues enable point-to-point communication. Components can post and read messages from queues.

Publish and subscribe. These so-called *pub-sub systems* enable applications to define several *topics*. Components can subscribe to specific topics so that messages published to matching topics are made available to them. Typically, a message posted to a given topic may be received by no more than one subscriber.

It is also worthwhile to explore the distinction between push and pull, which are the two main ways to program with messaging. Pull brings information into a receiver's context when the information is needed, but the sender must have it ready or be able to prepare it on demand; push offers information from a sender's context, but the receiver must be ready or be able to absorb it in a different context. Push interrupts the receiver: this can be intrusive, but also helpful. Therefore, push should be confined to control signals, while pull is used for data. However, push and pull are both low-level abstractions.

10.3   CORBA

This section gives a short, practical introduction to the Common Object Request Broker Architecture (CORBA), which was developed by the Object Management Group (OMG). It provides a good understanding of the basic mechanics of the CORBA architecture, overviews its components, and defines the relevant vocabulary.

An *object request broker* (ORB) is the distributed service (in the systems sense introduced in Section 1.4) that implements a request to a remote object. An ORB locates the remote object on the network, communicates the request to the object, waits for the results, and, when available, communicates those results back to the client. An ORB implements location transparency and programming language independence: the same request mechanism is used by the client and the CORBA object regardless of where the object is located, and the client and CORBA object can be written in different languages. CORBA also defines the Internet Inter-ORB Protocol (IIOP) for communications among ORBs.

The services that an object provides are given by its interface, defined in OMG's Interface Definition Language (IDL). In the OMG Object Model, clients request services from objects through this interface. A client accesses an object by issuing a request to the object. The request is a CORBA *event*, and it carries information about which operation is requested, the object reference of the service provider, and actual parameters, if any. The object reference is a name that identifies an object reliably. An object adaptor handles functions such as the generation and interpretation of object references, method invocation, security of interactions, object and implementation activation and deactivation, mapping references corresponding to object implementations, and registration of implementations. To do this, it obtains information about an object's location and operating environment from a database, called the "Implementation Repository."

As an example of how CORBA and its functionality might be used, the following is an IDL description of an object, a *Catalog*, and the services it provides:

```
module CatalogObjects {

  struct Item {
    string name;
    double price;
  };

  exception Unknown{};

  interface Catalog {

    // Returns the current catalog item
    Item get_item() raises(Unknown);

    // Sets the current catalog item
    void set_item(in Item catalog_item);
  };
};
```

An IDL compiler would convert the above module into its associated representation in a target programming language, such as Java. After compilation, a client written in Java could obtain a catalog item by the following code fragment:

```
Catalog theCat = ...
try {
    Item current_item = theCat.get_item();
}
catch (Throwable e) {
}
```

The IDL compiler would also generate a skeleton class for the object implementation. A skeleton is the entry point into the distributed object. It unmarshals incoming data, calls the method implementing the operation being requested, and returns the marshaled results. A developer must implement a Java class that extends the generated skeleton class and provide a method for each operation in the interface. In the example, the IDL compiler generates the skeleton class _CatalogImplBase for the Catalog interface. A possible implementation of the Catalog interface is:

```
public class CatalogImpl extends
      CatalogObjects._CatalogImplBase {

  private Item _item=null;

  public CatalogImpl(String name) {
    super();
  }

  public Item get_item() throws Unknown {
    if (_item == null) throw new Unknown();
    return _item;
  }

  public void set_item(Item item) {
    _item = item;
  }
}
```

The Java code below defines a server that when run makes the services of its objects available to clients. A server that will run with the Java ORB needs to do the following:

- define a main method;

- initialize the ORB;

- instantiate at least one object;

- connect each object to the orb;

- wait for requests.

```
public class theServer {
  public static void main(String[] args) {
    try {
      // Initialize the ORB
      org.omg.CORBA.ORB orb =
        org.omg.CORBA.ORB.init(args, null);

      // Create a catalog object
      CatalogImpl theCat = new CatalogImpl();

      // Let the ORB know about the object
      orb.connect(theCat);

      // Write stringified object reference
```

```
    PrintWriter out =
       new PrintWriter(new BufferedWriter(
               new FileWriter(args[0])));
    out.println(orb.object_to_string(theCat));
    out.close();

    // Wait for invocations from clients
    Object sync = new Object();
    synchronized (sync) {
       sync.wait();
    }
  } catch (Exception e) {
    System.err.println("Catalog error: " + e);
    e.printStackTrace(System.out);
  }
 }
}
```

The services that CORBA provides to objects are shown in Table 10.1. One of these, the CORBA *event service*, is a generalization of techniques for maintaining referential integrity. One way to maintain integrity constraints is to notify other objects of changes in a given object. The event service separates notification from an object's program logic, which decides how to accommodate the notification.

Table 10.1: CORBA Services

Service	Description
Object Life Cycle	Defines how CORBA objects are created, removed, moved, and copied
Naming	Defines how CORBA objects can have symbolic names
Event	Decouples the communication between objects
Relationship	Provides typed n-ary relationships between objects
Externalization	Coordinates the transformation of CORBA objects to and from external media
Transaction	Coordinates atomic access to CORBA objects
Concurrency Control	Provides a locking service for CORBA objects in order to ensure serializable access
Property	Supports the association of name-value pairs with objects
Trader	Supports the finding of CORBA objects based on properties describing the service offered by the object
Query	Supports queries on objects

ORB communications are of the main three kinds—synchronous, asynchronous, and deferred synchronous—described in Section 10.1. The CORBA event service architecture uses event channels to decouple communication. Each event from a supplier is sent to every consumer. The amount of storage for notifications affects the QoS of the channel, and is left to implementers.

CORBA allows different ORBs in different administrative domains to interoperate, but this requires a *bridging mechanism*. Different domains might have different security policies and different semantics. A bridging mechanism provides transformations that reconcile the differences.

The Object Management Group in late 2002 defined interoperability between CORBA, WSDL, and SOAP. Once developers have included SOAP and WSDL transports in their CORBA implementations, existing CORBA applications will be able to use a Web service transport without having to be changed.

10.4   Peer-to-Peer Computing

How much would you like to share files with another user without having to place them in an explicitly designated external location? Since the late 1990s, the successes of (and controversies surrounding) Napster, Gnutella, and FreeNet have drawn attention to peer-to-peer computing, a form of computing which allows precisely such client-to-client interactions. Consequently, peer-to-peer computing, or P2P for short, has become a popular topic. P2P, however, promises to go beyond file sharing or any other specific application. It can provide a general substrate for supporting distributed computation, for example, in Web services. This section briefly examines P2P and its main variants, both those that are popular and those that ought to be.

First, here is a brief motivation and definition. P2P can be defined most easily in terms of what it is not: it is not the client-server model, which is currently the most common model of distributed computing. In the client-server model, an application residing on a client computer invokes commands at a server. The client-server model is simple and effective but it has serious shortcomings. Each client interacts with the server independently of other clients. This property is codified in the traditional transaction model—great for isolation (as among bank accounts) but not so great for collaboration. Chapter 13 discusses transactions at length.

P2P is by no means a new idea. The distributed computing research community has studied it for decades. Networks themselves demonstrate P2P in action: Ethernet is nothing if not a P2P protocol, and network routing operates with routers acting as peers of routers to whom they send packets and from whom they receive packets. The difference prompting the recent attention seems to be that P2P finally has caught the imagination of people building practical systems at the application layer. And this is a significant expansion in its scope.

10.4.1 Going Beyond Client-Server

The centralization of information on servers makes for performance bottlenecks and makes the overall systems susceptible to failure at a single point. Cluster computing and even content networking are inspired by the idea of preserving the logical centrality of servers while replicating them to build sufficient redundancies into a system to sustain higher performance and to help the system's functionality degrade better under failure. The implicit claim behind P2P is that these redundancies are not enough. Let us examine the various motivations for this claim.

As long there is some centralization in a system it can be controlled from that central location. Arguably, Napster's original incarnation could be shut down simply because it had a server and the site running that server can be sued. P2P approaches that lack such a central server, such as FreeNet, would be much harder to shut down.

In client-server computing, the control rests entirely in the client; the server merely responds to requests. By requiring all control to reside on the client, the client-server model forces applications to be structured in such a way that coordination between their components is rigid. P2P can support richer models of interaction than client-server.

10.4.2 Models of P2P Computing

P2P computing models take on three main forms.

Symmetric client-server. Each party can query the other, thereby gaining power over the other at different times. The idea of symmetry is appealing in general, but this particular form, often touted as an explanation of P2P, is inherently limited, because it does not fundamentally look beyond client-server.

Asynchrony. In the client-server model, if the client needs to know of changes observed by the server, it must poll the server to learn of them. As peers, computations naturally communicate asynchronously. This is the original form of P2P. While the request-response paradigm corresponds to pull, asynchronous communication corresponds to push. Push, unfortunately, received bad publicity with applications that place their entire intelligence on the server (pushing) side. Pushing ads is viewed as spamming.

Federation of equals. When we take asynchrony but apply it in settings where the parties in principle have an equal share in and equal control over their joint computation, we can create applications of a performance and usability that is simply unattainable by client-server. This means symmetry not only at the level of communication, but also in terms of decision-making. This would require making the reasoning and policies of interaction explicit and enabling each party to take and concede the initiative as it sees fit. In simple terms, only intellectual equals can be true peers.

How we understand P2P depends on how we judge peerhood. The criterion that matters most for computer scientists is the model of computation supported (notwithstanding its political or marketing ramifications). And, it is only when we support flexible interactions that we achieve true P2P and realize the benefits that the Internet offers to computing. Chapter 15 discusses agents, which capture the above notion of P2P computing.

10.5   Jini

Jini is Sun Microsystems' system architecture for distributed computing among interconnected devices. Jini extends Java from one machine to a network of machines. It uses Remote Method Invocation (RMI) to move code around a network, and provides mechanisms for devices, services, and users to join and detach from the network. When a Jini device is plugged in, it announces itself to a special network service, a *matchmaker*. This is all it takes to enable the device to access and use other devices on the network and in turn to be usable by those devices.

A Jini announcement is a 512-byte packet that is broadcast to the network. The network replies with a description of itself so that the device can access its services. The device then sends a message registering its own capabilities with the network. Devices find each other through the matchmaker via a lookup process. Once they are matched, devices can interact directly. Jini is essentially a model for Web services (compare Figure 10.1 to Figure 2.1) within a local-area network, which might be implemented using Bluetooth or the IEEE 802.11x wireless LAN protocols. The Jini infrastructure for providing local-area Web

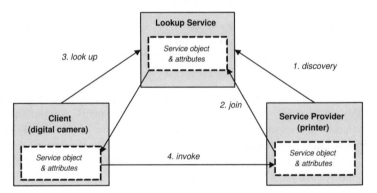

Figure 10.1: Jini services and protocols implement a local service-oriented architecture

services has the following positive and negative characteristics:

- Transactions support a two-phase commit protocol (positive).

- Clients lease services for specific durations (positive).

- Lookup services can be arranged hierarchically (positive).

- Services occupy nodes in tuple spaces, called JavaSpaces (positive).

- Lookup services require an exact match on the name of a Java class or its subclass (negative).

- Clients and servers exchange code *synchronously* via RMI (negative).

The development of and agreement on protocols for wireless local-area networks, such as Bluetooth and IEEE 802.11x provides a key enabler for Jini. However, its widespread deployment in practical settings also depends on (1) the ubiquity of wireless devices within homes, offices, and commercial establishments, and (2) a favored computing model based on synchronous invocations of individual services.

10.6 Grid Computing

The Grid was originally designed to be an environment with a large number of networked computer systems where computing and storage resources could be shared as needed and on demand. The development of standards, originally the Open Grid Service Architecture (OGSA) and more recently the Web Services Resource Framework (WS-Resource Framework), is leading the Grid toward an environment that is suited not only for computationally intensive applications, but also for typical distributed computing scenarios, such as information retrieval, multimedia environments, and ubiquitous computing.

Grid computing can be differentiated from other distributed computing paradigms, such as CORBA, P2P, and cluster computing, by the following characteristic: Grid computing involves the efficient utilization of an organization's heterogeneous, loosely coupled *resources* tied to workload management capabilities or information virtualization. The objective of workload management is to allocate resources to the most important applications.

To emphasize the differences from other paradigms, cluster computing involves a *static* number of *homogeneous* computing elements, and is concerned with computing, not storage. CORBA presumes an *object orientation*. Unlike P2P, Grid computing relies on centralized management and security, making it easier to manage but less scalable and less robust.

The purpose of the WSRF is to align Grid services with Web services. In this regard, a Grid service is a (potentially transient) Web service with specified interfaces and conventions: the interfaces address discovery, dynamic service creation, lifetime management, notification, and manageability, while the conventions address naming and upgradeability. As an example of transience, extra service instances supporting a Web server might be instantiated as needed to provide for constant user response time, i.e., capacity is added dynamically as workload increases. Other examples of transient service instances might be a query against a database, a data mining operation, a network bandwidth allocation, a running data transfer, and an advance reservation for processing capability.

The Globus Toolkit is an open-architecture set of services and libraries that supports Grid computing. It provides the

- Grid Resource Allocation and Management (GRAM) protocol and its gatekeeper (factory) service; these provide for the secure and reliable creation and management of arbitrary computations, termed transient service instances.

- Grid Security Infrastructure (GSI), which supports single sign on, delegation, and credential mapping. A two-phase commit protocol is used for reliable invocation.

- Meta Directory Service (MDS-2), which provides for information discovery through soft-state registration, data modeling, and a local registry.

The toolkit is being aligned with WSRF, so that its capabilities are expressed in Web service terminology.

The WS-Resource Framework, along with WS-Notification (another outgrowth of OGSA that deals with events), defines a common, standards-based infrastructure for business applications, Grid resources, and systems management. These specifications help in integrating heterogeneous resources and systems across and outside of an enterprise. The specifications provide the ability to utilize common Web services to support Grid and management-based solutions. The WS-Notification specification and the WS-Resource Framework provide a scalable pub-sub messaging model and the ability to model stateful resources using Web services.

Stateful resources are elements with state, including physical entities such as servers and logical constructs such as business agreements and contracts. Examples of components that may be modeled as stateful resources are files in a file system, rows in a relational database, and encapsulated objects such as Enterprise Java Beans. A stateful resource can also be a collection or group of other stateful resources. Access to these stateful resources enables customers to realize business efficiencies, including just-in-time procurement with multiple suppliers, systems outage detection and recovery, and Grid-based workload balancing. For example, a business might provide a Web service where customers can submit a purchase order. The business must maintain the state of the purchase in order to provide additional Web services whereby a customer can check on the status of the order or add an item to it.

Statelessness is desirable, however, because it can enhance reliability and scalability: a stateless Web service can be restarted without regard for its history; an arbitrary number of copies can be created to meet increased loads. To preserve these qualities without storing a static state within a Web service, dynamic state can be provided either within a request message (directly, by-name, or by-reference) or within other components with which the Web service can interact.

WS-Resource Framework provides the ability to (1) determine the type of the state and thus the specific message exchanges that may be supported, and (2) issue read, modify, and query requests against state components. It uses standard XML Schema global element declarations to define resource property elements. These elements are collected in a *Resource* properties document, which is associated with an interface by using an XML attribute on the WSDL 1.1 portType.

WS-Notification proposes to specify an agreed upon definition for events, as well as standardizing broker, publisher, subscriber, and consumer roles. It can automatically trigger an action in the IT infrastructure once certain criteria have been met. This can include, for example, suppliers automatically being notified to bid on replenishing inventory once current inventory drops to a set level. Several suppliers can be notified of this depletion in inventory and WS-Notification can be set up so that only the supplier with the best bid fills the order.

Besides describing how to model stateful resources with Web services, the WS-Resource Framework includes the following:

- *WS-Resource Properties* defines how data associated with a stateful resource can be queried and changed using Web service technologies. This allows clients to build applications that efficiently read and update data associated with resources, such as contracts, servers, and purchase orders.

- *WS-Resource Lifetime* enables a user to specify the period during which a resource definition is valid. WS-Resource Lifetime can, for example, automatically update suppliers from all systems once contracts or service-level agreements expire, or delete from inventory systems products that are no longer being manufactured.

Note that because a Web service can modify a stateful resource, a failed Web service might leave the resource in an inconsistent state. Provision must thus be made for restoring consistency and obeying the ACID properties (see Table 11.1). This can be done by using WS-Transaction (see Section 12.4) to govern use of the stateful Web service.

The problem being addressed by the Grid is resource sharing and coordinated problem solving in dynamic, multi-institutional virtual organizations. The architecture being defined specifies a protocol and service definitions for interoperability and resource sharing. Some implementors of Web services argue that Web services should not have state or instances, and some implementations do not accommodate dynamic service creation and destruction. WSRF makes an explicit distinction between a Web service and the stateful entities acted upon by that service. WSRF calls these entities WS-Resources and introduces an *implied resource pattern* to formalize the relationship between Web services and the stateful resources. As a result, the Grid is becoming an interesting and challenging environment supporting new services for cooperative applications. A research effort is needed not only to investigate innovative Grid infrastructures, but also to make the current Grid model suitable for the emerging usage scenarios.

10.7 Notes

Some useful articles on Grid computing are Foster [2002] and Foster *et al.* [2001]. The Grid community is also moving into a Web service-like model, where resources can be more easily accessed and bound [Tuecke et al., 2002].

10.8 Exercises

10.1. Which of the following are true about IIOP?

- It is the protocol that ORBs should use for talking to each other.
- It is an ORB implementation that is machine-independent.

A transaction is a program in execution. That is, a transaction is a computation, not the source or machine code that describes a computation. Typically, every time a program is run that would lead to a new transaction. Formally, a transaction is a set of database operations that are executed in some (partial) order. For our present purposes, a total or linear order is adequate. (Section 6.5 describes partial and total orders.) Figure 11.1 shows a simplified view of a central (single-site) database interfacing with applications. The applications begin transactions and request data items. The DBMS may give them the data items requested, possibly delaying the transactions arbitrarily. The application indicates its success by committing its transaction and failure by aborting its transaction. The DBMS would respect an abort request, but may deny a commit request by instead aborting a transaction. In fact, it is within the prerogative of the DBMS to abort a transaction unilaterally, if necessary to maintain database integrity or to break deadlocks among transactions.

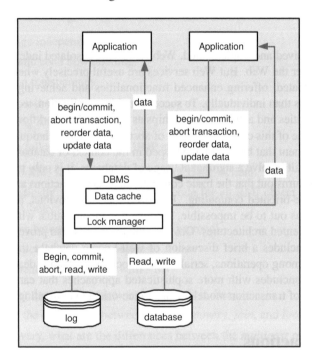

Figure 11.1: Simplified system architecture showing multiple applications running transactions on a single database. The applications could be hosted on one or more servers

The order of execution of the operations of a transaction depends on the source code, but the source code resides in the application and is not accessible to the DBMS. For example, we can imagine different programs to transfer money from one account in a bank to another account in the same bank. Mike's program may first perform a deposit and then a withdrawal, whereas Munindar's program may proceed in the reverse order. Here the deposit and withdrawal (of specified amounts and into or from specified accounts) are operations on the bank's database. Each program would presumably also perform some reasoning, e.g., to validate any user inputs and decide upon the amounts to be deposited and withdrawn. However, such reasoning is not visible to the database and, therefore, not included among the operations with which we model a transaction.

The essence of transactions is the following: when an ordered set of operations is modeled as a transaction, it is as if the set of operations were indistinguishable from *one logical operation*. The effects of a transaction appear to occur completely and become visible to other transactions if and when that single logical operation has concluded successfully; if the operation fails, then it is as if nothing happened. (Traditional transactions always terminate, meaning that their set of operations is finite.) Transactions thus provide a great approach for abstracting computations, because they help us map an arbitrary set of operations to a single operation. Each programmer need only worry about the logical operation that his or her source code implements and disregard the work of other programmers whose programs may be running as transactions on the same database at the same time.

At the elementary level for traditional databases, the transaction concept offers the following deal to the application programmers. If each programmer individually implements his or her transactions correctly, then the DBMS will ensure that any concurrently executing mix of those transactions will execute correctly. Given some assumptions about system recovery, the DBMS will ensure that the results of committed transactions are durable and the results of uncommitted transactions are not reflected in the database.

11.1.1   ACID Properties

Traditional database transactions satisfy the so-called ACID properties [Gray and Reuter, 1993]. A transaction happens entirely or not at all, does not violate consistency constraints, does not expose any partial results, and, if successful, has permanent results. Table 11.1 illustrates these properties.

Notice that because of atomicity and isolation, the above examples of Mike's and Munindar's programs are functionally equivalent. If the deposit and withdrawal operations both succeed or both fail, then their relative order is irrelevant. In general, though, there could be dependencies among the operations because of the flow of information among them or consequences on other decision making by the program logic. For this reason, a transaction is defined with a specific order in mind and the order of operations of a transaction cannot be modified arbitrarily.

Transactions are essential whenever we need definite and correct guarantees about access (viewing and modification) of data. But while the traditional transaction properties are desirable some of the time, they can also be a hindrance at other times. Further, setting up transactions across services is nontrivial, as examined in Chapter 13. In distributed settings, some sort of mutual commit (e.g., two-phase commit) is necessary to prevent violation of the ACID properties.

However, we can see that schedules S_1 and S_2 each yield a final value of 102 for z, but S_3 and S_4 each yield a final value of 101 for z. S_3 and S_4 illustrate a well-known problem of concurrency, known as the *lost update problem*. Our intuition is that S_1 and S_2 are correct whereas S_3 and S_4 are not.

Table 11.2 shows the conflicts for our operations. Any two operations of different transactions on the same data item conflict provided one of them is a write. Notice that other operations can be defined to obtain alternative sets of conflicts. The fewer the conflicts the better. However, conventionally, only read and write are considered in DBMS designs.

Table 11.2: Conflict matrix for transaction operations

	Read (r_j)	**Write** (w_j)
Read (r_i)	No conflict	If r_i precedes w_j, r_i retrieves a previously written value
Write (w_i)	If w_i precedes r_j, r_j retrieves the value written by w_i	The final value of the data item depends on whether w_i or w_j executes second

We define two additional database operations: increment (inc) and decrement (dec). The inc operation atomically increments the value of an integer data item by a given amount, and the dec operation atomically decrements the value of the data item. These operations correspond to a combination of read and write operations (with an intervening arithmetical step) but with one major difference. The inc and dec do not reveal the value of the data item to the transaction in which they are performed. Instead, if a transaction performs a read followed by write, it would become aware of the value of the data item as a consequence of the read. Notice that even if a transaction uses an inc or dec, it can still perform a separate read operation to obtain the value of the data item. The idea behind these operations is that inc and dec are common enough use cases and are often performed where the value of the data item is irrelevant to a transaction. Notice that inc and inc do *not* conflict. That is, if two transactions each increment the same data item, then the mutual order does not matter, because neither transaction reads the incremented value and the final value is the same regardless of the order in which the two increments take place. (For simplicity, let us neglect the possibility of integer overflow, because in case of such exceptions, all bets are off anyway.) Exercise 11.16 asks you to develop the complete conflict matrix incorporating these operations.

11.1.2.2 Serial Schedules

The classical definition of schedule correctness is based on the observation that in a correctly functioning DBMS, service requests, corresponding to different transactions, may come in slowly, one by one. In other words, not knowing the program logic of any transaction and assuming the correctness of the individual programs that the transactions are executions of, we must state that a schedule in which transactions occur one by one is correct. Such a schedule is termed a *serial* schedule. Serial schedules are correct but offer unacceptably poor performance and resource usage. In many settings, serial schedules simply will not work. For example, if an on-line bookstore were to run its sales transactions serially, then it would end up making most of its customers wait for excessively long periods. Indeed, if there is a confirmation step in the sales transactions that involves human input, a single sale would take several seconds to complete, say, one minute (to make up an estimate). This would limit the bookstore to $24 \times 60 = 1440$ sales per day. Likewise, a bank would be limited to a small number of transactions per day. As you can imagine, such limits are low; typical loads for successful businesses would easily be several orders of magnitude greater.

11.1.2.3 Serializable Schedules

Thus, for any realistic application scenario, we must allow concurrent transactions. Some concurrent executions of transactions correspond to schedules that are clearly incorrect, but some can be correct. The correct schedules are those that are equivalent to a serial schedule. The classical notion of equivalence is preservation of conflicts among operations. A schedule consists of a set of operations ordered in some manner. Let us consider two schedules consisting of the same set of operations, that is, executed by the same transactions and involving the same data items. Such schedules are said to be *conflict equivalent* provided that for any two conflicting operations, the orders of occurrence of those operation in the two schedules are the same. A *serializable* schedule is one that is conflict equivalent to a serial schedule. In essence, this means that the two schedules agree about what information does or does not flow from one transaction to another or from a transaction to a future transaction. Since a serial schedule is correct, so is a serializable schedule, because it may improve concurrency but does not alter the information flows or nonflows.

11.1.2.4 Strict and Rigorous Schedules

Besides ensuring correctness through serializability, a DBMS must also ensure that the transactions it runs are atomic and durable. Durability can be incompatible with consistency in the following way. Assume your bank account has a balance of $1 000. Assume a transaction deposits $1 000 into this account. A second transaction withdraws $1 500 from the same account. Based on the increased balance in your account, because of the first transaction, the second transaction can complete successfully. Say the money is handed out. Now if the first (deposit) transaction were to run into an error (or simply not receive a confirmation from the user), then that would mean it would only perform some of its operations. In other words, this would be a case of partial success. But atomicity requires all-or-none behavior. Since the deposit transaction is not succeeding, it is not taking the "all" branch. All that the DBMS can do is to cause it to be aborted or rolled back, i.e., to take the "none" branch. But if the DBMS does so, that means the deposit never took place. However, the withdraw transaction has used the results of the deposit transaction so it too is invalidated.

If the withdraw transaction has been recorded as having committed, (i.e., completed successfully)—meaning that the money was paid out—then the situation is past repair. In technical terms, the withdraw transaction *reads from* the deposit transaction. We should not have let the withdraw transaction commit prior to the deposit transaction. Such a schedule, where a transaction that reads from another transaction commits before the transaction being read from, is called a *nonrecoverable* schedule. This is the absolute no-no of database management, because it means that the data is corrupted and, further, because the second transaction is durable, the corruption is permanent.

An alternative would be to withhold committing the withdraw transaction until we knew whether the deposit transaction had committed. Now if the deposit transaction turns out to abort, we would abort the withdraw transaction as well. A schedule in which a transaction that reads from another transaction does not commit before the transaction being read from is called a *recoverable* schedule. Such a case avoids the above error of database corruption, but suffers from another potential problem. The problem is that the abort of one transaction can cause the aborts of other transactions, even if they had no internal reason to abort. Such *cascading aborts* cause a lot of work to be wasted. If only the transaction that would read from another transaction is made to wait until the transaction to be read from has committed or aborted, then such work would not be wasted.

For technical reasons to do with the implementations of logs, it is helpful if a transaction (say, T_2) that does not read from but overwrites the data items written by another transaction (say, T_1) is made to wait until T_1 has committed or aborted. The resulting schedule is called a *strict* schedule. In implementing this with locks, as discussed below, it is convenient to prevent a data item even from being written by T_2 after it has been read by T_1. The resulting schedule is called a *rigorous* schedule.

11.1.3 Locking

Conflict serializability is commonly used in commercial DBMSs. It is easy to realize through mechanisms such as the *two-phase locking protocol* (2PL). The mechanisms are largely transparent to the programmer. It is worthwhile to understand the functioning of 2PL in brief. Let us first review the basic concepts of locking. A *lock* is a data structure that controls access to a given data item. Locks are commonly used to prevent incompatible concurrent accesses to a data item. This is easy enough to ensure when data items are accessed through a particular software component, typically called a *lock manager*. When a computation needs to access a given data item, the lock manager assigns a lock on the data item to the given computation and gives it access to the data item. After the computation has accessed the data item, it relinquishes its lock on that data item. Notice that although we use the terms "acquire" and "relinquish" for locks, in a database environment these locks are generally not explicitly requested or released by a programmer, but are inferred by the DBMS based on the access requests made by the application that begins the given transaction and by an abort or commit generated by the application.

Lock types are often specific to the operation types that they allow. During the period that a transaction has a lock on a data item, the lock manager would not grant an incompatible lock on the same data item to another transaction. Since the operation types we are considering are read and write, it is appropriate to consider two kinds of locks: *shared* and *exclusive*. To ensure correctness, a lock manager cannot grant an exclusive lock if a shared lock or another exclusive lock has currently been granted and it cannot grant a shared lock if an exclusive lock has currently been granted. T_3 may obtain an exclusive lock on item x, which would enable it to read or write x. The exclusive lock would prevent another transaction from accessing x until the lock is released. Conversely, if T_4 obtains a shared lock on item y, it may read y. Another transaction, T_5, may also obtain a shared lock on y and may read y concurrently with T_6. However, if T_6 requests an exclusive lock on y, it would not be granted that lock until T_4 and T_5 have released their shared locks. This form of locking corresponds to the conflict matrix shown in Table 11.2. Two reads on a data item can happen concurrently, but a read and a write or two writes must happen one by one.

When a transaction requests a lock, either the lock is granted or a decision must be made about what to do with the transaction. Typically, the transaction is made to wait. This is reasonable based on the idea that the transaction cannot proceed without the data item it requested. However, alternative policies are considered in Section 11.2.2.

To continue with the transactions T_1 and T_2 introduced above, let us see how schedule S_1 may be realized. T_1 would first need to acquire an exclusive lock on z (because T_1 needs to both read and write z). T_1 would relinquish its lock upon committing. T_2 would then request an exclusive lock on z, acquire it, and proceed. Schedule S_1 could also be realized with a slightly different sequence of events. Again, T_1 would request and acquire a lock on z. Now T_2 could request a lock on z, but would be made to wait. In the meantime, T_1 would proceed. When T_1 relinquishes its lock, then the lock would be granted to T_2, which would resume at that point. The (observable) schedule would still be S_1, although T_2 requested its lock earlier than in the previous example. As you can verify, schedule S_2 may be realized through another sequence of lock acquisitions and releases.

11.1.3.1 Two-Phase Locking Protocol

Unfortunately, S_3 and S_4 can also occur provided one of the transactions acquires, releases, and acquires a lock again. In other words, undesirable schedules are not prevented by locking. Suppose we forbid such behavior. Would that prevent nonserializable schedules from being generated?

Unfortunately, no. A more interesting example can be constructed where two data items are involved. The above kind of locking applies to individual data items and ensures that each item is accessed serially whenever there is a conflict. However, transactions access multiple data items. If we are careless, the following kind of schedule over data items x and y may occur (we can infer the transactions T_7 and T_8 from this schedule):

- $S_5 = r_7(x); w_8(x); r_8(y); w_7(y); c_7; c_8$

Here T_7 reads x before T_8 writes x, but writes y after T_8 reads y. Such a schedule is clearly not serializable, because in any serial schedule of T_7 and T_8, one of the mutual orders of the conflicting operations on x and y, respectively, would be violated.

The above schedule can occur because, in essence, the locks are being released prematurely. To ensure that such is not the case, a simple approach considers each transaction to be separated into two main phases: a *growing phase* and a *shrinking phase*. Locks can only be acquired during the former phase and only be released during the latter phase. That is, all lock acquisitions precede all lock releases. It is simple to show that this protocol, aptly called two-phase locking (2PL), would not allow schedule S_5 to be obtained. Specifically, because T_7 would not relinquish its lock on x until after it has acquired a lock on y, this means that $w_8(x)$ cannot occur prior to T_7's having locked y, at which point T_8 would not be able to acquire a lock on y prior to T_8. In fact, $r_8(y)$ cannot occur until T_7 relinquishes its lock on y, which is only after $w_7(y)$. Thus, at best we would end up with the following schedule:

- $S_6 = r_7(x); w_8(x); w_7(y); r_8(y); c_7; c_8$

The following sequence of lock and unlock events shows how this schedule may be realized via 2PL: first T_7 locks both x and y; after $r_7(x)$, T_7 unlocks x; next T_8 locks x and performs $w_8(x)$; then T_7 does $w_7(y)$ and subsequently unlocks y; lastly T_8 locks y and performs $r_8(y)$.

In the above schedule, there are two conflicts between the operations on x and y, respectively, but both conflicts are in the same direction. Thus an equivalent serial schedule is where T_7 precedes T_8. More generally, 2PL can only produce a serializable schedule.

11.1.3.2   Variations of 2PL

Two variations of 2PL are worth mentioning. The first, *conservative 2PL* is like 2PL except that a transaction must acquire all of the locks that it might need prior to beginning. Thus, if a transaction has obtained one lock, then that means it has obtained all the locks it would need. Such a transaction would never have to wait for a lock during its computation. However, conservative 2PL is excessively pessimistic. A transaction would be delayed from starting because all of the locks were not available at the outset. Further, a DBMS cannot determine the entire set of locks a transaction might need, so the programmer would need to request the locks explicitly.

The second variation of 2PL, *strict 2PL*, is one in which the locks acquired by a transaction are held until the transaction has ended, i.e., committed or aborted. This is conceptually straightforward to implement: a DBMS lets the transaction keep its locks until receiving a commit or abort notification from the transaction (or aborting the transaction, if necessary). Strict 2PL ensures that the schedules realized using it are *strict* and even *rigorous*. Such schedules make it easier to maintain the integrity of the database while reducing the number of transactions the DBMS would have to abort to ensure integrity. For this reason, strict 2PL is common in practice.

11.1.4   Distributed Transactions

The transaction concept as introduced above applies to an individual DBMS in which all transactions are executed. It offers guarantees for the integrity of the information on the DBMS provided the individual transactions are correct. This is a powerful feature and one we would like to ensure for transactions distributed over more than one database, information system, or, in general, information resource.

In the above, a transaction, also called a *flat transaction*, consists of a set of operations, all executing on a single database. For distributed settings, we need to put together a transaction whose constituents are also transactions. Such a transaction is called a *distributed transaction* or a *nested transaction*. Table 11.3 describes the ACID properties as applied to distributed transactions.

Table 11.3: The ACID properties for closed nested distributed transactions

Property	Meaning	Example
Atomicity	All or nothing	All component transactions of a composed transaction complete successfully or none does, e.g., if paying to purchase a product, either the product ships and is paid for, or neither happens, meaning there are no unpaid shipments and no extra payments
Consistency	Integrity preserving	If each constituent transaction is individually correct, then the composed transaction with its constituents executing concurrently is also correct
Isolation	Hidden partial results	Different composed transactions are unaware of one another, e.g., a business deal with one customer is either entirely before or entirely after a business deal with another customer
Durability	Permanent committed results	The composed transactions are durable, e.g., once the inventory and payment databases are modified, the modifications are durable

Figure 11.2 illustrates a nested transaction. For the same reasons as for flat transactions, we would like nested transactions to support the ACID properties. However, the fact that we now have distributed databases causes a complication. The data items on which the operations are performed reside on different databases. Thus a single lock manager cannot ensure that all accesses to the various data items satisfy a protocol such as 2PL.

It is customary to separate the ACID properties into two main aspects. One aspect emphasizes atomicity and durability, and ensures the all-or-none behavior of a transaction. The other aspect deals with isolation and consistency and ensures that the concurrent execution of distributed transactions is correct. Traditionally, the first aspect is termed *recovery* and the second is termed *concurrency control*. A conventional DBMS provides both recovery and concurrency control. However, for distributed settings, the two aspects often behave differently. It is easier and more common to ensure atomicity than to ensure isolation.

So what happens when a CT fails? First, each service will ensure the atomicity and durability of its local subtransactions. Second, with 2PC, the CTM can guarantee that all or none commit. Otherwise, only weaker atomicity and durability are possible.

11.2.2.2 Compositional Serializability

For traditional databases, serializability corresponds to correctness of execution. For this reason, DBMSs are designed to ensure the serializability of the transactions running on them. One might require that transactions throughout the system should be serializable.

In Figure 11.6, let us assume that each of the transaction managers involved ensures the serializability of the transactions that run through it. Thus, the CTM ensures that the composed (upper-level) transactions are serializable; LDB1 ensures that all transactions executing on LDB1 are serializable; and LDB2 ensures that all transactions executing on LDB2 are serializable. In terms of our example, this means that Expedia, United Airlines, and Sheraton each ensures the serializability of the transactions that it runs.

11.2.3 Difficulty with Compositional Serializability

Interestingly, exemplary behavior by each of the parties involved *does not guarantee compositional serializability*, because of indirect conflicts. As an example from our travel scenario consider the following situation, which is formalized using the notation of Section 11.1.2:

- Alice wishes to determine the availability of rooms at a specific hotel (say, a specific room such as the presidential suite in the Manhattan Sheraton on June 15) and the availability of seats on a specific flight (say, UA 1 on June 15). The CTM performs T_1: $r_1(a); r_1(c)$ to serve Alice's needs.

- Bob wishes to determine the availability of rooms at a specific hotel (say, a specific room such as the presidential suite in the Manhattan Sheraton on June 16) and the availability of seats on a specific flight (say, UA 1) on June 16. The CTM performs T_2: $r_2(b); r_2(d)$ to serve Bob's needs.

- Vladimir goes to the Sheraton site directly and changes his booking for the presidential suite in the Manhattan Sheraton from June 15 to June 16. In other words, LDB1 performs T_3: $w_3(a); w_3(b)$. Notice that the data item a corresponds to the booking of June 15 and b to June 16; both are affected.

- Wlodek goes to the United site directly and changes his booking for UA 1 from June 15 to June 16. In other words, LDB2 performs T_4: $w_4(c); w_4(d)$.

- Since T_1 and T_2 are read-only, they have no conflicts with each other. Hence, any schedules involving just T_1 and T_2 are automatically serializable. This stands to reason: two different customers asking about two different dates should have no conflicts with each other.

- Let us assume that the operations take place on Sheraton in the following order. That is, LDB1 sees the schedule $S_1 = r_1(a); w_3(a); w_3(b); r_2(b)$.

- Let us assume that the operations take place on United in the following order. That is, LDB2 sees the schedule $S_2 = w_4(c); r_1(c); r_2(d); w_4(d)$.

- Thus, each LDB has a serializable schedule

- Yet the schedule at LDB1 puts T_1 *before* T_2 and the schedule at LDB2 puts T_1 *after* T_2, which is inconsistent. The schedule is *not* compositionally serializable.

In other words, *correct local actions by all parties yield incorrect outcomes*. This should be thought of as a major impossibility result. Below are the main approaches that have been proposed to fix this problem.

11.2.4 Achieving Compositional Serializability

Achieving compositional serializability presupposes the composed transaction manager has some level of control over the local services that it invokes. In particular, if the services execute fully autonomously, it is not possible to ensure compositional serializability, because each service could unilaterally take a decision (about serialization) and the decisions of the services may be incompatible with each other. Section 11.2.1 describes the skeletons that the local services should support so as to expose their precommit states. Given such skeletons, the CTM could serve as the coordinator for an execution of the 2PC protocol (as described in Section 11.1.4) among the composed services.

However, the point of the above impossibility result is that the CTM simply would not know whether the observed schedule was serializable or not. Compositional serializability fails because of local conflicts that the CTM does not see. So we need to find a way to work around this challenge.

11.2.4.1 Tickets

It is not possible to make the local conflicts visible without seriously compromising the autonomy of the local services. However, the problem of hidden local conflicts can be corrected by always causing visible conflicts among the composed transactions.

The idea is that whenever two composed transactions execute at a site, they must conflict there. Assuming that each LDB allows only serializable schedules, this means that the schedules produced by the LDB are compatible with the new conflicts. In other words, if there were any previously hidden local conflicts, they would have to be consistent with the newly introduced conflicts. Consequently, the newly introduced conflicts would provide sufficient information to the CTM.

A simple way to generate such visible conflicts is to require that each composed transaction takes a *ticket* at each site. A ticket is simply implemented as a counter, which is read and modified by each transaction. Because the counter is a data item, and each transaction writes

11.2.4.3 Performance of Compositional Serializability

Ticketing causes all subtransactions of a transaction to go through a local hotspot. The composed transactions are serialized, but at a huge price: there is a high probability of the transactions being aborted. Exercise 11.10 asks you to perform a simple analysis of the chances of a composed schedule not suffering any aborts due to ticketing.

Moreover, system-wide deadlocks are also possible when the local sites use pessimistic, i.e., lock-based, serialization approaches. The example of Section 11.2.2.1 applies unchanged in the present case as well.

Strict scheduling, being a pessimistic approach, is not subject to as many aborts after all the work of a schedule has been performed. However, it requires transactions to be held up until they can be committed in the correct order. In effect, strict scheduling causes sites to be held up until all of them are ready to commit, which is essentially like a two-phase commit (2PC) approach. For our travel example, the hotel, airline, and car rental agency would all have to keep a reservation open until all were ready to commit. That is, United Airlines might have to wait for Sheraton to reach an agreement with the traveler. Further, this approach is also susceptible to system-wide deadlocks of the kind explained above.

11.3   Limitations of Traditional Transactions

Traditional transactions are highly effective in homogeneous and centralized databases. In the above, we discussed why the ACID properties of transactions are difficult to attain. More importantly, the ACID properties may not even be desirable for service-oriented architectures, which are geared toward open environments consisting of autonomous and heterogeneous components. Table 11.4 reviews the limitations of traditional transactions in terms of the ACID properties.

Table 11.4: The ACID properties reviewed in the context of service-oriented architectures

Property	Meaning	Undesirable when
Atomicity	All or nothing	Legacy systems or nonterminating processes are involved
Consistency	Integrity preserving	Integrity conditions cannot be defined or data values expire
Isolation	Hidden partial results	Collaboration is desired
Durability	Permanent committed results	Backing out is necessary

ACID transactions are applicable for brief, simple activities (involving just a few updates that occur within a few seconds) on centralized architectures. By contrast, open environments require tasks that

- Run forever or for long durations (e.g., months rather than seconds for many business processes). Isolation requires locking data items at one partner based on the behavior of another, remote partner.

- Are prone to failure. Rolling back is usually impossible but proceeding in the face of error may appear to violate database integrity.

- Update data across systems with consistency requirements that might be quite subtle.

- Are cooperative, i.e., involve several applications and humans, which are *not isolated* from one another.

- Execute over autonomous and heterogeneous partners. Atomicity requires the component databases to expose their internal control states and operations, violating their autonomy and heterogeneity.

11.4 Relaxing Serializability

One of the potential problems caused by removing isolation is that inconsistencies may be introduced.

The notion of serializability can potentially be relaxed depending on the application domain. This relaxation can be valid based on the data types that the application uses. As a classical example, consider the case of a bank. Assume that there are is data type called Account and operators on accounts to deposit, withdraw, and transfer money, and to check the balance. The first three operators each involve a write operation in the sense of Section 11.1.2; the last involves a read operation. Therefore, deposit, withdraw, and transfer would be taken to conflict with each of the other operators. However, the operators are specific combinations of reads and writes, and may not truly conflict.

For instance, deposit would not conflict with deposit, because the outcome is the addition of the sum of the deposited money to the account (regardless of the order in which the operators are performed). Deposit would conflict with withdraw and transfer, because the money being deposited may determine whether a withdrawal or a transfer can occur. However, the conflict is "partial" in the sense that if a withdrawal followed by a deposit executes without exceptions, then clearly the deposit followed by the withdrawal would also execute without exceptions.

In a similar spirit, we can consider an airline flight reservation data type representing the bookings made on a given flight, and supporting operators such as book and cancel, which involve reads and write on the underlying database. The cancel of a reservation on a flight does not conflict with the cancel of another reservation on the same flight, because the seats are simply released by each cancellation. If we assume that the specific seats do not matter (what matters is if you can get on the flight), then book may still conflict with book. However, if the second book operator does not throw an exception, then the order does not matter. Likewise, if a book succeeds prior to a cancel, then the book will definitely succeed after a cancel.

Thus, if the database underlying such a service were aware of the data types being accessed and manipulated by the transactions, it may be able to avoid some possible conflicts. Consequently, a larger number of schedules would be allowed, meaning that there would be more composed schedules that were acceptable.

A downside of such an approach is that it requires careful, advance consideration of the data types involved. Unless the operations on the various data types are carefully formalized, it is difficult to take advantage of such improvements. Moreover, the formalized behavior of the data types would need to be exposed as part of the interface of a service. We can imagine that such behaviors may be standardized in specific industries. However, in general settings, that is not the case, and data type considerations are not used in current practice.

When no cross-service constraints apply, local serializability may be adequate. For example, if there were no relationships between the bookings you make on a hotel and an airline, it might be satisfactory (for a given application) to simply ensure that each site is locally consistent. We may still need a protocol such as 2PC to ensure atomicity, but we may not care about violations of compositional serializability.

When cross-service constraints apply, data is split into local data and shared data, such that each LDB controls its local data and the CTM controls shared data. For the shared data, this approach allows local reads, but data can be written only via the CTM. For independently existing services, this would be an onerous restriction, and not likely to be feasible in practice. The disadvantage of an approach based on cross-service constraints is that it does not work in all cases: because all shared data are managed through a special service, only the most trivial compositions are feasible. Such approaches are not found in practice.

11.5   Extended Transaction Models

One of the main motivations behind services is that they naturally accommodate the heterogeneity and autonomy of business partners. However, we still need higher-level abstractions to capture the complexity and cooperative nature of the tasks.

The extended models seek to capture weaker specifications of the ACID properties that would be acceptable in different application scenarios and yet are easier to implement. Ideally, various mathematical properties of these specifications may also be proved.

In simple terms, approximations to the atomicity requirements of transactions may be ensured by following a strategy based on one or a combination of the following:

Redo. Rerun the writes from a transaction log, meaning that the specific values written previously are attempted to be written again. This would work only if the given data items were not accessed in the meanwhile. For example, in a travel setting, you would get the same seat on the same flight as you would have gotten originally.

Retry. Rerun all of a subtransaction, meaning that the program is reapplied from scratch. Different data values might be read, different conditional branches may be taken, and different results may be written or presented to users. For example, you may get a different seat or a reservation on a different flight.

Contingency. Define contingency transactions for selected, typically crucial, subtransactions. When one such subtransaction fails, run one of the contingency transactions instead. There could be multiple layers of contingency transactions. A classical example in the travel domain is to try for bookings on a second airline if the first airline is sold out, and on a third if the second airline is also sold out.

Compensation. Semantically undo all subtransactions that completed before the failure, meaning that if some subtransactions completed even as the main transaction failed, they must be undone so that there are no residual effects on the data. For example, if a traveler booked a hotel, a car, and a flight, but the airline seat assignment failed, then all bookings would be cancelled instead of retrying to find another seat.

Compensation is different from the effects of undoing an operation from the database log, because the previous value (as recorded on the log would, in general, be invalid). For example, you can compensate for an erroneous withdrawal from a bank account by depositing the erroneously withdrawn amount back into the same account. Say, there were $1 000 in the account and you withdrew $700 in error. The database log would say that the value of the given account prior to the operation was $1 000. When the error is detected, it would be risky simply to restore the account to the amount in the log, because there could have been intervening transactions, which would thus be lost. For instance, after you withdrew $700 in error, say, someone deposited $400 into the same account. To compensate for the erroneous withdrawal, you would have to change the account value to $1 400, not to $1 000.

Vital subtransaction. Given a transaction consisting of one or more subtransactions, the vital subtransactions are identified as being critical to the successful completion of the upper-level transaction. In other words, the vitality of a subtransaction is considered as key to the atomicity of the upper-level transaction: no execution of it can be considered "all" if one of the vital subtransactions fails.

Numerous extended transaction models have been proposed that relax the ACID properties in different ways. These are built on combinations of the above. They consider features such as the following:

- The type of nesting allowed: traditional distributed transactions require the nesting to be *closed* (ACID), while newer models permit it to be *open* (non-ACID).

- The constraints that must be enforced among subtransactions, such as commit dependencies and abort effects.

- Atomicity variations, in the form of the contingency procedures needed to ensure an effect similar (under various assumptions) to the all-or-none of traditional transactions.

- The procedures that are needed to restore consistency, e.g., compensation techniques.

Several extended transaction models were defined in the late 1980s and the early 1990s. These are now drawing attention again for modeling Web service transactions. There is also earlier relevant work on generic schemes for capturing transactions that is being adapted, e.g., in CORBA Activity Services. We describe some of this work below, and relate it to problems that arise in Web service transactions.

11.5.1 Sagas

A *saga* is a sequence of steps, where each step is a transaction. These transactions are executed in sequence: one transaction commits before the next one begins. The motivation behind sagas is to improve concurrency by reducing isolation, while still assuring atomicity. Thus the results of a subtransaction are not locked until the saga ends. However, a saga satisfies the all-or-none atomicity requirement in the following manner. If all the subtransactions of a saga commit, then the saga succeeds and clearly we have the "all" branch. On the other hand, if one of the subtransaction fails, we must undo the effects of the previous subtransactions (which we know have committed, because otherwise, the given subtransaction would not have been started). To undo those subtransactions, we must simply run their compensation transactions. To ensure that the compensates apply properly, they are run in the *reverse* order of the original execution.

Thus a saga presupposes that compensation transactions are defined for all its subtransactions except the last. Further, it presupposes that the compensates are guaranteed to succeed. A variation is when compensates are defined for the first several subtransactions and it is guaranteed that the remaining subtransactions will succeed after a finite number of retries. In such a case, once enough subtransactions have succeeded, we can guarantee that the saga will take the "all" branch; otherwise it takes the "none" branch.

Typically, the subtransactions of a saga should be as independent as possible so there are few data dependencies between them. As a result, there would be fewer problems in exposing the results of a subtransaction, which would be analogous to the partial results of the saga. Not sharing the data item does not entail that there would be no problems, because there could be constraints among the data items accessed by the different subtransactions. For example, if a saga consists of a hotel booking followed by an airline booking, it is conceivable that a hotel booking may appear to have succeeded even though the airline booking fails (and thus the saga is undone).

11.5.2 Flex Transactions

The *Flex transaction* model assumes that we are given functionally equivalent subtransactions for achieving some specified tasks. If one subtransaction fails, then an alternative subtransaction may be used to accomplish the same task.

The flex model relies upon rules for processing under different circumstances. Each rule is in restricted syntax and is termed a *dependency*. The dependencies may involve either the failure or success of a subtransaction, or can involve external events.

11.5.3   DOM Transactions

The distributed object management (DOM) transaction model goes beyond the flex model in considering a richer variety of constraints among transactions. It also considers details specific to distributed objects, which are not germane here. At the outermost level, DOM defines a long-running activity, which is called a *multitransaction*. A multitransaction consists of a set of *top transactions*, some of which are declared vital. The top transactions may optionally be assigned a mutual order.

Each top transaction is a nested transaction and may have any number of subtransactions. The subtransactions would generally have contingency and compensation transactions—also part of the same top transaction.

DOM allows two main kinds of precedence dependencies. It may be stated that a transaction may not begin before the commit of another transaction, or that a transaction may not commit before the commit of another transaction. The former causes the transactions to run serially; the latter allows additional ordering.

The commit of a multitransaction entails the commit of all the vital top transactions. The abort of a multitransaction entails the abort of all its top transactions. If a component subtransaction has committed, then it must be compensated (in reverse order of the order of commitment).

11.6   Notes

The discussion of compositional transactions is based on previous work on distributed transactions, which were originally studied in the context of distributed, heterogeneous database systems. The impossibility result for ensuring serializability is discussed in Breitbart *et al.* [1994]. Tickets are discussed in Georgakopoulos *et al.* [1994].

Extended transaction models developed in the late 1980s and the early 1990s have gradually converged to those that can be expressed using dependencies. The saga model was introduced in Garcia-Molina and Salem [1987] and open nested transactions in Traiger [1983]. Davies [1978] presents an early account of transactions.

11.7   Exercises

11.1. What is the difference between a serial schedule and a serializable schedule?

11.2. Verify that the strict 2PL scheduling of composed transactions as well as local transactions prevents the failure of compositional serializability in the example of Section 11.2.

11.3. Verify that the strict 2PL scheduling of composed transactions as well as local transactions prevents the failure of compositional serializability *in general*.

We want to define a transaction that attempts to perform a delete on R_1 by doing the following three things concurrently:

Start the delete on R_1
Start the decrement on R_2
Ask the user for confirmation

Complete the table below such that the consistency of R_1 and R_2 is maintained. Also, please list the key assumption necessary to guarantee eventual consistency.

User Says:	Delete on R_1	Decrement on R_2	Actions
		Abort	
OK	Abort	Commit	
		Abort	
	Commit	Commit	
		Abort	None
Cancel	Abort	Commit	
		Abort	
	Commit	Commit	

11.16. Produce a conflict matrix involving the operations read, write, inc, and dec. That is, consider all possible pairs of these operations.

11.17. Specify the following supply chain scenario where assembly is understood as a saga (hence, its atomicity is required).

- An assembly saga consists of transactions for procuring hoses, valves, and coupling sleeves.
- The valves are not needed without hoses.
- The coupling sleeves are not needed without valves.
- Assume that the valves are not needed for the saga to complete.

Describe what compensation transactions are needed (and what they would compensate) for the above to work. State any assumptions you need to make about the compensation transactions. Order the above subtransactions into an appropriate saga.

11.18. Extend the scenario of Exercise 11.17 to be a multitransaction as in DOM.

- When an assembly multitransaction starts (because of an order being received), subtransactions for procuring hoses, valves, and coupling sleeves are started concurrently.

- The subtransaction for procuring hoses is *vital* to the assembly multitransaction.

- The coupling sleeves are not needed without valves.

Write the start, commit, and failure dependencies for the above cases.

11.19. Recall that the inc database operation atomically increments the value of an integer data item by a given amount (for simplicity, let us take this to always be 1). Tickets could thus possibly be implemented using the inc operation on the ticket data item. Based on the conflict matrix entry for inc with itself, what would be the main advantages or disadvantages of doing so? In particular, consider whether this approach will work and, if so, how. Describe briefly (in up to five sentences).

11.20. Implement a simple TP monitor. Your TP monitor will not only manage distributed transactions, but also simulate transaction failures and recoveries. For this, you will need to implement two related databases: one database will contain base-level data and the other will contain summary data. For example, a travel office might have one database containing sales transactions and a second database containing a summary of the performances of the sales force. Your TP monitor can use whatever transaction semantics you prefer. Simulate the failure of an update on either of your two databases and implement a recovery mechanism. For example, a salesman might record a sale in a sales-transaction database, but the update to the salesman's monthly sales totals in the second database might fail. If possible, use two different database management systems, such as Microsoft Access and Oracle. For this exercise, you may assume that the TP monitor has exclusive control of the underlying databases.

11.21. Imagine a distributed information system for the Widget Manufacturing Co. consisting of an on-line database server containing records of widget sales and sales representatives who use PDAs to enter new sales at WiFi "hot spots." When the sales representatives are visiting a customer, the PDAs are not connected to the server. Discuss the architecture of this information system in terms of the CAP Principle.

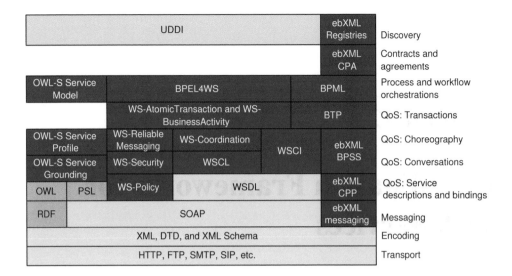

Figure 12.1: The relationships among the different proposed standards and methodologies for describing activities and business processes

- The parties must agree on the documents to be exchanged, and the semantics of the documents as defined by XML, XML Schema, RDF, or OWL.

- The parties must agree on the protocol used to transmit a document (such as SOAP-RPC, asynchronous SOAP, or ebXML transport), the routing for the transmission, and the packaging of the document.

- The parties must know each other's location.

- An ordering of the documents to be transmitted must be specified, in the form of a *conversation*.

Note that the conversation does not specify any internal implementation or mapping to back-end applications within the various enterprises that are interacting.

12.1   WSCL: Web Services Conversation Language

The Web Services Conversation Language, allows the business level conversations or public processes supported by a Web service to be defined. The W3C Web Service Choreography (WS-Chor) Working Group's emerging specifications such as the Conversation Definition Language are slated to replace WSCL and WSCI (discussed next), but it is worth studying WSCL and WSCI to get an understanding of the basic challenges involved.

WSCL specifies the *sequencing* of XML documents—as well as specifications for the documents themselves—being exchanged between a Web service and a user of that service.

WSCL conversation definitions are themselves XML documents and can therefore be interpreted by Web service infrastructures and development tools. When used in conjunction with a service description language such as WSDL, WSCL can provide protocol binding information for abstract interfaces or can specify the abstract interfaces supported by a concrete service. Unlike BPEL4WS (see Section 13.4.1), which describes the connections and interactions among more than one service, WSCL describes the interactions with just one service. As a standard, WSCL is no longer being pursued, but is useful for the purposes of illustration, because it provides a simple example of how conversations can be encoded and applied.

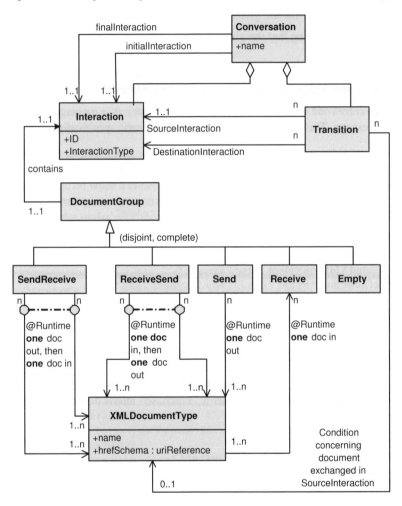

Figure 12.2: Concepts in WSCL

A WSCL specification, using the concepts shown above in Figure 12.2, consists of the

```
            id="InvalidLoginRS" />
  </Interaction>
        ...
  <Interaction interactionType="Empty" id="Start" />
  <Interaction interactionType="Empty" id="End" />
</ConversationInteractions>
```

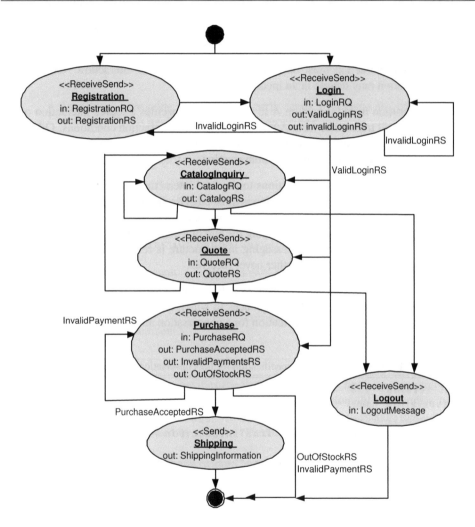

Figure 12.3: A WSCL definition for a conversation about on-line purchasing

Listing 12.2: A WSCL specification (cont.)

```
<ConversationTransitions>
  <Transition>
    <SourceInteraction href="Start"/>
    <DestinationInteraction href="Login"/>
  </Transition>
    ...
  <Transition>
    <SourceInteraction href="Login"/>
    <DestinationInteraction href="Registration"/>
    <SourceInteractionCondition href="InvalidLoginRS"/>
  </Transition>
  <Transition>
    <SourceInteraction href="Logout"/>
    <DestinationInteraction href="End"/>
  </Transition>
</ConversationTransitions>
</Conversation>
```

WSCL has several limitations for describing the interactions that can occur between the participants in a conversation. Most significantly, the conversation is limited to two participants. Additional conversations can occur concurrently among more than two participants, but there is no support for aligning or reconciling the conversations. Other limitations of WSCL are

- It provides excellent graph primitives for describing control flows, but does not have specific constructs for iteration or recursion (i.e., for conditionally terminating iterations).

- Conversations are modeled as scripted procedures (graphs), but the procedures are not flexible.

- Cooperation is not supported.

- Exception handling is done only at a low level, and it is syntactic, not semantic.

12.2 WSCI: Web Service Choreography Interface

The Web Service Choreography Interface, an interface description language for business processes, provides a global message-oriented view of the choreographed interactions among a collection of Web services. It describes the flow of messages exchanged by a Web service that is interacting with other services according to a choreographed pattern. By capturing the temporal and logical dependencies among the messages, WSCI characterizes the externally observable behavior of the Web service, not its internal operation. WSCI is a common denominator to BPML and BPEL4WS (described in Chapter 13), and offers interoperability across these two languages, as well as ebXML's BPSS and WfMCs XPDL.

WSCI is an enhancement to WSDL, and a WSCI specification is intended to be part of a WSDL document describing a Web service. Listing 12.3 builds on the WSDL description of an example stock-quotation Web service that is specified in Exercise 2.8. Listing 12.3 includes the WSCI snippet that specifies the order in which the log-in and stock-quote operations must proceed.

Listing 12.3: An example WSDL document for a stock-quotation Web service, enhanced by WSCI

```
<!— Assume the default namespace is
  xmlns="http://www.w3.org/2002/07/wsci10" —>
<!— WSCI Selectors —>
  <correlation name="quotationCorrelation"
    property="tns:quotationID">
  </correlation>
  <interface name="StockQuoteWS">
    <process name="ProvideStockQuote" instantiation="message">
      <sequence>
        <action name="ReceiveLogin"
          role="tns:StockQuoteWS"
          operation="tns:QuoteToUser/LogIn"/>
        <action name="ReceiveStockQuoteRequest"
          role="tns:StockQuoteWS"
          operation="tns:QuoteToUser/ProvideQuote">
            <correlate correlation="tns:quotationCorrelation"/>
            <call process="tns:LookupPrice"/>
        </action>
        <action name="ReceiveLogout"
          role="tns:StockQuoteWS"
          operation="tns:QuoteToUser/LogOut"/>
      </sequence>
    </process>

    <process name="LookupPrice" instantiation="other">
      <action name="QueryNYSE"
        role="tns:StockQuoteWS"
        operation="tns:QuoteToUser/QueryNYSE"/>
    </process>
  </interface>
</wsdl:definitions>
```

The static (i.e., non-WSCI) portion of Listing 12.3 describes the data types supported by the stock-quotation service, the messages it can send or receive, and the operations it provides. The dynamic (i.e., WSCI) portion uses its action construct to make each operation (login, stock-quote request and response, and then logout) into an atomic unit of work. Using sequence, these are then specified to be part of an ordered sequence with a user. Notice that the "ReceiveStockQuoteRequest" action invokes a separate Web service with the NYSE. Although not used in this example, choices among actions and repeated actions are allowed in WSCI interfaces as complex activities.

A service can participate in several different conversations at the same time, with the correlate element used to manage them and associate messages with the proper conversation. WSCI also supports both atomic transactions and open-nested transactions, the latter of which can be compensated when exceptions occur, as shown in the partial specification of Listing 12.4. This example also shows a repeated action in the form of a while-loop.

Listing 12.4: An example of WSCI specifying a transaction, its compensation, and a while-loop

```
<sequence>
  <context>
    <transaction name="buyStock" type="atomic">
      <compensation>
        <action name="NotifyUnavailable"
          role="NYSE"
          operation="tns:NYSEtoBroker/NotifyUnavailable"/>
      </compensation>
    </transaction>
  </context>
  <action name="BuyShare"
    role ="Broker"
    operation="tns:BrokerToNYSE/BuyShare"/>
  <while name="BuyShares">
    <condition>defs:fundsRemain</condition>
    <action name="BuyShare"
      role ="Broker"
      operation="tns:BrokerToNYSE/BuyShare">
        <correlate correlation="defs:buyingCorrelation"/>
    </action>
  </while>
</sequence>
<!-- Compensating Behavior for the Above Transaction -->
<exception>
  <onTimeout property="tns:expiryTime"
    type="duration"
    reference="tns:BuyShares@end">
    <compensate transaction="tns:buyStock"/>
  </onTimeout>
</exception>
```

12.3   WS-Coordination: Specifying Coordination

Multiparty business processes whose components are Web services require the Web services to operate in a consistent context and to be coordinated. One way to achieve this is to provide a coordination service, i.e., a service whose job it is to coordinate the activities of the Web services that are part of the business process. WS-Coordination is a specification for just such a coordination service. The service might be provided by a dedicated coordinator, or by one of the component Web services that would assume the role of coordinator in addition to providing its own service. Using WS-Coordination, multiple participants can hide their proprietary protocols and reach agreement on the outcome of their activities. It is helpful to define the following terminology:

Protocol. A set of well-defined messages that are exchanged between the Web services participating in a process. The WS-Coordination specification describes a general, extensible, framework for defining such coordination protocols, which then can support a variety of activities. For example, WS-AtomicTransaction and WS-BusinessActivity, described in Section 12.4, use WS-Coordination to define coordination types for short-running atomic transactions and long-duration business transactions, respectively (the long durations might be caused by business or network latencies or having to wait for user inputs).

Context. A uniquely identified, conceptually coherent activity that includes the services being coordinated. A context identifier is passed from one application-level computation to the next to ensure that they remain in the scope of the same context.

Application. An executing program instance at one site. This could be a service viewed as a proactive entity that can initiate coordination requests and cause a coordination context to be launched.

12.3.1   Coordination Service

As shown in Figure 12.4, a *coordination service* (or *coordinator*) is an aggregation of the following services:

- An *activation service*, which defines a CreateCoordinationContext operation that allows a CoordinationContext to be created. The exact semantics is defined in the specification for the coordination type. An activation service is optional.

- A *registration service*, which defines a Register operation that allows a Web service to register its participation in a coordination protocol. A coordinator must support a registration service.

- A *coordination protocol service* for each supported coordination type. These are defined in the specification for the coordination type.

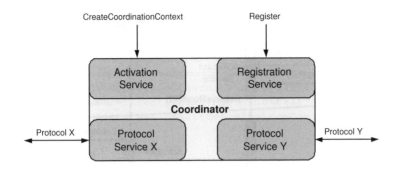

Figure 12.4: A coordination service for Web services

Once an application creates or acquires a coordination context, it can send the context to another application. The context contains the necessary information for a receiving application to register itself into the activity. The application can choose either the registration service of the original application or one that is specified by another coordinator, which can then be interposed into the coordination. The application then follows the specified coordination behavior.

A possible reason for a service to prefer its own coordinator is that it might not trust the other party's coordinator. This is like bringing your own lawyer to a tough negotiation. When the coordinator is an independent authority, it can help ensure the compliance of the other party to the selected protocol and, outside of the protocol, to the business deal that emanates from the protocol.

Suppose a travel agency would like its Web service (WStravel) for arranging travel packages to interact with a hotel's Web service (WShotel) for room reservations. Because they interact with many other Web services, WStravel and WShotel each make use of their own coordinators, CoordinatorT and CoordinatorH, respectively. WStravel and WShotel need coordination, because WStravel expects to receive an ACK for each message it sends, but WShotel does not send or expect to receive ACKs.

As shown in Figure 12.5, an interaction might proceed as follows. It begins with WStravel asking its coordinator to create a coordination context for an ACID-transaction type of coordination. WStravel does this by sending CreateCoordinationContext for coordination type ACIDTransaction to the activation service of CoordinatorT. CoordinatorT returns a context, C_t, with an activity identifier A_1 and the PortReference where its registration service, RS_t, can be found.

WStravel then sends a *ReserveRoom* message to WShotel containing the context C_t. WShotel prefers CoordinatorH, so it uses CreateCoordinationContext with C_t as an input to interpose CoordinatorH. CoordinatorH creates its own CoordinationContext C_h that contains the same activity identifier and coordination type as C_t, but with its own registration service RS_h. WShotel determines the coordination protocols supported by the coordination type ACIDTransaction and then registers for the coordination protocol TwoPhaseCommit at CoordinatorH, sending PortReferences for WShotel and receiving the protocol service one-way messaging. The CoordinationType provides the identifier for the desired coordination type for the activity, in this case a URI (http://schemas.xmlsoap.org/ws/2002/08/wstx) to the ACIDTransaction coordination type.

The CreateCoordinationContextResponse message returns the coordination context that was created. Here is an example response message sent from the travel agency's coordinator:

```
<wsc:CreateCoordinationContextResponse>
  <wsc:RequesterReference>
    <wsa:Address>
        http://TravAgency.com/WStravel
    </wsa:Address>
  </wsc:RequesterReference>
  <wsc:CoordinationContext>
    <u:Identifier>
        http://CoordinatorT.com/context1234
    </u:Identifier>
    <wsc:CoordinationType>
        http://schemas.xmlsoap.org/ws/2002/08/wstx
    </wsc:CoordinationType>
    <wsc:RegistrationService>
        <wsa:Address>
            http://CoordinatorT.com/registration
        </wsa:Address>
        <WSt:PrivateInstance>
            1234
        </WSt:PrivateInstance>
    </wsc:RegistrationService>
  </wsc:CoordinationContext>
</wsc:CreateCoordinationContextResponse>
```

The RequesterReference tag provides the port reference of the caller that invoked CreateCoordinationContext. The resultant CoordinationContext is given the URI http://CoordinatorT.com/context1234.

12.3.3   Registration Service

The Registration service definition requires a port type on the coordinator side for the request and on the requester side for the response. The coordinator's Registration service is defined as:

```
<wsdl:portType name="RegistrationCoordinatorPortType">
  <wsdl:operation name="Register">
    <wsdl:input message="wsc:Register"/>
  </wsdl:operation>
</wsdl:portType>
```

The requester's Registration service is defined as:

```
<wsdl:portType name="RegistrationRequesterPortType">
  <wsdl:operation name="RegisterResponse">
    <wsdl:input message="wsc:RegisterResponse"/>
  </wsdl:operation>
  <wsdl:operation name="Error">
    <wsdl:input message="wsc:Error"/>
  </wsdl:operation>
</wsdl:portType>
```

The Register message enables

- A participant to select and register for a particular coordination protocol of the type supported by the coordination service.

- A participant and the coordinator to exchange port references.

Participants can register for multiple coordination protocols by issuing multiple Register operations. The following is an example registration message:

```
<wsc:Register>
  <wsc:RegistrationService>
    <wsa:Address>
        http://CoordinatorT.com/registration
    </wsa:Address>
    <WSt:TravCode>T9876</WSt:TravCode>
  </wsc:RegistrationService>
  <wsc:RequesterReference>
    <wsa:Address>http://TravAgency.com</wsa:Address>
  </wsc:RequesterReference>
  <wsc:ProtocolIdentifier>
    http://schemas.xmlsoap.org/ws/2002/08/wstx/2PC
  </wsc:ProtocolIdentifier>
  <wsc:ParticipantProtocolService>
    <wsa:Address>
        http://TravAgency.com/travel2PCservice
    </wsa:Address>
  </wsc:ParticipantProtocolService>
</wsc:Register>
```

The RegistrationService tag provides the registration port reference. The RequesterReference tag is the port reference where WStravel wants the registration service CoordinatorT to return status information. The ProtocolIdentifier is the URI that identifies the coordination protocol selected for registration. The ParticipantProtocolService tag is the port reference that WStravel wants CoordinatorT to use for the coordination protocol.

7. CoordA tells WSairline to commit T. When WSairline receives the COMMIT message (9), it commits.

8. When the Committed message returns (10, 11, and 12), the 2PC protocol has ended.

If WSairline had been unable to prepare to commit, the flows would be the same, except for the following:

1. The PREPARED notification messages (5 and 6) would be replaced by ABORTED notification messages.

2. Because WSairline reported ABORTED, the phase two 2PC messages and replies (messages 8 through 11) would be unnecessary. The coordinator would send ABORT notification messages to other participants who voted PREPARED (none in this example).

3. Message 12 to WStravel would be replaced by ABORTED.

If a Web service is not a resource manager but wants to vote on the outcome of a transaction, it can register for the 2PC protocol and respond to a phase one PREPARE with either ReadOnly (vote for commit) or Aborted (vote for abort).

The problem of deciding that all the actions requested as part of a transaction have completed is not part of the WS-AtomicTransaction specification. Instead, it is the responsibility of the application to determine this prior to attempting to commit or rollback the transaction.

12.5   BTP: Business Transaction Protocol

The Organization for the Advancement of Structured Information Standards (OASIS) has developed the Business Transaction Protocol to automate and manage long-running, Web-based, collaborative business applications. BTP is designed to support interactions that cross application and administrative boundaries, thus requiring extended transactional support beyond the classical ACID properties. BTP relaxes the ACID properties via two sub-protocols: (1) *atoms*, where isolation is relaxed, and (2) *cohesions*, where both isolation and atomicity are relaxed.

Figure 12.7 shows an example of a cohesion transaction in which an investment manager tries to satisfy the goal of a balanced portfolio, in this case defined as at least two different kinds of investment. A BTP composer service coordinates the processing of the transaction and ensures that it reaches a successful conclusion, even when some of the participating services cancel unilaterally or are cancelled by the composer.

After establishing a context for the transaction with the composer, the investment manager begins the transaction by notifying the potential services of its interest in making an investment. Some of these services agree to participate by enrolling in the transaction and returning their price. Based on the prices, the manager chooses the services to be used and the services to be cancelled. The composer then handles the termination of the business transaction by obtaining acknowledgments of either confirmation or cancellation.

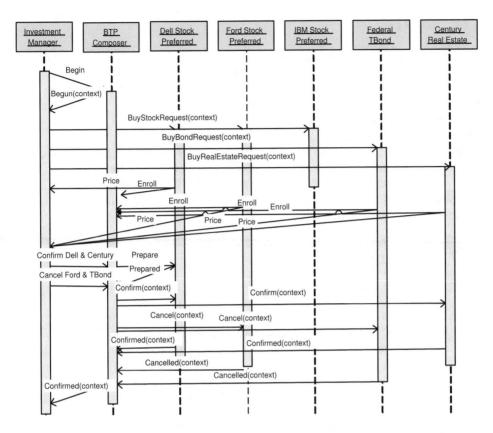

Figure 12.7: An example of the Business Transaction Protocol for *cohesions* among an investment manager and several different kinds of financial instruments

Compared to WS-Transaction, BTP is more suitable for loosely coupled applications, because it relaxes isolation. For example, a service providing travel packages (flight, hotel, and car) could take advantage of relaxed isolation by getting a confirmed airline seat and hotel room while negotiating for a car; depending on the location of the hotel, the car might not be needed. All reservations could be cancelled later if an acceptable rate for a hotel room is not obtained. BTP provides slightly finer-grained control over which parts of an overall transaction commit and which are aborted, and over the timing of these actions. WS-Transaction is dependent on Web services, whereas BTP is an XML message protocol with a SOAP binding. Both can be used to support business process execution environments, such as BPEL4WS, WSFL, WSCI, and BPML.

12.6 Notes

A paper by Francisco Curbera and colleagues [2003] provides a nice description of WS-Coordination and WS-Transaction in the context of evolving standards for composed Web services. BTP is described in Dalal *et al.* [2003] and Little *et al.* [2003]. Specifications for WS-Coordination and WS-Transaction can be found at
http://www-106.ibm.com/developerworks/library/ws-coor/ and
http://www-106.ibm.com/developerworks/webservices/library/ws-transpec/,
respectively. There is currently industry pressure to unify the BTP and WS-Coordination specifications, because they address similar problems and in a largely similar manner.

WSCL, currently found at http://www.w3.org/TR/wscl10, and WSCI, currently found at http://www.w3.org/TR/wsci, are both designed to specify conversations among services. We expect that they will be unified soon.

12.7 Exercises

12.1. How does a conversation as defined in WSCL differ from a CORBA IDE or a Java interface?

12.2. Study the WSCL and WSCI specifications. Construct a comparison that considers, for example, their features for describing: abstract conversations, conversation instances, control flow, nested conversations, multiparty conversations, exceptions in conversations, more than one simultaneous conversation among some of the same parties, explicit timing of conversational steps, a formal semantics for conversations, and transactional support in conversations.

12.3. Assume that a student has a personal software agent that behaves like a Web service and that helps in enrolling for courses on-line at NSCU. The university provides a coordinator (CoordNSCU), as defined by WS-Coordination, that ensures an ACID transaction between the student's agent and the NSCU course enrollment Web service. Show in a UML sequence diagram (similar to Figure 12.5) the coordination between the student's agent and the Web service for enrollment.

12.4. For the course enrollment scenario in Exercise 12.3, construct the WSCL specifications for the two conversations between the student's agent and the coordinator, and between the coordinator and the Web service for enrollment.

12.5. Draw a UML sequence diagram for the example of WS-Transaction shown in Figure 12.6, assuming that all operations commit successfully.

12.6. Draw a UML sequence diagram for the example of WS-Transaction shown in Figure 12.6, assuming that DB is unable to commit and all participants need to abort.

Chapter 13

Process Specifications

No service is an island. The key point about service-oriented computing is that it involves extended, loosely coupled activities among two or more autonomous business partners. Such activities can be thought of as (business) processes that engage several services in a manner that brings about the desired (business) outcome. The previous chapter described the underpinnings of processes from the perspective of transactions. This chapter covers process specifications, discussing their modeling and enactment, as well as key emerging standards.

13.1 Processes

A process is an activity. Generally a process would be a composite activity and be geared to serve some purpose. Depending on the specific process, its tasks could be some combination of services that correspond to queries, transactions, applications, and administrative activities. These services may be distributed within or across enterprises and would be coordinated by constraining control and data among them. The services may themselves be composite, i.e., implemented as processes. The following discussion emphasizes business processes consisting of services, but the concepts developed could apply equally well to scientific computing and other settings. Examples of settings where processes apply include *intraenterprise* environments (i.e., within an enterprise), such as production scheduling and inventory control, and *interenterprise* environments (i.e., across enterprises), such as supply-chain management and purchase negotiation. Clearly, intraenterprise and interenterprise processes need to correlate with each other, because intraenterprise activities are needed to support interenterprise interactions.

Processes present a number of technical challenges. First, we must be able to model a process, incorporating correctness of executions with respect to the model, and respecting the constraints of the underlying services and their resources. The normal executions of a process are often easy, since they can be as simple as a partial order of the activities in the process. By contrast, the exception conditions can be difficult to model and handle. More importantly, because interesting business processes are often long running, their mutual interactions are nonatomic, leading to the prospect that the information they take as input may be subject to revision and thereby causing their own results to be invalidated. Exceptions and revisions are the main sources of complication in the modeling of a process.

Second, we must be able to interface a process to underlying functionalities. In the case of database systems, these would include a suitable model of transactions that incorporates constraints on the concurrency control and recovery mechanisms of a DBMS. A transaction model provides the necessary abstractions and shields process models from the implementational details of DBMSs.

Because processes are used in a number of places by an enterprise to support its internal functioning as well as its interactions with its business partners, processes can end up being modeled in several different ways, typically based upon process representations that are proprietary to the software vendors and applications involved. For example, if production scheduling software employs a different modeling formalism than purchase order processing software, then the enterprise's participation in a supply chain may be adversely affected. However, interoperation among processes, while clearly an important need in practical settings, is nearly impossible without some kind of translator among process models. The challenges of heterogeneity that Section 5.1 discussed in the context of information sharing apply equally to process model interaction.

Before we get into the details, it is worth describing the main perspectives we can have on processes and the distinctions between them.

Orchestration. This takes the view of a process as a program or a partial order of operations that need to be executed. This view is logically centralized in that it views a process from the perspective of one "orchestrating" engine. It is as if the process specification is being executed under the control of or on behalf of a specific party. Orchestration corresponds best to the workflow representations discussed in Section 13.3 and to process languages such as BPEL4WS. Representations such as OWL-S (introduced in Section 15.5.2) enable the right orchestrations to be produced, given the requirements for a desired process and the functionalities of the available services.

Choreography. This takes the view of a process as being a set of message exchanges between participants. The message exchanges are constrained to occur in various sequences and may be required to be grouped into various transactions. Choreography corresponds best to languages such as WSCL and WSCI.

Collaboration. This takes the view of a process as a collaboration among business partners. The business partners not only send messages to one another, but also enter into business relationships such as contracts and obligations. They generate flexible message exchanges depending on the evolving circumstances and their local policies, e.g., to handle business exceptions. Collaboration is emerging as a serious approach for carrying out large-scale business processes.

13.2 Describing Dynamics with UML

UML provides graphical constructs that can be used to describe (1) actions and activities, and (2) their temporal precedences and flows of control. The allowable control constructs are:

- *Sequence*, which is a transition from one activity to the next in time.

- *Branch*, which is a decision point among alternative flows of control.

- *Merge*, where two or more alternative flows of control rejoin.

- *Fork*, which is a splitting of a flow of control into two or more concurrent and independent flows of control.

- *Join*, which is a synchronization of two or more concurrently executing flows of control into one flow.

These control constructs are a sufficient set for describing an arbitrary process or workflow. As such, they can also describe a composite Web service. A particular process is shown on an activity diagram, an example of which is shown in Figure 13.1.

A UML sequence diagram is used to show the interactions among concurrently existing objects or concurrently executing threads and process instances. It focuses on the time ordering of the messages between such entities. Figure 12.5 is an example of a sequence diagram.

13.3 Workflows

A *workflow* is an activity that addresses some business need by carrying out specified control and data flows among subactivities that involve information resources and possibly humans. A classic example of a workflow is loan processing: when you apply for a loan, you fill out a form, a clerk reviews it for completeness, an auditor verifies the information, and a supervisor invokes an external credit agency or uses a credit risk assessment tool. Each person in the loan process receives information concerning your application, modifies or adds to it, and forwards the results.

There is a fine line between processes and workflows. Some research seems to treat these as isomorphic. A lot of the research on processes is based on previous research on workflow. For example, one of the inputs to the current leading standard for processes, BPEL4WS (discussed in Section 13.4.1), is the Web Services Flow Language (WSFL), which was closely based on IBM's Flowmark workflow product.

For the purposes of this book, workflows are a narrower concept than processes. Processes may be realized through workflows, but possibly through other means as well, e.g., business protocols or conversations among agents, which are introduced in later chapters. The above definition of workflows emphasizes the control and data flow among subactivities that are the essence of workflows. These flows are necessary to realize the desired processes.

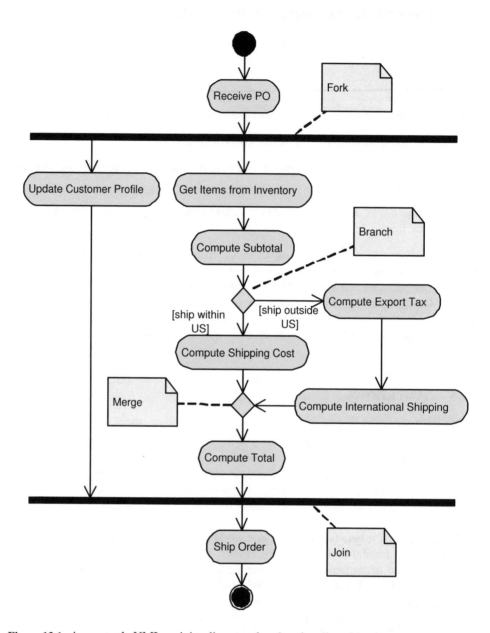

Figure 13.1: An example UML activity diagram showing the allowable control constructs that can be used to describe workflows and processes

Ultimately, no matter how you specify a process, control and data flows will occur when it is enacted. The key point is that in workflow technology, such control and data flows are directly specified from a logically central perspective. This modeling assumption accounts for both the strengths and the weaknesses of workflow technology. Service providers can manage workflows that are used to implement the given service. An implementation based on workflow techniques can help manage potential exception conditions better than a traditional application, which would hide the necessary reasoning. However, workflows too have their limitations as discussed below.

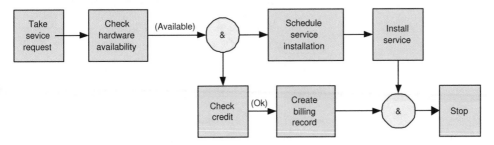

Figure 13.2: A workflow for processing a telecommunications service order

Figure 13.2 illustrates an example of a workflow that is executed when you order a service from a telecommunications provider. You initiate the order by interacting with a sales representative from the provider, who fills out a form on your behalf. The sales representative checks with a provisioning database to determine whether the necessary hardware is in place. If it is, you receive an estimate of when the service will be ready for your use. A local service installer is dispatched to install your service, while the telecommunications provider checks your credit history.

13.3.1 Exceptions

If all goes well, the installer successfully installs the service, the auditors find your credit history acceptable, the billing department is notified to begin charging you, and the workflow concludes successfully. However, things do not always go that smoothly. For example, in checking whether you already have an account, the telecommunications provider might discover that you have an unpaid and overdue balance—or that someone else previously at the same address has an unpaid balance. Such discoveries would raise a red flag.

Perhaps the service installer for your area calls in sick, requiring a revision in the installation schedule. Or the installer might discover that the available hardware is unusable and must be replaced. Each of these situations can lead to modified behavior, as illustrated in Figure 13.3. Such modifications might lead to an additional change in schedule or possibly even cause you to cancel the order altogether because you do not want to wait indefinitely.

These occurrences are instances of exceptions that can arise during process execution. The number of possible exceptions is extremely large; their scope and the great variety of

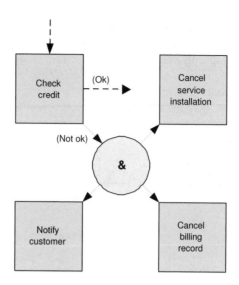

Figure 13.3: Exceptions—unexpected occurrences that interrupt and possibly alter a workflow—can arise during workflow execution

possible contexts make it practically impossible to specify all exceptions statically and in advance. Unfortunately, the only sure thing about exceptions is that they are far from exceptional. As a consequence, most natural processes are inherently incomplete.

Exceptions are not just alternative flows of control; indeed, the two are conceptually distinct. Attempting to include all exceptions is not only futile, but also would clutter the workflow so much as to render it incomprehensible. For the same reasons that programming languages such as Java treat exceptions separately, it is preferable to think of exceptions as parasitic on the main workflow. Of course, if some exceptions occur often enough to become almost routine, they will be incorporated as explicit alternatives within the workflow, as illustrated in Figure 13.4.

In many cases, multiple workflows can arise and interact with each other. For example, in a telecommunications setting, a channel assignment workflow must wait until enough channels have been created by another workflow. Workflow interactions necessarily occur when business partners collaborate. By design, these interactions are intended to be useful, although some might be pernicious in that one workflow could cause the failure of another. The challenge is to identify the potential interactions and to control them appropriately.

13.3.2 Workflow Interoperability

Workflows interact by sharing data or functionality. An interaction can occur (1) directly, (2) via message passing, (3) through a gateway that translates protocols, or (4) by mutual use of a common repository. For each of these means of interaction, there are three primitive patterns

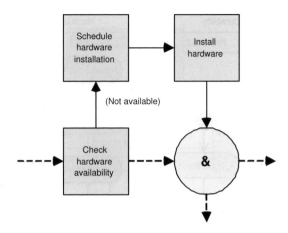

Figure 13.4: An exception that occurs often enough to be considered routine can be incorporated into the workflow as an alternative flow of control

for the interoperability—chained, nested, and synchronized—as depicted in Figure 13.5.

In a *chained pattern*, one process triggers the creation and enactment of another. The triggering process either terminates at this point or continues independently and concurrently with the second. In a *nested pattern*, a triggering process creates and enacts the other, and then waits for the other to return results and terminate. The triggering process can also execute concurrently with the other process, and receive its results at a later step [Jung et al., 2004]. In a *synchronized pattern*, two concurrently executing processes synchronize at a specified point in their respective executions. Only after both reach that point do they continue independently.

13.3.3   A Metamodel for Workflow

The following terms and meanings are defined and advocated by the Workflow Management Coalition (WfMC):

- A *process*, typically a business process, is a collection of tasks organized into a graph. This reflects the workflow view of processes.

- A *task* is an atomic work item.

- A *service* implements a task and may be implemented.

- An *actor* is a human or machine that performs a task by fulfilling a service.

- A *role* abstracts a set of tasks.

- A *workflow* is an instance of a process that binds and consumes resources in fulfilling the tasks of a process.

ation among the different resources while satisfying the resources' local constraints.

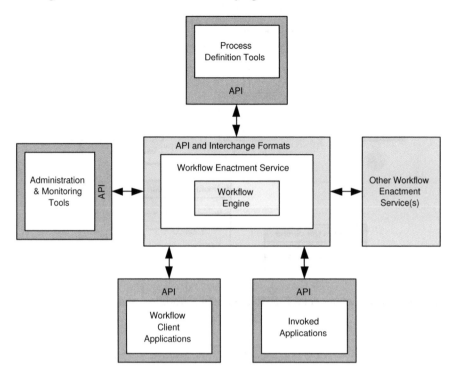

Figure 13.8: A reference model for workflow management systems, showing their major components and interfaces [WfMC]

Another, more profound, kind of interoperation occurs among different workflows. A workflow represents a meaningful unit of processing that affects a number of people and information resources. Clearly, multiple units must interact with each other, because some people participate in more than one, and the units inevitably share resources. Workflow designers must understand, model, and manage these interactions properly. If they do not, all manner of chaos may ensue—and indeed often does. For example, one workflow of the above communications provider might be underway to upgrade wiring with a view to discarding the old wiring, while another workflow might treat the old wiring as freely available and be actively assigning new telephone circuits to it.

This requires an ability to communicate and negotiate. Such coordination benefits from standards that enable workflows modeled and managed by tools from different vendors to be related. One standard is the Simple Workflow Access Protocol (SWAP) announced by WfMC and the IETF. SWAP governs both the control and monitoring of workflows. Control means instantiating the workflow, starting it, stopping it, being informed of exceptions, being informed of completions, and obtaining the results. Monitoring means checking on the current status of the workflow and obtaining its history. SWAP's protocol for basic interaction is:

- The client invokes createProcessInstance command on the workflow server.

- The server returns the URI of the workflow instance.

- The client sends its own URI to the instance.

- When it is done, the workflow instance invokes the completed command on the client.

Other commands can be invoked by both the client and server during execution to provide status, exception, and result information. The resultant protocol is lightweight and, although it has largely been superseded by business process protocols such as BPEL4WS, it is representative of the capabilities needed to describe, control, and monitor a workflow.

13.3.5 State of the Art

There are many workflow tools available in the marketplace—at least 100, or by some counts as many as 250. Each tool provides some type of modeling mechanism coupled with an execution framework. In general, the metamodels underlying most workflow tools are based on a variant of activity networks, which show different activities as nodes and use links to represent various temporal and exception dependencies among the nodes. Figures 13.2 to 13.4 reflect this general idea.

System analysts design workflows on the basis of their understanding of the given organization and the abstractions the chosen workflow tool supports. Once designed, the workflow can be executed automatically by the tool, typically resulting in improved efficiency. For example, when workflows involve human workers, the workers can be automatically informed of the tasks they should be performing and given the resources they need to complete the tasks, thereby reducing idle time.

13.3.6 Challenges Facing Workflow Technology

Workflow technology is not universally acclaimed, and many CIOs are not convinced of its capabilities and benefits. One problem is that current workflow technology is often too rigid. Because workflows are constructed prior to use and are enforced by some central authority, this rigidity is inevitable. However, the lack of freedom accorded to human participants causes workflow management systems to appear unfriendly. As a result, workflows are often ignored or circumvented.

This rigidity also causes productivity losses by making it harder to accommodate the flexible, *ad hoc* reasoning that is the strong suit of human intelligence. This need for flexibility is most apparent when an exception occurs and rigid workflow management tools behave incorrectly. In our earlier example, if the credit bureau is unresponsive, a poorly designed workflow might just wait indefinitely, whereas a flexible one would let a human make a decision based on available information.

- *flow*, for parallel execution. Within activities executing in parallel, execution order constraints are indicated by using service links.

- The control structures relate the following atomic actions:

  - invoke, invoking a specific Web service;
  - receive, a server waiting to receive a message from a client, which would invoke the server's service;
  - reply, generating the response to an invocation request;
  - wait, waiting either for a deadline or some amount of time;
  - assign, assigning a value, which might have come from a received message, to a variable;
  - throw, indicating that something went wrong;
  - terminate, terminating an entire service instance;
  - empty, doing nothing.

In modeling a business protocol as an abstract process, BPEL4WS describes just public aspects of the protocol. For example, in a supply-chain protocol, BPEL4WS would describe the roles of a buyer and a seller as abstract processes, with their relationship modeled as a service link. Abstract processes are restricted to manipulation of values contained in message properties, and use nondeterministic values to reflect the results of hidden private behavior.

In modeling an executable business process, BPEL4WS does not necessarily define a partner's individual implementation completely, but it does define a portable execution format for business processes. Such processes execute and interact with their partners in a consistent way regardless of the supporting platform or the programming model used by a particular implementation.

The result of using BPEL4WS to model an executable business process is a new Web service composed of existing services. The interface of the composite service is a collection of WSDL portTypes, just like any other Web service. Figure 13.10 illustrates this external view of a BPEL4WS process.

13.4.1.1 Transaction Flow

BPEL4WS provides a compensation protocol that is a variant of earlier work on sagas and open nested transactions (see Section 11.5). It enables flexible control of rollbacks and reversals through application-specific definitions for fault handling and compensation, resulting in a Long-Running (Business) Transaction (LRT).

An LRT can be undone by reversing individual operations, using business rules that typically depend on the application. Scope elements delineate the parts of a behavior that are allowed to be reversible by a compensation handler. Scopes can be nested to an arbitrary depth.

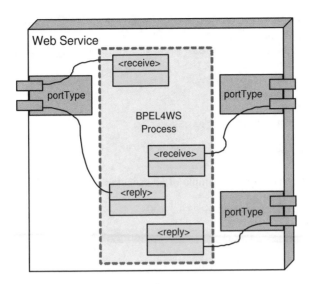

Figure 13.10: A BPEL4WS process is a composite Web service with an interface that is a collection of WSDL portTypes, just like any other Web service

An LRT occurs within a single business process instance, and there is no distributed coordination among the partners regarding an agreed-upon outcome. Achieving distributed agreement is outside the scope of BPEL4WS, to be solved by using protocols described in WS-Transaction (see Section 12.4). In essence, WS-Transaction uses WS-Coordination to extend BPEL4WS to provide a context for transactional agreements between services. Different agreements may be described in an attempt to achieve consistent, desirable behavior while respecting service autonomy.

13.4.1.2   Implementing a BPEL4WS Web Service

A BPEL4WS process is an algorithm expressed as a flow chart, where each step is an activity. Information is passed between activities through data containers and the use of ⟨assign⟩ statements. For example, a customer's address would be copied from a purchase order to a shipping request by the following:

```
<assign>
  <copy>
    <from container="PO" part="customerAddress"/>
    <to container="shippingRequest" part="customerInfo"/>
  </copy>
</assign>
```

A service link is used to define the relationship between two partners and the role that each partner plays. For example, a service link between a buyer and seller might be

```
<serviceLinkType name="BuySellLink"
 xmlns="http://schemas.xmlsoap.org/ws/2002/07/service-link/">
  <role name="Buyer">
    <portType name="BuyerPortType"/>
  </role>
  <role name="Seller">
    <portType name="SellerPortType"/>
  </role>
</serviceLinkType>
```

The following is a complete example of an executable BPEL4WS process for the implementation of a stock quoting service:

```
<!ENTITY BPEL
  "http://schemas.xmlsoap.org/ws/2002/07/business-process"
<process name="simple"
  targetNamespace="urn:simple:stockQuoteService"
  xmlns:tns="urn:simple:stockQuoteService"
  xmlns:sqp="http://tempuri.org/services/stockquote"
  xmlns=&BPEL;/>

  <containers>
    <container name="request"
               messageType="tns:request"/>
    <container name="response"
               messageType="tns:response"/>
    <container name="invocationRequest"
               messageType="sqp:GetQInput"/>
    <container name="invocationResponse"
               messageType="sqp:GetQOutput"/>
  </containers>

  <partners>
    <partner name="caller"
             serviceLinkType="tns:StockQuoteSLT"/>
    <partner name="provider"
             serviceLinkType="tns:StockQuoteSLT"/>
  </partners>

  <sequence name="sequence">
    <receive name="receive" partner="caller"
             portType="tns:StockQuotePT"
             operation="wantQuote" container="request"
             createInstance="yes"/>
    <assign>
      <copy>
```

```
            <from container="request" part="symbol"/>
            <to container="invocationRequest" part="symbol"/>
        </copy>
      </assign>
      <invoke name="invoke" partner="provider"
              portType="sqp:StockQuotePT"
              operation="getQuote"
              inputContainer="invocationRequest"
              outputContainer="invocationResponse"/>
      <assign>
        <copy>
            <from container="invocationResponse" part="quote"/>
            <to container="response" part="quote"/>
        </copy>
      </assign>
      <reply name="reply" partner="caller"
             portType="tns:StockQuotePT"
             operation="wantQuote" container="response"/>
    </sequence>
</process>
```

This process is a simple five-step sequence that begins when a request for a quote is received from the caller. The request is copied to an invocation container, the getQuote operation is invoked with the parameters of the request, the result is copied to a result container, and a reply is returned to the requester.

13.4.1.3 UML to BPEL4WS Translation

The Unified Modeling Language (UML) is a popular representation and methodology for characterizing software and information processes, so we consider its use here for describing business processes. BPEL4WS processes are stateful and can have instances, so the appropriate UML construct for modeling them is a class with stereotype «Process» and whose attributes are the state variables of the process. The behavior of the class is described using an activity diagram. Other aspects of a mapping from UML to BPEL4WS are shown in Table 13.1.

13.4.2 BPML

The Business Process Modeling Language (BPML) and BPEL4WS share similar roots in Web services (SOAP, WSDL, and UDDI), take advantage of the same XML technologies (XPath and XML Schema), and are designed to leverage other specifications (WS-Security and WS-Transaction). Beyond these areas of commonality, BPML supports the modeling of real-world business processes through its support for advanced semantics, such as nested processes and complex compensated transactions. BPML builds on the foundation of WSCI for expressing public interfaces and choreographies.

Table 13.1: UML to BPEL4WS mappings

UML Construct	BPEL4WS Concept
≪*process*≫ *class*	BPEL process definition
Activity graph on a ≪process≫ class	BPEL activity hierarchy
≪*process*≫ *class attributes*	BPEL variables
Hierarchical structure	BPEL sequence and flow activities
Control flow	BPEL sequence and flow activities
≪*receive*≫ activities	BPEL activities
≪*reply*≫ activities	BPEL activities
≪*invoke*≫ activities	BPEL activities

13.4.3   ebXML

The Electronic Business Extensible Markup Language (ebXML) has been established by the United Nations CEFACT (Centre for Trade Facilitation and Electronic Business) and the OASIS (Organization for the Advancement of Structured Information Standards) group to provide specifications for defining standard business processes and trading agreements among different organizations. It also specifies the business messages that are exchanged as part of a business process. The objective is for ebXML to be a global standard for governmental and commercial organizations of all sizes to find business partners and interact with them.

Suppose you are the owner of a disk-drive manufacturing company that sells its disk drives to the computer industry, and you decide that your company should receive purchase orders electronically. To implement this as an ebXML business process, you could follow the typical three-step procedure described in Figure 13.11.

According to this procedure, the recommended way for you to design an ebXML process is to first construct a model of one of your business processes, using a process modeling language. For example, you might use the UN/CEFACT Modeling Methodology (UMM), which is a qualified UML notation for business processes, or you might use the Process Specification Language (PSL). Based on your process model and using the ebXML Business Process Specification Schema (BPSS), you would then extract and format the set of elements necessary to configure an ebXML runtime system that will be able to execute the required set of ebXML business transactions. The result is an ebXML Process-Specification Document, which might be a RosettaNet Partner Interface Process (PIP) as introduced in Section 13.4.4. The following example describes a transaction whereby a customer (buyer) issues a request for a purchase order (PO) and your company, the seller, confirms the purchase order.

Listing 13.1 is the Process-Specification Document corresponding to a well-known RosettaNet PIP for purchase orders and acknowledgments. RosettaNet has given this PIP an identifier 3A4, hence the use of that string in the document.

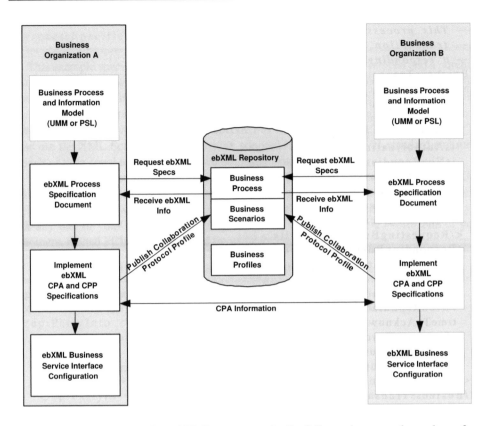

Figure 13.11: The design of an ebXML system typically follows the steps shown here, from modeling a business process to constructing the CPP and CPA specifications

Listing 13.1: An example ebXML Business Process Specification Schema document

```
<ProcessSpecification
  xmlns="http://www.ebxml.org/BusinessProcess"
  name="PIP3A4RequestPurchaseOrder">

<!-- The request document and its XML Schema -->
 <BusinessDocument name="PO_Request"
   nameID="Pip3A4PORequest"
   specificationLocation="PurchaseOrderRequest.xsd"/>

<!-- The confirmation document and its XML Schema -->
 <BusinessDocument name="PO_Confirmation"
   nameID="Pip3A4POConfirmation"
   specificationLocation="PurchaseOrderConfirmation.xsd"/>
```

```
                cp:isNonRepudiationRequired="true"
                cp:isSecureTransportRequired="true"
                cp:isAuthorizationRequired="true"
                cp:timeToAcknowledgeReceipt="PT2H"
                cp:timeToPerform="PT1D"/>
            </cp:ThisPartyActionBinding>
          </cp:CanSend>
        </cp:ServiceBinding>
      </cp:CollaborationRole>
    </cp:PartyInfo>
</cp:CollaborationProtocolProfile>
```

Listing 13.3 provides a feel for the additional practically important, but conceptually trivial details that must be worked out in a collaboration protocol profile to make it effective. These details involve message delivery, transport protocol (the following is a wordy way of specifying HTTPS), reliable messaging, and security, including nonrepudiation.

Listing 13.3: Additional details to include within the partyInfo element of Listing 13.2 to make it complete

```
<cp:DeliveryChannel cp:channelId="asyncChannelA1"
    cp:transportId="transportA2"
    cp:docExchangeId="docExchangeA1">
    <cp:MessagingCharacteristics
        cp:syncReplyMode="none"
        cp:ackRequested="always"
        cp:ackSignatureRequested="always"
        cp:duplicateElimination="always"/>
</cp:DeliveryChannel>
<cp:Transport cp:transportId="transportA2">
    <cp:TransportSender>
        <cp:TransportProtocol cp:version="1.1">
            HTTP
        </cp:TransportProtocol>
        <cp:TransportClientSecurity>
            <cp:TransportSecurityProtocol cp:version="3.0">
            SSL
            </cp:TransportSecurityProtocol>
        </cp:TransportClientSecurity>
    </cp:TransportSender>
</cp:Transport>
<cp:DocExchange cp:docExchangeId="docExchangeA1">
    <cp:ebXMLSenderBinding cp:version="2.0">
        <cp:ReliableMessaging>
            <cp:Retries>3</cp:Retries>
            <cp:RetryInterval>PT2H</cp:RetryInterval>
            <cp:MessageOrderSemantics>
```

```
            Guaranteed
          </cp:MessageOrderSemantics>
       </cp:ReliableMessaging>
       <cp:SenderNonRepudiation>
         <cp:NonRepudiationProtocol>
            http://www.w3.org/2000/09/xmldsig#
         </cp:NonRepudiationProtocol>
         <cp:HashFunction>
            http://www.w3.org/2000/09/xmldsig#sha1
         </cp:HashFunction>
         <cp:SignatureAlgorithm>
            http://www.w3.org/2000/09/xmldsig#dsa-sha1
         </cp:SignatureAlgorithm>
       </cp:SenderNonRepudiation>
       <cp:SenderDigitalEnvelope>
         <cp:DigitalEnvelopeProtocol cp:version="2.0">
            S/MIME
         </cp:DigitalEnvelopeProtocol>
         <cp:EncryptionAlgorithm>
            DES-CBC
         </cp:EncryptionAlgorithm>
       </cp:SenderDigitalEnvelope>
     </cp:ebXMLSenderBinding>
   </cp:DocExchange>
```

The CPP specifies the buyer role for this customer in a RosettaNet PIP, with a *service binding element* specifying the customer's ability to send a purchase order request. A *delivery channel element* defines characteristics of the business transaction and the messaging. The *transport element* defines the buyer's network communication capabilities. Together, the CPA and CPP agreements serve as configuration files (e.g., messaging headers) for ebXML Business Service Interface software.

To summarize, the vocabulary used for an ebXML specification consists of the following three parts:

1. A Process-Specification Document describing the activities of the parties in an ebXML interaction.

2. A Collaboration Protocol Profile (CPP), which describes an organization's profile, i.e., which business processes it supports, its roles in those processes, the messages exchanged, and the transport mechanism for the messages.

3. A Collaborative Partner Agreement (CPA), which is an intersection of two CPP's, representing a technical agreement between two or more partners, and potentially negotiated as shown in Figure 13.12. It may have legal binding.

- One or more REQUIRED Certificate elements that identify the certificates used by this Party in security functions.

- One or more REQUIRED DeliveryChannel elements that define the characteristics of each delivery channel that the Party can use to receive Messages. It includes both the transport level (e.g., HTTP) and the messaging protocol (e.g., ebXML Message Service).

- One or more REQUIRED Transport elements that define the characteristics of the transport protocol(s) that the Party can support to receive Messages.

- One or more REQUIRED DocExchange elements that define the message-exchange characteristics, such as the messaging protocol, that the Party can support.

Listing 13.5: The PartyInfo field for an ebXML Collaboration Protocol Agreement

```
<PartyInfo>
    <PartyId type="...">  <!—one or more—>
        ...
    </PartyId>
    <PartyRef xlink:type="...", xlink:href="..."/>
    <CollaborationRole>  <!—one or more—>
        ...
    </CollaborationRole>
    <Certificate>  <!—one or more—>
        ...
    </Certificate>
    <DeliveryChannel>  <!—one or more—>
        ...
    </DeliveryChannel>
    <Transport>  <!—one or more—>
        ...
    </Transport>
    <DocExchange>  <!—one or more—>
        ...
    </DocExchange>
</PartyInfo>
```

Listing 13.6: The CollaborationRole field for an ebXML Collaboration Protocol Agreement

```
<CollaborationRole id="N11">
    <ProcessSpecification name="BuySell" version="1.0">
        ...
    </ProcessSpecification>
    <Role name="buyer" xlink:href="..."/>
    <CertificateRef certId="N03"/>
```

```
<!—primary binding with preferred DeliveryChannel—>
<ServiceBinding name="aProc"
  channelId="N02" packageId="N06">
  <!—override default DeliveryChannel—>
  <Override action="OrderAck"
    channelId="N05" packageId="N09"
    xlink:type="simple" xlink:href="..."/>
</ServiceBinding>
<!— the first alternate binding —>
<ServiceBinding channelId="N04" packageId="N06">
  <Override action="OrderAck"
    channelId="N05" packageId="N09"
    xlink:type="simple" xlink:href="..."/>
</ServiceBinding>
</CollaborationRole>
```

Based on the above CPA and CPP documents, the following would be an example of a message header for sending a Purchase Order Request document from a buyer to a seller.

Listing 13.7: An example SOAP message header for sending a Purchase Order Request document

```
<SOAP:Envelope
  xmlns:SOAP="http://schema.xmlsoap.org/soap/envelope/">
  <SOAP:Header
    xmlns:eb="http//www.ebxml.org/msg-header-2_0.xsd">
    <eb:MessageHeader id="123" eb:version="2.0"
      SOAP:mustUnderstand="1">
      <eb:From><eb:PartyId>123456</eb:PartyId></eb:From>
      <eb:To>
        <eb:PartyId eb:type="someType">987654</eb:PartyId>
        <eb:Role>
          http://rosettanet.org/processes/3A4.xml#seller
        </eb:Role>
      </eb:To>
      <eb:CPAId>uri:companyA—and—companyB—cpa</eb:CPAId>
      <eb:ConversationId>987654321</eb:ConversationId>
      <eb:Service eb:type="anyURI">
        bpid:icann:rosettanet.org:3A4v2.0
      </eb:Service>
      <eb:Action>Purchase Order Request Action</eb:Action>
      <eb:MessageData>
        <eb:MessageId>UUID—2</eb:MessageId>
        <eb:Timestamp>2000—07—25T12:19:05</eb:Timestamp>
        <eb:RefToMessageId>UUID—1</eb:RefToMessageId>
      </eb:MessageData>
      <eb:DuplicateElimination/>
    </eb:MessageHeader>
```

EDI messages.

We saw above how ebXML's BPSS could be used to define a RosettaNet PIP. An important distinction between RosettaNet PIPs and ebXML BPSS is that PIPs define specific processes (like a purchase-order process), whereas BPSS is a *language* for defining processes. ebXML as such does not define processes and RosettaNet does not provide a process modeling or definition language.

13.5   The Process Specification Language

The Process Specification Language (PSL) is designed for describing or exchanging information among models of discrete processes, i.e., processes consisting of individually distinct events, tasks, or service invocations [Gruninger, 2003]. Examples of such processes are production scheduling, resource planning, workflows, and project management. PSL is not appropriate for continuous processes, whose behavior might be more appropriately described by differential equations.

PSL is intended to be a process representation that is common to all business and manufacturing applications, and powerful enough to represent the processes in any given application. This representation would facilitate interoperation by serving as an interlingua (i.e., a *lingua franca*) for process models. To achieve this, PSL has a formally defined semantics in the language of first-order logic and represented using the Knowledge Interchange Format (KIF). (KIF is now included in the proposed ISO standard called Common Logic.) The semantics consist of a set of KIF definitions that enable PSL statements about processes to be understood. For example, the KIF statement

$$\text{(between ?task1 ?task2 ?task3)}$$

is given semantics by the definition

$$\text{(defrelation between (?a ?b ?c)} \equiv$$
$$\text{(and (before ?a ?b)(before ?b ?c)))}$$

which defines *between* in terms of *before*. The semantics of *before* is provided by axioms, such as

$$\text{(forall (?x) (not (before ?x ?x)))}$$

which states that nothing can be before itself.

A first-order semantics for PSL has several advantages. First, we can specify and implement inference techniques that are sound and complete with respect to models of the theories, i.e., a theory is consistent if and only if there exists a model that satisfies the axioms of the theory. Second, a process ontology with a first-order axiomatization can be more easily integrated with other ontologies. Third, a first-order semantics allows a simple characterization of incomplete service specifications.

There are six basic things that are important to consider for an ontology of business and manufacturing processes, such as PSL:

- Objects, which are concepts in the world that have identity. An example is Mike's credit card.

- ActivityOccurrences, which are actions or events that have a temporal extent and involve specific objects. An example is checking Mike's credit rating beginning at 10:00 a.m. and ending at 11:00 a.m. on October 24.

- TimePoints, which are instances that separate discrete states. An example is the instant between Mike's account having a balance of $1 000 and a balance of $900.

Each of these can be typed, i.e.,

- the type of an Object is a Class;

- the type of an ActivityOccurrence is an Activity;

- the type of a TimePoint is Time.

Next, we can define relationships between some of the pairs of these six basic things. Ignoring Time and TimePoint for the moment, out of the ten possibilities for domain-independent relationships among the remaining four things, the following seven are meaningful:

1. instanceOf between Object and Class.

2. subclass between Class and Class.

3. subclass between Activity and Activity.

4. occurrenceOf between ActivityOccurrence and Activity.

5. partOf between Object and Object.

6. subactivityOf between ActivityOccurrence and ActivityOccurrence.

7. participatesIn between Object and ActivityOccurrence.

A relationship between Activity and Class is not meaningful, e.g., between all hammering *ActivityOccurrences* and all hammers. Similarly, relationships between Object and Activity and between Class and ActivityOccurrence are not meaningful. Note that the binary relations *subclass*, *partOf*, and *subactivityOf* are partial orders (transitive, antisymmetric, and reflexive) as described in Section 6.5.

Each of the seven basic relationships, shown in Figure 13.15 leads to a form of inference: classification and instantiation for things related by *instanceOf*, subsumption for things related by *subclass*, aggregation for things related by *occurrenceOf*, *subactivityOf*, and *partOf*, and association for things related by *participatesIn*.

When Time and TimePoint are included, there are 21 possible binary relationships. In addition to the seven above, the following are meaningful:

1. existsAt between Object and TimePoint.

2. existingFor between Object and Time.

3. occursAt between ActivityOccurrence and TimePoint.

4. occurringFor between ActivityOccurrence and Time.

5. subset between Time and Time.

6. instanceOf between TimePoint and Time.

7. equality, lessThan, and greaterThan between TimePoint and TimePoint.

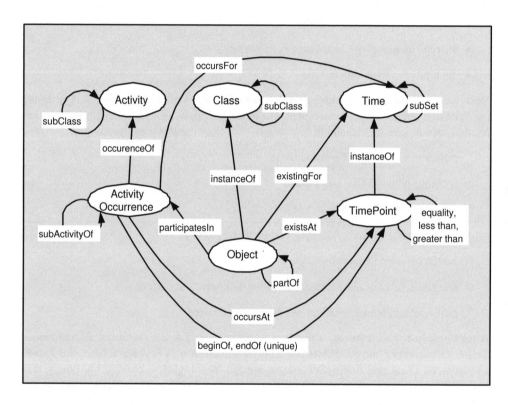

Figure 13.15: The key concepts and relationships of PSL

Class and Time, Class and TimePoint, Activity and Time, and Activity and Timepoint do not participate in meaningful relationships. Finally, PSL provides the two functions beginOf and endOf, which return the TimePoints that define the temporal extent of an ActivityOccurrence.

In the PSL conceptual model, time is understood as discrete. This is based on the intuition that measurable time is essentially discrete, although mathematically approximated via a continuous set, such as the real numbers. A Time can then be represented by a linearly ordered set of integers. Hence, the correct relationship between TimePoint and Time is *instanceOf*, not *partOf*.

PSL consists of the core set of concepts listed above and several extensions. The PSL core includes axioms specifying the semantics of the core concepts. An example axiom is that everything is either an ActivityOccurrence, an Object, or a TimePoint, and that each of these are distinct.

There are five defined extensions: durations, activities and duration, temporal ordering relations, reasoning about state, and interval activities. Within these extensions are definitions for:

Ordering. ActivityOccurrences can take place in sequences delimited by TimePoints.

Concurrency. ActivityOccurrences can take place at the same time, i.e., during the same time interval.

Resource. A Resource is an Object that is used or consumed during an ActivityOccurrence.

PSL also provides support for you to define your own extensions for specific domains or applications. More importantly, PSL can be used to define translations among different process ontologies. For example, we can specify that the *composedOf* property in OWL-S is equivalent to the *subactivity* relation in PSL by the following KIF statement:

```
(forall (?activity1 ?activity2)
    (iff (composedOf ?activity1 ?activity2)
        (subactivity ?activity2 ?activity1)))
```

For another example, a *CompositeProcess* in OWL-S (see Section 15.5.2) is a PSL *activity* that is not *primitive*:

```
(forall (?activity)
    (iff (CompositeProcess ?activity)
        (and (activity ?activity)
            (not (primitive ?activity)))))
```

Via PSL, specifications for Web services in workflows, BPEL4WS (see Section 13.4.1), OWL-S, and others can be given a sound and complete axiomatization, and interoperation among services specified in different formalisms can be facilitated.

13.6 Notes

In 2003, BPEL4WS was submitted to the OASIS Web Services Business Process Execution Language (WSBPEL) Technical Committee. The committee is working on refining the specification for the language.

Piccola is an experimental composition language that has been developed recently [Achermann and Nierstrasz, 2001]. It is based on the π-calculus, and represents an attempt to define a formal execution semantics for workflows composed from Web services.

Information on RosettaNet is available from http://www.rosettanet.org.

13.7   Exercises

13.1. Construct a UML activity diagram for a process with which you are familiar, such as withdrawing money from an ATM, paying for a purchase with a credit card, registering for classes, paying tuition and fees, or obtaining a student or visitor visa. Imagine that each is a two-party interaction: you and the other participant (bank, merchant, university registrar, university cashier, foreign consulate, respectively).

13.2. For the scenario described in Exercise 13.1, construct an equivalent BPEL4WS description.

13.3. Repeat Exercise 13.1 but model three parties for each case. For example, add in the ATM network, credit card company, university department, financial aid office, and host (who provides you with an invitation letter for a visa), respectively.

13.4. For the scenario described in Exercise 13.3, construct an equivalent BPEL4WS description.

13.5. A bank manager must be able to monitor a customer's account to make sure the customer has not transferred any money to terrorists. For this use-case, the detailed interactions between the manager and the system are:

> manager starts the monitoring system;
> monitoring system asks for customer ID;
> manager enters customer ID;
> monitoring system retrieves customer information from account database;
> monitoring system requests table of terrorist names from FBI database;
> FBI database sends table of terrorist names to monitoring system;
> monitoring system compares customer information with table from FBI;
> monitoring system notifies manager if there is a match;
> manager halts the monitoring system.

Construct a UML activity diagram for this scenario.

13.6. For the scenario described in 13.5, construct an equivalent BPEL4WS description.

13.7. Consider the workflow for recording student registration in Figure 13.16. As shown, some of the tasks require operations on databases. Assume that each database management system implements the two-phase commit protocol for transactions. When student Bob registers, Task #2 is a check with the Graduate Coordinator to verify that he has completed the necessary prerequisites for the courses for which he is registering. Assume that Tasks #3, #4, and #5 succeed, but that Task #2 fails.

- As the system administrator, what operations would you have to perform in order to restore consistency to your system?

- How would you modify the workflow in order to prevent problems such as this from occurring in the future?

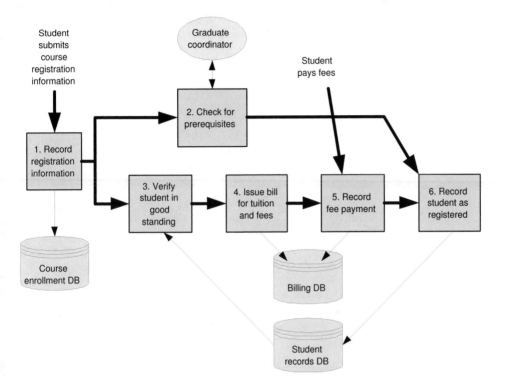

Figure 13.16: An example workflow for student registration

13.8. Compare the procedure for designing an ebXML system in Figure 13.11 with the design procedure for a system based on a service-oriented architecture, as described in Section 5.2.

13.9. Develop an OWL representation of PSL's conceptual model as captured in Figure 13.15.

13.10. PSL: Two time intervals meet if the end time of the first equals the start time of the second. This can be represented by the following defrelation:

$$(\text{defrelation meet } (?\text{task1 } ?\text{task2}) \equiv$$
$$(\text{equal } (\text{endOf } ?\text{task1})(\text{beginOf } ?task2)))$$

Assuming that the temporal predicate *before* has not already been defined in PSL, define *before* in terms of *meet*.

13.11. PSL has a foundation in the *situation calculus*, which has the primitives *situation*, *action*, and *fluent* (a fluent is a relation that might vary over time). It has been claimed that if a situation involves more than one process and if information about the exact timings of the steps in the processes is unavailable, then a situation-calculus reasoner will fail. Discuss why this claim might or might not be true.

Chapter 14

Formal Specification and Enactment

Services are most valuable when composed in novel ways. Two key aspects of realistic scenarios are that the services are inherently autonomous and their compositions are often long-lived. For example, a long-lived interaction occurs in e-business when you try to change an order because of some unexpected conditions or try to get a refund for a faulty product. Further, specific configurations may impose their own requirements. Even short-lived settings may require flexibility, e.g., routing an order differently in some cases or checking if the service requester is authenticated and properly authorized before accepting its order.

Given the autonomy of the partners and the long-lived exception-prone nature of their interactions, hard-coded, procedural abstractions prove quite limiting, because they are inflexible. Thus they impinge upon the autonomy of the partners. A declarative approach would have the advantage of flexibility. But how expressive should the declarative approach be? Its expressiveness will determine how easy it is to use and to implement.

Based on the foregoing chapters, we can reasonably consider two bodies of work. One body of work deals with process specification. Leading process approaches such as BPEL4WS (introduced in Section 13.4.1) specify compositions of services in the form of flows. Typical process approaches such as BPEL4WS capture the flows procedurally in terms of branch and join primitives indicating the specific orderings of the different tasks. BPEL4WS specifications orchestrate invocations of services from a central engine, but do not readily accommodate the kinds of flexibility described above.

The other body of work deals with extended transaction models, as discussed above. These provide a means to organize the tasks of different partners in a manner that offers a level of atomicity. Whereas the earliest extended transaction models considered the semantics of the operations on the data items in detail, this body of work soon converged into a set of abstractions dealing with the existence and ordering of various subtransactions. Clearly, there are considerations of business logic that come into play in defining contingency and compensating transactions and determining whether a certain subtransaction is vital for a given transaction. However, once suitable services exist, the transaction models come down to stating whether or not a given transaction is a contingency procedure for another, a given transaction compensates another, or a given transaction is a vital subtransaction of another.

Recall from Section 13.1 that orchestration views a process from a central perspective; it involves specifying the steps that the different partners may perform. By contrast, choreography involves specifying how the partners may interact. Interestingly, the two approaches come together in the approach of this chapter. Because orchestration takes a central perspective, it can yield more natural specifications. And, because choreography takes a distributed perspective, it can yield more natural execution, where each partner need know and consider only its own interactions.

The approach in this chapter can capture the flow aspects of the process models, as well as the dependencies of the extended transaction models, which are also akin to flow constraints. This approach has two key features. First, the desired service compositions are specified declaratively and enacted automatically by an engine. Second, the engine itself can be dissolved into the partners themselves. In this manner the specification is like an orchestration, but the enactment is like a choreography. Formal specifications yield greater flexibility, which enhances the autonomy and heterogeneity of the partners.

Because of the inherent nature of service-oriented computing, it relies more upon narrow interfaces (to maximize heterogeneity) and flexible descriptions of behavior (to maximize autonomy). Both these aspects emphasize the importance of formal specifications. Successful practitioners of service-oriented computing will need a deeper grounding in formal methods than do those interested in more conventional programming. This chapter serves the pedagogical purpose of helping you refresh and build on your knowledge of logic.

14.1 Scheduling with Dependencies

The basis for scheduling is an expression of the requirements on the occurrence and ordering of various tasks and transactions. It is not always appropriate to specify activities simply as black boxes. Although black box specifications can be appropriate in some cases, it is generally helpful to provide more structure for the tasks in question. For example, a pure black box specification would model a task merely in terms of its start and finish. Simple process models would find such a characterization adequate to be able to chain various activities. But it is almost always essential to distinguish between successful and unsuccessful terminations of tasks. In other circumstances it is necessary to model tasks in even greater detail.

By providing more detailed models of task structures, we are able to coordinate the tasks better. However, the more details that are required, the less the heterogeneity. There is a trade-off between reducing heterogeneity and enabling complex coordination. Reducing the heterogeneity excessively would be inappropriate for an open environment and would violate the basic tenets of service-oriented architectures.

However, in many applications, it is quite reasonable that a small amount of structure be revealed. Such structure is represented quite simply through a set of what are called *signif-*

icant events. Examples of significant events are the start, commit, or abort of a transaction; or the receipt of a purchase order by a service; or the sending of an update notification by a service. "Significant" means that the events are relevant for coordination. Thus a complex activity may be reduced to a single state and termination of that activity to a significant event. Workflows can then be modeled in terms of dependencies among the significant events of their tasks. For example, if a transaction for booking a hotel room fails, then a compensating transaction for the billing transaction would be scheduled.

Thus, significant events can be used to capture succinctly various kinds of flow primitives as needed by the process models or dependencies as needed by the extended-transaction models. For example, we can capture the following kinds of requirements quite naturally in terms of dependencies:

- Chaining tasks in a process: state that the second task is started after the commit of the first task (and not otherwise).

- Joining multiple flows in a process: state that the following task is started only after all tasks preceding it have occurred.

- Failure dependency in a transaction model: state that the start of the dependent transaction follows the abort event of the first transaction.

- Vital subtransactions: the abort of a vital subtransaction causes the abort of the containing transaction.

The significant events can be thought of as being organized into a finite state *skeleton*. Figures 11.3 and 11.7 show two examples of skeletons; the former for a transaction that can participate in 2PC and the latter for an OS task that resembles the black box mentioned above. The edges of these skeletons are the significant events. Skeletons can be used as abstractions to guide the interactions of desired services. Alternatively, they can be used to decide how to structure the implementations of the services based on the skeletons that they must present to others.

Section 14.2 describes the specification language. Section 14.5 shows how flows are enacted, and the enactment is formalized via a series of sophisticated and provably correct strategies in Section 14.6. Section 14.7 discusses relevant literature and research directions.

14.2 Specifying Service Composition

Let us now discuss how to compose services by specifying constraints on events (described below) of the different services. We must begin with a language in which specifications can be expressed. Such a language would for practical purposes be given an XML rendition. However, our present interest is in understanding the concepts that underlie the syntax. For this, a more succinct, conventional logical notation is preferable.

To this end, let us consider $\mathcal{I}$, an event-based linear temporal logic. $\mathcal{I}$ is interpreted as dealing with the occurrences of events. It is temporal, because it includes support for

M_1. $\tau \models e$ if and only if $(\exists i : \tau_i = e)$

> A run τ satisfies a dependency consisting of a single event literal provided the specified event occurs somewhere on the run. For example, if the dependency requires that the hotel service commits, then precisely those computations will satisfy it on which the hotel service does in fact commit.

M_2. $\tau \models I_1 \vee I_2$ if and only if $\tau \models I_1$ or $\tau \models I_2$

> A run satisfies a disjunctive dependency provided it satisfies at least one of the disjuncts. For example, if our dependency is that the hotel service must start or the airline service must start, then at least one of the two services must start.

M_3. $\tau \models I_1 \wedge I_2$ if and only if $\tau \models I_1$ and $\tau \models I_2$

> A run satisfies a conjunctive dependency provided it satisfies both of the conjuncts. For example, if our dependency is that the hotel service must commit and the airline service must abort, then the first must commit and the second must abort.

M_4. $\tau \models I_1 \cdot I_2$ if and only if $(\exists i : \tau_{[0,i]} \models I_1$ and $\tau_{[i+1,|\tau|]} \models I_2)$

> A run satisfies an ordering dependency provided it satisfies both parts and in the correct order. For example, if our dependency is that the hotel service must commit prior to the airline service committing, then both must commit and the hotel should commit first; otherwise, it is a violation.

The semantics of individual events has no temporal component. The idea is that the specifier does not care about when an event occurs except for the restrictions specified using the $\cdot$ operator.

The denotation of a dependency I is the set of runs that satisfy I. The purpose of specifying a dependency is to discriminate between the computations that satisfy it and the computations that do not. The denotation characterizes the good computations precisely. Formally we write $[\![I]\!] = \{\tau : \tau \models I\}$. Thus we can define equivalence of two dependencies as $D \equiv E$ if and only if $[\![D]\!] = [\![E]\!]$.

14.2.1   Coordination Relationships

As running examples, we use two dependencies due to Klein [1991]. In Klein's notation, $e < f$ means that if both events e and f happen, then e precedes f. In other words, f disables e. Also in Klein's notation, $e \rightarrow f$ means that if e occurs then f also occurs (before or after e). That is, e requires f. The reason these dependencies are important is that $<$ orders events without any presumption of occurrence and $\rightarrow$ asserts the conditional occurrence without any presumption of ordering. However, although Klein's work was pioneering, the above notation becomes cumbersome and confusing: it lacks a formal semantics, for instance. (This chapter does not follow Klein's notation.) The following examples formalize these dependencies.

Example 1 Let $D_< \triangleq \overline{e} \vee \overline{f} \vee e \cdot f$. To understand this expression, let us consider a possible run that would satisfy it. Let $\tau \in \mathbf{U}_{\mathcal{I}}$ satisfies $D_<$. If τ satisfies both e and f, then e and f both occur on τ. Thus, neither $\overline{e}$ nor $\overline{f}$ can occur on τ. Hence, τ must satisfy $e \cdot f$, which requires that an initial part of τ must satisfy e and the remainder must satisfy f. In other words, if e and f both occur on τ, then e must precede f on τ.

Example 2 Let $D_\rightarrow \triangleq \overline{e} \vee f$. In the same vein as the above, let $\tau \in \mathbf{U}_{\mathcal{I}}$ satisfy $D_\rightarrow$. If τ satisfies e, then e occurs on τ. Thus, $\overline{e}$ cannot occur on τ. Hence, f must occur somewhere on τ. In other words, the occurrence of e must be accompanied by the occurrence of f. There is no implication about the mutual ordering of the two events.

Now that we have worked through a couple of the dependencies in some detail, we can more easily see how the above approach would capture a variety of other coordination requirements as dependencies. These are summarized below. You are encouraged to study these to be convinced that the formulations are correct.

D_1. e feeds or enables f. f requires e to occur before: $e \cdot f \vee \overline{f}$

D_2. e conditionally feeds f. If e occurs, it feeds f: $\overline{e} \vee e \cdot f \vee \overline{f}$

D_3. Guaranteeing e enables f. f can occur only if e has occurred or will occur: $e \vee \overline{e} \wedge \overline{f}$

D_4. e initiates f. f occurs if and only if e precedes it: $\overline{e} \wedge \overline{f} \vee e \cdot f$

D_5. e and f jointly require g. If e and f occur in any order, then g must also occur (in any order): $\overline{e} \vee \overline{f} \vee g$

D_6. g compensates for e failing f. If e happens and f does not, then perform g: $(\overline{e} \vee f \vee g) \wedge (\overline{g} \vee e) \wedge (\overline{g} \vee \overline{f})$

The above dependencies are mostly self explanatory. Dependency D_6 is an interesting one, however. It captures requirements such as that if e occurs, but is not matched with f, then g must occur, and g must not occur otherwise. This is a typical requirement in information applications with data updates, where g corresponds to an action to restore the consistency of the information (potentially) violated by the success of e and the failure of f. Hence the need to run a *compensation* transaction for e if f does not occur.

14.2.2 Example Scenario

Armed with the above background, we can now attempt to formalize a more realistic situation. For this purpose, consider the following simple scenario inspired by supply chains. Here an assembly service composes three services that supply hoses, valves, and elbow joints. The assembly service orders a matching hose and valve to create a requested assembly. For simplicity, each service (A, V, H, E) can be started and might complete successfully or might fail. The elbow-joint service supports cancellation (undo), which always succeeds. Thus, the events defined are A_s, A_c, V_s, V_c, H_s, H_c, E_s, E_c, and E_u (the subscripts s, c, and u indicate start, successfully complete, and undo, respectively), and their complements. The failure of a service is the complement of its successful completion. For example, the failure of the valve service is given by $\overline{V_c}$.

- If (and only if) an assembly is started, start the valve and hose services: $(\overline{A_s} \vee V_s \wedge H_s) \wedge (\overline{V_s} \vee A_s) \wedge (\overline{H_s} \vee A_s)$.

- As soon as the hose service completes successfully, start the elbow-joint service, except that if the valve service has failed before the elbow-joint service is started, do not start the elbow-joint service: $(\overline{H_c} \vee E_s \wedge V_c \vee \overline{V_c} \cdot \overline{E_s} \vee E_s \cdot \overline{V_c})$.

- If the valve service has failed, but the elbow-joint service has completed successfully, then and only then undo the elbowjoint service: $(V_c \vee \overline{E_c} \vee E_u) \wedge (\overline{E_u} \vee \overline{V_c}) \wedge (\overline{E_u} \vee E_c)$.

14.3   Residuation

Given a specification as a set of dependencies, we must ensure that the right events occur, each at the right time, and that the wrong events do not occur. To this end, imagine a scheduler that somehow causes events to satisfy all the stated dependencies. Notice that the partners whose significant events we are considering are autonomous, so no scheduler may be able to cause events to occur or not occur. Section 14.6.3 returns to this consideration, but for now let us assume that the events can in fact be controlled by a scheduler.

Let us attempt to model the scheduler as a state machine. In practical terms, we can characterize the state of the scheduler by the runs it can allow. The scheduler's changing state determines which events may or may not occur from now on. As remarked above, a dependency is satisfied when a run in its denotation is realized. Therefore, initially, the allowed runs are given by the stated dependencies. As events occur, the set of allowed runs is progressively narrowed. If the set of allowed runs should ever become empty, that means we have hit an inconsistency: there is no way to satisfy the dependencies now. Thus, the scheduler should take care never to enter such an unsatisfiable state.

Intuitively, two questions must be answered for each event under consideration: (a) can it happen now? and (b) what will remain to be done later? The answers can be determined from the stated dependencies and the history of the system. One can examine the runs allowed by the original dependencies, select those compatible with the actual history, and infer how to proceed. However, the present approach achieves this effect symbolically, *without* examining the runs. This is important, because it makes our reasoning depend on the finite specifications, not on the potentially infinite runs.

The dependencies stated in a flow fully describe the initial state of the scheduler; successive states are computed symbolically. Figure 14.1 shows how the states and transitions of the scheduler may be captured symbolically. The state labels give the corresponding obligations, and the transition labels name the different events. An event that would make the scheduler obliged to 0 cannot occur.

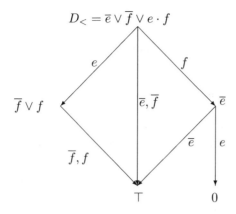

$$D_< = \overline{e} \vee \overline{f} \vee e \cdot f$$

Figure 14.1: Scheduler states and transitions for $D_<$

For a source state given by an expression and transition labeled with event, the target state is given by an operation termed *residuation*. For an expression D and event e, the residuation is written D/e. The key property is that residuation yields the *largest* set of runs satisfying the given expression after the given event has occurred. Importantly, the result of residuating an expression by an event is also an expression.

M_5. $\nu \in [\![D/e]\!]$ if and only if $(\forall \upsilon : \upsilon \in [\![e]\!]) \Rightarrow (\upsilon \nu \in \mathbf{U}_\mathcal{I} \Rightarrow \upsilon \nu \in [\![D]\!]))$

Example 3 (Figure 14.1) If $\overline{e}$ or $\overline{f}$ happens, then $D_<$ is necessarily satisfied: this is why the resulting state in each case is $\top$. If e happens, then either f or $\overline{f}$ can happen later, either of which would take the state to $\top$. But if f happens, then only $\overline{e}$ must happen afterwards (e cannot be permitted any more, since that would mean that f precedes e).

14.4 Symbolic Calculation of Residuals

As discussed above, M_5 characterizes the evolution of the state of a scheduler, but offers no suggestions about how to determine the transitions. Fortunately, a set of equations exists using which the residual of any dependency can be computed symbolically. Importantly, dependencies not mentioning an event have no direct effect on it. Consequently, the reasoning with respect to different dependencies can be performed modularly.

The symbol $\doteq$ is used to indicate that the equations legitimize our simplifying the expression on the left with the expression on the right. Of course, the $\doteq$ would correspond to $=$ so that the equations are sound.

The equations below seek to capture the independence of the scheduler's behavior with respect to an event that does not feature in a given dependency. For this purpose, it helps to define Γ_D as the set of literals mentioned in a specific D and their complements. For example, e is a trivial dependency consisting of just the event literal e, and thus $\Gamma_e = \{e, \overline{e}\}$.

E_1. $0/e \doteq 0$

    No events are allowed by an impossible dependency.

E_2. $\top/e \doteq \top$

    All events are allowed by the $\top$ dependency.

E_3. $(E_1 \wedge E_2)/e \doteq ((E_1/e) \wedge (E_2/e))$

    Residuation distributes over $\wedge$, meaning that e is allowed by a conjunction if it is allowed by each conjunct.

E_4. $(E_1 \vee E_2)/e \doteq (E_1/e \vee E_2/e)$

    Residuation distributes over $\vee$, meaning that e is allowed by a disjunction if it is allowed by at least one disjunct.

E_5. $D/e \doteq D$, if $e \notin \Gamma_D$

    A dependency in which e or $\bar{e}$ do not occur has no bearing on whether e is allowed. However, the dependency itself still remains obligated, hence, we still have D in the result.

E_6. $(e \cdot E)/e \doteq E$, if $e \notin \Gamma_E$

    For a sequence expression beginning with e, if e occurs first, then the rest of the sequence must still materialize.

E_7. $(e' \cdot E)/e \doteq 0$, if $e \in \Gamma_E$ (e' is any event literal)

    For a sequence expression not beginning with e (assume $e \neq e'$), but where e occurs sometime later, e is not allowed now because that would violate the stated order. If the sequence begins with e and e occurs in the rest of it as well, then we would have multiple occurrences of e, which are not allowed anyway. Thus the result of the residuation is 0 in either case.

E_8. $(\bar{e} \cdot E)/e \doteq 0$

    For a sequence expression that begins with the complement of an event, that event cannot occur.

Example 4 The reader can verify that the above equations yield the transitions shown in Figure 14.1.

    The scheduler can take a decision to accept, reject, or trigger an event only if *no* dependency is violated by that decision. There are several ways to apply the above algebra. The relationship between the algebra and the scheduling algorithm is similar to that between a logic and the proof strategies for it. For scheduling, the system accepts, rejects, or triggers events to determine a run that satisfies all dependencies. Equations E_1–E_8 can be proved to be sound and complete.

14.5 Distributed Scheduling

One requirement for a service-oriented architecture is that the agents act as autonomously as possible, constrained only by their coordination relationships. This presupposes that the decisions on events be taken based on local information. Further, distribution is attractive because it promises greater scalability and reliability by placing decision-making functionality right where the decision needs to be made. We can potentially implement such distributed processing using a reliable messaging framework such as message queues.

To enable sound local decisions, we place a *guard* on each event. The guard on an event is a condition such that when the guard is true, it is correct to let the event happen. The guards depend on the dependencies that have been specified. We want the guards to be as general as possible. Moreover, as some events occur, other events can become enabled or disabled, i.e., the guards of the latter events can become true or false. This means that the guards of events can be modified based on messages from other events.

In other words, this approach requires (a) initially determining the guards on each event, (b) arranging for the relevant information to flow from one event to another, and (c) modifying the guards to assimilate the information received from other events.

14.5.1 Temporal Logic for Internal Reasoning

The guard on an event is the weakest condition whose truth guarantees correctness if the event occurs. Guards must be temporal expressions, so that decisions taken on different events can be sensitive to the state of the system. The guards are compiled from the stated dependencies; in practice, they are quite succinct.

$\mathcal{T}$ is a temporal language used for expressing guards. $\Box E$ means that E will always hold; $\Diamond E$ means that E will eventually hold (thus $\Box e$ entails $\Diamond e$); and $\neg E$ means that E does not (yet) hold. $E \cdot F$ means that F has occurred preceded by E. For simplicity, we assume the following binding precedence (in decreasing order): $\neg$; $\cdot$; $\Box$ and $\Diamond$; $\land$; $\lor$. The syntax of $\mathcal{T}$ is given in BNF with T as the start symbol.

L_6. $T \longrightarrow conj \mid conj \land T$

L_7. $conj \longrightarrow disj \mid disj \lor conj$

L_8. $disj \longrightarrow bool \mid \Box\, seq \mid \Diamond\, seq \mid \neg\, event$

The semantics of $\mathcal{T}$ is given with respect to a run (as for $\mathcal{I}$) *and* an index into that run (unlike for $\mathcal{I}$). In addition, we need an auxiliary notion of semantics, which requires two indices. The semantics given next characterizes progress along a given computation to determine the decision on each event. For $0 \le i \le k$, $u \models_{i,k} E$ means that E is satisfied over the subsequence of u between i and k. For $k \ge 0$, $u \models_k E$ means that E is satisfied on u at index k—implicitly, i is set to 0. A run u is *maximal* if and only if, for each event, either the event or its complement occurs on u. The universe, $\mathbf{U}_{\mathcal{T}}$, is the set of maximal runs.

events on ρ up to e have occurred in the right sequence, and (2) the events of ρ after e have not occurred, but will occur in the right sequence.

We define a series of operators to calculate guards as $\mathsf{G} : \mathcal{I} \times \Gamma \mapsto \mathcal{T}$. $\mathsf{G}_b(\rho, e)$ denotes the guard on e due to path ρ (b stands for *basic*). $\mathsf{G}_b(D, e)$ denotes the guard on e due to dependency D. To compute the guard on an event relative to a dependency D, we sum the contributions of different paths in D. $\mathsf{G}_b(\mathcal{F}, e)$ denotes the guard due to flow $\mathcal{F}$ and is abbreviated as $\mathsf{G}_b(e)$ when $\mathcal{F}$ is known. This definition redundantly repeats information about the entire path on each event. Below, we remove this redundancy to obtain a semantically equivalent, but superior, solution.

Definition 1 $\mathsf{G}_b(\rho, e) \triangleq$ if $e = e_i$, then $\Box(e_1 \cdot e_2) \wedge \ldots \wedge \Box(e_{i-2} \cdot e_{i-1}) \wedge \neg e_{i+1} \wedge \ldots \wedge \neg e_n \wedge \Diamond(e_{i+1} \cdot e_{i+2}) \wedge \ldots \wedge \Diamond(e_{n-1} \cdot e_n)$, else 0.
$\mathsf{G}_b(D, e) \triangleq \bigvee_{\rho \in \Pi(D)} \mathsf{G}_b(\rho, e)$.
$\mathsf{G}_b(\mathcal{F}, e) \triangleq \bigwedge_{D \in \mathcal{F}} \mathsf{G}_b(D, e)$.

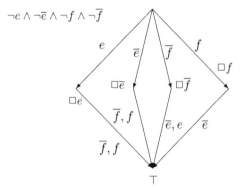

Figure 14.2: Guards with respect to $D_< = \overline{e} \vee \overline{f} \vee e \cdot f$

Figure 14.2 illustrates the above procedure for the dependency of Example 1. The figure implicitly encodes all paths in $\Pi(D_<)$ (here, for simplicity, $\Gamma = \Gamma_D$). The initial node is labeled $\neg e \wedge \neg \overline{e} \wedge \neg f \wedge \neg \overline{f}$ to indicate that no event has occurred yet. The nodes in the middle layer are labeled $\Box e$, etc., to indicate that the corresponding event has occurred. To avoid clutter, labels like $\Diamond e$ and $\neg e$ are not shown after the initial state.

Example 6 Using Figure 14.2, we can compute the guards for the events in $D_<$. Each path on which e occurs contributes the conjunction of a $\Box$ term (what happens before e) and a $\neg$ and a $\Diamond$ term (what happens after e).

- $\mathsf{G}_b(D_<, e) = (\neg f \wedge \neg \overline{f} \wedge \Diamond(\overline{f} \vee f)) \vee (\Box \overline{f} \wedge \top)$. But $\Diamond(\overline{f} \vee f) \cong \top$. Hence, $\mathsf{G}_b(D_<, e) = (\neg f \wedge \neg \overline{f}) \vee \Box \overline{f}$, which reduces to $\neg f \vee \Box \overline{f}$, which equals $\neg f$.

- $\mathsf{G}_b(D_<, \overline{e}) = (\neg f \wedge \neg \overline{f} \wedge \Diamond(\overline{f} \vee f)) \vee (\Box f \wedge \top) \vee (\Box \overline{f} \wedge \top)$, which reduces to $\top$.

- $\mathsf{G}_b(D_<, \overline{f}) = \top$.

- $\mathsf{G}_b(D_<, f) = (\neg e \wedge \neg \overline{e} \wedge \Diamond \overline{e}) \vee \Box e \vee \Box \overline{e} \cong \Diamond \overline{e} \vee \Box e$.

Thus $\overline{e}$ can occur at any time, e can occur if f has not yet happened (possibly because f will never happen), $\overline{f}$ can occur any time, but f can occur only if e has occurred or $\overline{e}$ is guaranteed.

14.5.3 Scheduling with Guards

To execute an event e, check if its guard is $\top$ (execute e), 0 (reject e), or neither (make e wait). Whenever an event e occurs, notify all events depending on e that $\Box e$ now holds, thereby causing their guards to be updated.

Example 7 Using the guards from Example 6, if e is attempted and f has not already happened, e's guard evaluates to $\top$. Consequently, e is allowed and a notification $\Box e$ is sent to f (and $\overline{f}$). Upon receipt of this notification, f's guard is simplified from $\Diamond \overline{e} \vee \Box e$ to $\top$. Now if f is attempted, it can happen immediately.

If f is attempted first, it must wait because its guard is $\Diamond \overline{e} \vee \Box e$ and not $\top$. Sometime later if $\overline{e}$ or e occurs, a notification of $\Box \overline{e}$ or $\Box e$ is received at f, which simplifies its guard to $\top$, thus enabling f. The guards of $\overline{e}$ and $\overline{f}$ equal $\top$, so they can happen at any time.

The above development shows how we can compute the semantics of $\mathcal{T}$, i.e., realize the appropriate runs, incrementally. But in some situations potential race conditions and deadlocks can arise. To ensure that the necessary information flows to an event when needed, the execution mechanism should be more astute in terms of recognizing and resolving mutual constraints among events. This reasoning is essentially encoded in terms of heuristic graph-based reasoning. Although these heuristics can handle many interesting cases, they are not claimed to be complete.

14.6 Formalization

There are two main motivations for carrying out a formalization of an approach for scheduling (as for any other purpose). Formalization can help in proving the correctness of an approach and in justifying improvements in efficiency, e.g., in updating guards incrementally as messages are exchanged and to simplify guards prior to execution.

Correctness is a concern when (a) guards are compiled, (b) guards are preprocessed, and (c) events are executed and guards updated. Correctness depends on how the guards are used to yield actual computations. That is, correctness depends on the *evaluation strategy*, which determines how events are scheduled. We formalize evaluation strategies by stating what initial values of guards they use and how they update the guards. We begin with a strategy that is simple but correct and produce a series of more sophisticated, but semantically equivalent (hence correct), strategies.

The idea is that an evaluation strategy incrementally generates the given run. At any index in the run, an event may take place if its guard is true at the preceding index in the run. That is, an event may be allowed only at a specified index in the run. A given partial run may be completed in various ways, all of which would respect the stated dependencies.

Although the guard is verified at the designated index on the run, its verification might involve future indices on that run. That is, the guard may involve $\diamond$ expressions that happen to be true on the given run at the index of e's occurrence. Because generation looks into the future, it is more abstract than execution.

14.6.1 Evaluating Guards

At run time, a guard equal to $\top$ means the given event can occur; a guard equal to 0 means the given event cannot occur. For all guards in between, we would potentially have to modify them in light of messages about other events. The operator $\div$ captures the processing required to assimilate different messages into a guard. This operator embodies a set of "proof rules" to reduce guards when events occur or are promised. Table 14.1 defines these rules. Because the sequences are limited to two literals, we do not need to consider longer sequences. This leads to an improved guard definition, which creates only two-sequence guard expressions, because it works straight from the dependency syntax.

Table 14.1: Assimilating messages

Old Guard (G)	Message (M)	New Guard ($G \div M$)
$G_1 \vee G_2$	M	$G_1 \div M \vee G_2 \div M$
$G_1 \wedge G_2$	M	$G_1 \div M \wedge G_2 \div M$
$\Box e$	$\Box e$	$\top$
$\Box \bar{e}$	$\Box e$ or $\diamond e$	0
$\diamond e$	$\Box e$ or $\diamond e$	$\top$
$\diamond \bar{e}$	$\Box e$ or $\diamond e$	0
$\Box(e_1 \cdot e_2)$	$\Box(e_1 \cdot e_2)$	$\top$
$\Box(e_1 \cdot e_2)$	$\Box(e_2 \cdot e_1)$	0
$\Box(e_1 \cdot e_2)$	$\Box \bar{e_i}$ or $\diamond \bar{e_i}$	0
$\diamond(e_1 \cdot e_2)$	$\Box(e_1 \cdot e_2)$ or $\diamond(e_1 \cdot e_2)$	$\top$
$\diamond(e_1 \cdot e_2)$	$\Box(e_2 \cdot e_1)$ or $\diamond(e_2 \cdot e_1)$	0
$\diamond(e_1 \cdot e_2)$	$\Box \bar{e_i}$ or $\diamond \bar{e_i}$	0
$\neg e$	$\Box e$	0
$\neg \bar{e}$	$\Box e$ or $\diamond e$	$\top$
G	M	G, otherwise

When the dependencies involve sequence expressions, the guards may also have sequence expressions, which indicate ordering of the relevant events. In such cases, the information that is assimilated into a guard must be consistent with that order. For this reason, the updates in those cases are more complex.

The repeated application of $\div$ to update the guards corresponds to a new evaluation strategy. This strategy permits possible runs on which the guards are initially set according to the original definition, but *may* be updated in response to expressions verified at previous indices. It does not require that every M that is true be used in reducing the guard. This enables us to accommodate message delay, because notifications need not be incorporated immediately. This is because when $\Box e$ and $\Diamond e$ hold at an index, they hold on all future indices.

It can be established that the evaluation of the guards according to $\div$ is sound and complete. All runs that could be generated by the original guards are generated when the guards are updated (completeness) and any runs generated through the modified guards could be generated from the original guards (soundness).

The main motivation for performing guard evaluation as above is that it enables us to collect incrementally the information necessary to make a local decision on each event. However, executability requires, in addition, that we can take decisions without having to look into the future. The above theorem does not establish that the guards for each event will be reduced to $\Box$ and $\neg$ expressions (which require no information about the future). That depends on how effectively the guards are processed.

14.6.2 Simplification

Computing the guards as given in Definition 1 is cumbersome and involves looking at the paths at a lot of detail. Fortunately, the guards can be computed in a much more straightforward and efficient manner. Let us define a symbolic calculation for guards as below. These cases cover all of the syntactic possibilities of $\mathcal{I}$. Importantly, this definition distributes over $\wedge$ and $\vee$: using the above normalization requirement, each sequence subexpression can be treated separately. Thus the guards are quite succinct for the common cases, such as the relationships of Section 14.2.1.

Definition 2 The guards are given by the operator $G : \mathcal{I} \times \Xi \mapsto \mathcal{T}$:

(a) $G(D_1 \vee D_2, e) \triangleq G(D_1, e) \vee G(D_2, e)$

(b) $G(D_1 \wedge D_2, e) \triangleq G(D_1, e) \wedge G(D_2, e)$

(c) $G(e_1 \cdot \ldots \cdot e_i \cdot \ldots \cdot e_n, e_i) \triangleq \Box e_1 \wedge \ldots \wedge \Box e_{i-1} \wedge \neg e_{i+1} \wedge \ldots \neg e_n \wedge \Diamond(e_{i+1} \cdot e_{i+2}) \wedge \ldots \wedge \Diamond(e_{n-1} \cdot e_n)$

(d) $G(e_1 \cdot \ldots \cdot e_n, e) \triangleq \Diamond(e_1 \cdot \ldots \cdot e_n)$, if $\{e, \overline{e}\} \not\subseteq \{e_1, \overline{e_1}, \ldots, e_n, \overline{e_n}\}$

(e) $G(e_1 \cdot \ldots \cdot e_i \cdot \ldots \cdot e_n, \overline{e_i}) \triangleq 0$

(f) $G(0, e) \triangleq 0$

(g) $G(\top, e) \triangleq \top$

Example 8 We compute the guards for the events in $D_<$:

- $\mathsf{G}(D_<, e) = (\Diamond \overline{f} \vee (\neg f \wedge \Diamond f)) \cong \neg f$

- $\mathsf{G}(D_<, \overline{e}) = \top$

- $\mathsf{G}(D_<, f) = \Diamond \overline{e} \vee \Box e$

- $\mathsf{G}(D_<, \overline{f}) = \top$

Thus $\overline{f}$ and $\overline{e}$ can occur at any time. However, f can occur only if e has occurred or never will. Similarly, e can occur only if f has not yet occurred (it may occur in the future).

14.6.2.1 Eliminating Irrelevant Guards

It is fairly straightforward to show that the guard on an event e due to a dependency D in which e does not occur can be set to $\top$, provided D is entailed by the given flow—an easy test of entailment is that D is explicitly a member of the given flow. Thus dependencies in the flow that do not mention an event can be safely ignored for that event. This makes sense, because the events mentioned in D will ensure that D is satisfied in any generated run. Thus, at all indices of any generated run, we will have $\Diamond D$ anyway.

Consequently, we can establish that the guard on an event e due to a conjunction of dependencies is the conjunction of the guards due to the individual dependencies that mention e. Thus, we can compile the guards modularly and obtain expressions that are more amenable to processing.

14.6.3 Formalizing Event Classes

Not all events are alike. We consider four classes of events, which have different properties with respect to coordination. Because the significant events are visible, the service always knows of their occurrence (some interesting results deal with distributing this knowledge requirement over the agents). When an agent is willing to delay or omit an event, it is on the basis of constraints provided by the service, which in turn involve the occurrence or nonoccurrence of other events. The event classes are

- *Flexible*, which the agent is willing to delay or omit.

- *Inevitable*, which the agent is willing only to delay.

- *Immediate*, which the agent performs unilaterally, i.e., is willing neither to delay nor to omit.

- *Triggerable*, which the agent is willing to perform if requested.

The first three classes are mutually exclusive and exhaustive; each can be conjoined with triggerability. We do not have a category where an agent will entertain omitting an event, but not delaying it, because unless the agent performs the event unilaterally, there must be some delay in receiving a response from the service. This is because the above event classes apply to the interface between an agent and a logical coordinator.

Event classes do not replace the dependencies, which specify the constraints among different agents. For example, it is possible that a postal service agent may offer to deliver at two addresses in a particular order, but let the residents at the two addresses decide whether delivery should be made there at all. Thus at the conceptual level, it is possible that an agent may appear to another agent to be willing to cancel an action but not to delay it. However, this requirement is captured through dependencies. If p_1 and p_2 are the postal deliveries and a_1 and a_2 are the addressees' permissions, then, following Examples 1 and 2, we would have $\overline{p_1} \vee \overline{p_2} \vee p_1 \cdot p_2$ (if both deliveries occur, they are ordered), and $\overline{p_1} \vee a_1$ and $\overline{p_2} \vee a_2$ (deliveries are only made if permitted).

For inevitable and immediate events, the dependencies must be strengthened. The basic idea for strengthening is to eliminate paths whose prefixes lead to a state where an inevitable event may have to be denied, or an immediate event denied or delayed. An algorithm to derive strengthened dependencies proceeds by iteratively removing unsafe paths; it is iterative because removing one path to make a dependency safe for one event can make it unsafe for another.

The strengthened dependencies are then used in all reasoning, e.g., in computing the guards. Below, we show revised representations for some relationships.

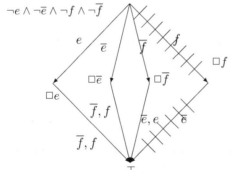

Figure 14.3: Guards from $D_<$ assuming e is inevitable

Example 9 Figure 14.3 shows the dependency of Figure 14.2. The path $\langle f\bar{e} \rangle$ is deleted because if f occurs first, e must not occur. We can verify that $G_b(D_<, e)$ is unchanged but $G_b(D_<, f)$ is stronger: since e cannot be rejected, we cannot let f happen unless e or $\bar{e}$ has already happened. Figure 14.3 still holds when e is immediate. Thus the same guards are obtained as before.

Example 10 Starting with $\bar{e} \vee f$, if e is inevitable, the path $\langle f\bar{e} \rangle$ must be deleted (because if

Describe the skeletons for assembly, hose, and valve tasks, specifying the significant events of each. For each event, state its event class and briefly explain why (in ≤ 1 sentence each). Next state the dependencies in the language $\mathcal{I}$.

14.5. Specify the following supply-chain scenario. This supply chain functions as follows:

- When an assembly is ordered, matching valves and hoses are ordered.
- If exactly one of the hose and valve requests is aborted, cancel the one that committed.
- If the hose and valve requests both commit, but the assembly finds that the hoses and valves fail to match, return the hoses and order new hoses.

Describe the skeletons for assembly, hose, and valve tasks, specifying the significant events of each. For each event, state its event class and briefly explain why. Next state the dependencies in the language $\mathcal{I}$.

14.6. Describe all legal traces involving the events e, f, g, and their complements that satisfy the following dependencies:

- $\bar{e} \vee f \vee e \cdot \bar{f} \cdot g$
- $\bar{e} \vee e \cdot f \vee g \cdot f$
- $\bar{e} \vee e \cdot f$

14.7. We are given two dependencies as follows. Formalize them to determine if both can be satisfied together.

- If e and f both occur, then e precedes f.
- If e and f both occur, then f precedes e.

14.8. Compute the guards for the events (and their complements) that feature in the following dependencies:

- $\bar{e} \vee f \vee e \cdot \bar{f}$
- $\bar{e} \cdot f \vee e \cdot f$
- $\bar{e} \cdot f \vee e \cdot f \vee f \cdot e$

14.9. Compute the revised dependencies corresponding to the dependency: $\bar{e} \vee f \vee e \cdot \bar{f}$ under the following conditions:

- e is immediate
- f is immediate
- e is inevitable

14.10. Compute the revised dependencies corresponding to the dependency: $\bar{e} \cdot f \vee e \cdot f \vee f \cdot e$ under the following conditions:

- e is immediate
- f is immediate
- e is inevitable

14.11. Compute the revised dependencies corresponding to the dependency: $\bar{e} \vee e \cdot \bar{f}$ under the following conditions:

- e is immediate
- f is inevitable

14.12. Compute guards for the events (and their complements) that feature in the following dependency. Simplify the guard expressions.

- $\bar{e} \cdot f \vee f \cdot \bar{e}$

14.13. Compute the revised dependencies corresponding to the following dependency: $\bar{e} \cdot f \vee e \wedge \bar{f}$ under the following conditions:

- e is immediate
- f is inevitable

Chapter 15

Agents

Given that we are now able to to build and use individual services, it is only natural that we should want to put services together to create more sophisticated and valuable services. Service composition is about putting together services to create new functionality. Often, the composed functionality would itself be exposed as a new service with a standard interface.

Let us consider some challenges faced while composing services. Imagine that a merchant would like to enable a customer to be able to track the shipping of a sold item. Currently, the best that the merchant can do is to point the customer to the shipper's Web site, and the customer can then go there to check on delivery status. If the merchant could compose its own production notification system with the shipper's Web services, the result would be a customized delivery notification service by which the customer—or the customer's agents—could find the status of a purchase in real time.

The foregoing chapters have referred informally to agents and multiagent systems. The next sections introduce agents at a technical level. They describe the intellectual background behind agents, the principles that underlie them, and the technologies to realize them.

It is helpful to distinguish three fundamental aspects of agents: (1) individual agents, (2) systems of agents and their interactions, and (3) the environment in which agents operate. This chapter describes individual agents and their environments, and discusses how to implement and apply them. Multiagent systems and their architectures are deferred to the next chapter.

15.1 Agents Introduced

The term *agent* in computing covers a wide range of behavior and functionality. In general, an agent is an active computational entity that

- has a persistent identity;

- can perceive, reason about, and initiate activities in its environment;

- can communicate with other agents, including humans.

Agents act with varying levels of autonomy, depending on environmental constraints and their ongoing interactions, e.g., as reflected in their previous commitments. Because services are best modeled as being autonomous and heterogeneous, they can be naturally associated with agents. Agents make it possible to capture the interactions among the services and the creation of new services as subtle compositions of others.

Agents can balance cooperation with self-interest. Agents also have the property of persistence, which is necessary to carry out business transactions, handle exceptions, and to build a history of interactions, itself necessary to establish trust. Moreover, agents typically interact via the exchange of declarative messages.

The currently popular way of thinking about Web services is based on procedural abstractions using the interaction facilities of CORBA, .NET, or RMI. The history of computing has shown that declarative approaches are ultimately favored because of their benefits of productivity and requirements gathering. For example, databases are queried declaratively through the Structured Query Language (SQL), instead of through procedures as they previously were, and formal grammars are specified through notations such as the Backus Naur Form (BNF), instead of through procedural code. More obviously, literally hundreds of XML-based markup languages exist and continue to be proposed to specify not only documents or knowledge, but also functions as diverse as network administration and data center management. In each case, the development of a declarative (and preferably standardized) language leads to clearer specifications, easier exchange of models, enhanced tools, and improved productivity in programming and administration. The same considerations yield a trend that is inherently biased in favor of declarative agent languages.

Agents are not a panacea, however. But, applying agents enables us quite naturally to (1) capture deeper constraints on what services are willing to offer, thereby capturing richer requirements for service composition, (2) discover trustworthy services, (3) negotiate within teams of providers, and (4) judge the compliance of service providers with their contracts regarding specific compositions.

15.2   Agent Environments

Agents, as well as services, do not exist and operate in isolation, but rather in some physical or computational *environment*. There are an unlimited number of environments, but they can be described in terms of the following six characteristics [Russell and Norvig, 2003]:

Observability. An environment is *fully observable* by an agent if its sensors can detect all aspects that are relevant to its choice of action; it is *partially observable* otherwise.

Determinism. An environment is *deterministic*, from the point of view of an agent, if its next state is completely determined by the current state and the agent's action; otherwise, it is *stochastic*.

History Freedom. An episode is a single cycle of an agent perceiving its environment and taking an action. If the choice of action depends only on the episode itself and not previous episodes, then the environment is *episodic*. If the current decision affects future decisions, as in deciding on a move in chess, then the environment is *sequential*.

Dynamism. An environment is *dynamic* if it can change while an agent is deciding on the action it should take; otherwise, it is *static*.

Continuity. From the point of view of an agent, an environment is *discrete* if the agent perceives it as being in one of a finite number of distinct states, if the agent has a finite number of possible actions, and if there is a distinct set of time points at which it is perceived or actions are taken. If the perceived variables can have a continuous range of values, then the environment is *continuous*.

Multiagent. From the point of view of an agent, if there are other agents that can affect its environment and of which the agent is aware, then the environment is considered to be *multiagent*.

From an implementation standpoint, environments for agents consist of: a communication infrastructure and protocols for interaction, security services for authentication and authorization, remittance services for billing and accounting, and operations support for logging, recovery, and validation.

For an agent to act properly in an environment, some combination of its data structures and program must reflect the information it has about its environment. Because this information would reflect the state of the environment according to the agent, it can be termed its *knowledge* or a set of its *beliefs*. (The distinction between knowledge and beliefs is stronger in ordinary language than it is in the literature about agents, where knowledge is usually treated simply as "true belief." Some researchers represent the relationship between the two using additional attributes, such as justifications, but we will just accept the simpler definition.) An agent's *desires* correspond to the state of the environment the agent prefers. And, an agent's *intentions* correspond to the state of the environment the agent is trying to achieve, which should be a consistent subset of the agent's desires and directly connected to the agent's actions.

Notice that it is the human designer who determines the agent's beliefs, desires, and intentions in an environment. However, to make sense, these beliefs, desires, and intentions must be related to the agent's perceptions and actions. The relationship can be captured in an agent architecture such as the one shown in Figure 15.1.

The relationship is mediated by the agent's reasoning system. For simplicity and as is customary, we assume that the agent's desires are given. The cognitive concepts can then be used in two ways:

- *Means-ends reasoning.* The agent must decide what intentions to adopt or revise, and what actions to perform.

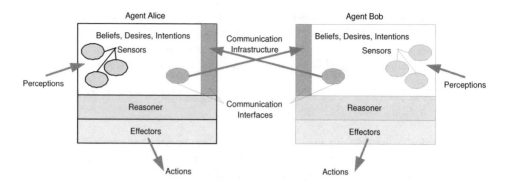

Figure 15.1: An architecture for an agent that captures the beliefs, desires, and intentions ascribed to the agent, and relates them to the agent's perceptions and actions in an environment

- *Plan recognition.* The agent must infer the beliefs, desires, and intentions of other agents in order to cooperate or compete with them.

For example, suppose Agent Alice desires both ice cream and soup, but given the cold weather (and based on beliefs not mentioned here) Alice intends to have only soup. Means-end reasoning causes Alice to get soup from the pantry and heat it in the microwave oven. Next, Alice observes Bob perform actions or hears Bob's statements indicating that Bob is opening the refrigerator door. From this action, Alice uses plan recognition to infer that Bob is about to get ice cream. Knowing Bob to be "rational," Alice decides that Bob does not believe that it is cold outside. Since Alice is a helpful agent, she informs Bob that it is cold outside.

Although the example describes only two agents in a kitchen, the agents are reasoning about each other's actions, which could be actions to access information when the agents are in a Web environment.

A somewhat limited, but widely employed, class of logical models, called truth maintenance systems (TMS), involves the single abstraction of belief. The models capture relationships among the beliefs of an agent, which enables maintaining consistency of the beliefs. Section 16.5 describes a distributed version of a TMS.

15.3   Agent Descriptions

Agents can be described at three different levels of abstraction:

- *Knowledge Level (or Epistemological Level)*—the agent is described by saying what it knows.

- *Logical Level*—the level at which the knowledge is encoded into sentences.

- *Implementation Level*—the level that runs on the agent architecture; this level is important only for efficiency.

An agent can be built by *telling* it what it needs to know, which is a declarative approach. This presupposes a suitable reasoning engine that can operationalize what the agent is being told. An agent can also be built by giving it an ability to interact with its environment and a mechanism for learning, and then placing it in an environment where it can learn for itself, which is an inductive approach. With either approach, the agent needs a means for representing its knowledge, i.e., a *knowledge-representation language*. Such a language has two aspects:

- *Syntax*—how each sentence appears or is recorded in the agent's memory.

- *Semantics*—how the agent's sentences correspond to (facts in) the real world. For example, if x and y are expressions denoting numbers, then $x \geq y$ syntactically is a sentence about natural numbers in the language of arithmetic expressions. The semantics of the language says that $x \geq y$ is false when y is a bigger number than x, and true otherwise.

If the syntax and semantics are defined precisely, then the language is called a *logic*. A knowledge-representation language imposes a systematic relationship between sentences and facts. Most such languages have a *compositional semantics*, so that the meaning of a sentence is a function of the meaning of its parts. Since the syntax is typically given via a context-free grammar using a notation such as BNF, a compositional semantics is readily generated by giving a separate meaning rule for each production in the grammar.

15.3.1 Reasoning

A knowledge base (KB) is a set of sentences expressed in a knowledge representation language. Reasoning is the process of constructing new sentences from existing ones in the KB of an agent. If the existing sentences are true, the process of generating new ones that are necessarily true is called *entailment*, written

$$\text{KB} \models \alpha$$

A reasoning, or *inference* procedure, can do either of the following:

- given a KB, generate new sentences that are entailed by the KB;

- given a KB and a sentence, determine whether or not the sentence is entailed by the KB.

There are two important properties that inference procedures can have. *Soundness* means that the given procedure generates only entailed sentences. *Completeness* means that the given procedure can generate all entailed sentences. The record of operation of a sound inference procedure is called a *proof*. The key to soundness is to have the procedure generate only new sentences that correspond to facts that follow from the facts represented by the KB. Sound reasoning is also called logical inference or *deduction*. A description of an agent at the knowledge level is shown in the following listing:

```
function: KB-Agent (percept)
   static:   KB - a description of the current world
             t - a counter, initially 0, indicating time
   Tell(KB, Make-Percept-Sentence(percept, t))
   action <- Ask(KB, Make-Action-Query(t))
   Tell(KB, Make-Action-Sentence(action, t))
   t <- t+1
   return action
```

The function Tell adds new sentences to the KB, and the function Ask is used to query the KB.

15.3.2   Internal Architectures

Agents are implemented according to a few common architectures. Agent features relevant to implementation are unique identity, proactivity, persistence, autonomy, and sociability. An agent inherits its *unique identity* simply by being an object. To be *proactive*, an agent must be an object with an internal event loop similar to that possessed by an object in a derivation of the Java thread class. Here is simple pseudocode for a typical event loop, where events result from sensing the environment:

```
Environment e;
RuleSet r;
while (true) {
   state = senseEnvironment(e);
   a = chooseAction(state, r);
   e.applyAction(a);
}
```

This is an infinite loop, which also provides an agent with *persistence*. Persistence is a prerequisite for agents to participate in long-term interactions, such as long-lived business transactions, and to detect and handle exceptions that may occur long after a traditional transaction would be over. Ephemeral agents would find it difficult to converse, making them, by necessity, asocial. Additionally, persistence makes it worthwhile for agents to learn about and model each other. To benefit from such modeling, they must be able to distinguish one agent from another, hence the need for unique identities.

Agent *autonomy* is akin to human free will and enables an agent to choose its own actions. For an agent constructed as an object with methods, autonomy can be implemented by declaring all of the methods private. With this restriction, only the agent can invoke its own methods, under its own control, and no external object can force the agent to do anything it does not intend. Other objects can communicate with the agent by creating events or artifacts (especially messages) in the environment that the agent can perceive and react to.

Enabling an agent to converse with other agents achieves *sociability*. The conversations, normally conducted by sending and receiving messages, provide opportunities for agents to coordinate their activities and cooperate, if so inclined. We can achieve sociability by generalizing the input class of objects an agent might perceive, as shown in Figure 15.2. Events serving as input are simply reminders the agent sets for itself. For example, an agent wanting to wait 5 minutes for a reply would set an event to fire after 5 minutes. If the reply arrives before the event, the agent can disable the event. If it receives the event, then it knows it did not receive the reply in time and can proceed accordingly.

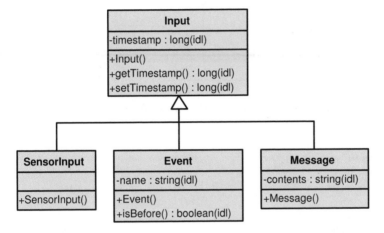

Figure 15.2: An agent's input can be a piece of sensory information, a message from another agent, or an event defined by the agent

A reactive agent is the simplest kind to build, since it does not maintain information about the state of its environment, but simply reacts to its current perceptions. A *reactive architecture* is shown in Figure 15.3.

A belief-desire-intention (BDI) architecture includes and uses an explicit representation for an agent's beliefs, desires, and intentions. The *BDI architecture* shown in Figure 15.4 uses a voluntary multitasking method whereby the environment thread constantly checks to make sure the current intention is applicable. If it finds that it is not, it will tell the intention to stop itself, which the intention does by calling stopCurrentPlan(). This method in turn will call stopExecuting(). Thus the plan is responsible for stopping itself and cleaning up. By giving each plan this capability, we eliminate the possibility of a deadlock resulting from the plan's having some resource reserved when it was stopped. The pseudocode in Listing 15.1 illustrates the two main loops, one for each thread, of our BDI architecture.

Listing 15.1: Pseudocode for voluntary multitasking in the BDI architecture

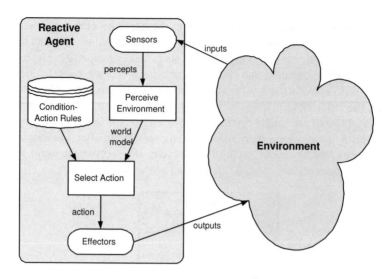

Figure 15.3: An architecture for a reactive agent, which simply reacts to the current observable state of its environment

```
Agent::run () {
Perception p;
p.run (); // start perception system in its own thread

while (true) {
  I = getBestPlan ();
  if (I.execute ()) // true if goal was achieved
    D.remove (I.goal);
}

Perception::run (){
while (true) {
  a.B.incorporateNewObservations (getInput (w));
  if (! a.currentPlanIsApplicable ())
    a.stopCurrentPlan ();
  sleep (someShortTime);
}
}
```

   The agent's run method consists of finding the best applicable plan and executing it to completion. If the plan returns true, it means the goal was achieved, so the goal is removed from the desire (goal) container. If the thread for perceiving the environment finds that an executing plan is no longer applicable and calls for a stop, the plan will promptly return from the execute() call with a false. Notice that the perception thread modifies the agent's set of beliefs. The belief container needs to synchronize these changes with any changes the

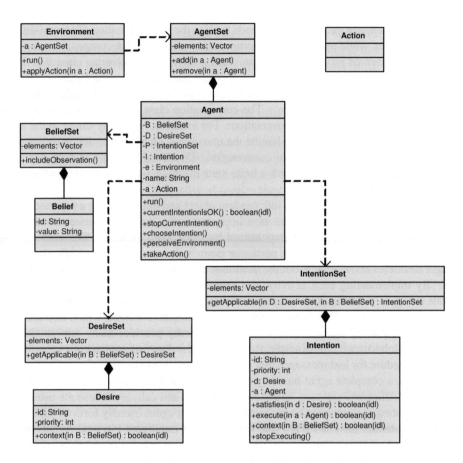

Figure 15.4: A belief-desire-intention (BDI) architecture for an agent, which enables the agent to plan or deliberate about its present and future actions in order to achieve its desires

plans make to the set of beliefs. Finally, the perception thread's sleep time can be modified, depending on the system's real-time requirements. If we do not need the agent to change plans rapidly when its perception of the environment changes, the thread can sleep longer. Otherwise, a short sleep will make the agent check the environment more frequently, using more computation. A more efficient callback mechanism could easily replace the current run method if the agent's input mechanism supported it.

Most popular agent architectures include a set of behaviors and a method for scheduling them. A behavior is distinguished from an action in that an action is an atomic event, while a behavior can span a longer period of time. In multiagent systems, we can also distinguish between physical behaviors that generate actions, and conversations between agents. We can consider behaviors and conversations to be classes inheriting from an abstract activity class. We can then define an activity manager responsible for scheduling activities.

Method invocation may be appropriate for closed systems, but services are inherently autonomous and often to be used in long-lived interactions. For example, a long-lived interaction occurs in e-commerce when you try to change an order because of some unexpected conditions or try to get a refund for a faulty product. Even short-lived settings involve certain kinds of protocols, e.g., checking if the service requester is authenticated and properly authorized before accepting its order. Consequently, interesting service compositions will often involve subtle constraints on interactions among the participating services. The services will not simply be invoked as methods by an aggregator, but will engage each other using rich protocols for potentially long-lived interactions.

Composing services clearly involves overcoming the challenges of semantic disparities, as well as of invoking services correctly. The previous chapters have examined these topics in some detail. Further challenges involve ensuring that the autonomous interests represented by the services can be suitably captured and respected, that the services are able to collaborate, comply with their collaboration, and confirm their collaboration, that exception conditions can be detected and resolved, and lastly that the services can be selected (or can select each other) so that the right behavior is obtained and trust is engendered.

15.5.1   Representing and Reasoning about Action

Several formalisms for representing and reasoning about action have been developed in computer science. The recent representations of services that are intended to support automatic planning of the execution of individual or composed services are typically based on these formalisms. The execution of a service corresponds to performing an action; when a service is composite, so is the corresponding action.

Some of the most common action languages are the imperative programming languages. In particular, some of the primitives that originated with Algol-60, and reflected in modern languages such as Java and C#, are still relevant. These include the classical notions of sequence (;), conditional branching (if then else), and conditional iteration (while), and the well-known constructs for parallel execution (fork and join).

To illustrate a practical application of the above ideas, we consider the OWL-S language next. OWL-S develops richer representations than previous approaches and its expected usage would exploit the ability of agents to reason flexibly and plan. However, although it is placed in this chapter, it does not use the collaborative aspects of agents, which the following chapter elaborates.

15.5.2   OWL-S

OWL for Services makes use of prior work done in workflow management systems, artificial intelligence approaches to planning, formal process models, multiagent planning, and description logic. The objectives for the development of OWL-S are to enable reasoning about Web services, planning compositions of Web services, and automating the use of services by software agents. The goal is to make Web services unambiguously interpretable by a computer.

OWL-S provides an OWL ontology for describing Web services. Using OWL-S, a Web service can advertise its functionality to potential users. A request for a service would then be matched against the Web service's advertisement via a matchmaking process (which OWL-S does not specify).

An OWL-S description for a service consists of three components—a *service profile*, a *service model*, and a *service grounding*—as shown in the upper ontology for services in Figure 15.5.

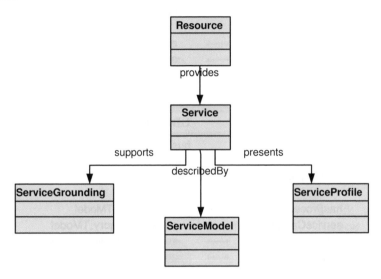

Figure 15.5: An upper ontology for services, describing a service in terms of what it does, how it works, and how it can be accessed

A service profile is an abstract characterization of what a service can do. It is similar to the yellow-page entry for a service and is used for entries in directories and service registries, from where the service can be discovered and matched to requirements. It relates and builds upon the type of content in UDDI, describing properties of a service necessary for automatic discovery, such as what the service offers, and its inputs, outputs, preconditions, and side-effects. From the profile, which presents information about the provider, the functionality, and the functional attributes of the service, service advertisements and service requests can be constructed. Table 15.1 compares the OWL-S service profile to UDDI.

The functionality is specified by the *inputs*, *outputs*, *preconditions*, and *effects* of the service. These are together known as the IOPEs of the given service. For example, an online bookstore could have as inputs the book title, customer's address, and customer's credit card number, with the precondition being the validity of the credit card. The output would be an electronic receipt, and the effects would be a debiting of the credit card and the shipping of the book with a corresponding transfer of ownership. The functional attributes might describe the quality of the service, such as the bookstore's average time-to-ship and the fee for an order

Table 15.2: The IOPEs for an example bookstore service

Process	Inputs	Outputs	Preconditions	Effects
Bookstore Composite	Name, Password, ISBNs, Credit card type, Credit card no.	Order number, Amount, OK?	Valid credit card	Account debited, Ship books, Transfer ownership
Login Selection	Name, Password, Address		Valid address	Account exists
Order Iteration	Title	Order number, Amount		Inventory decreased
Create Account	Name, Password, Address		Account does not exist, Valid address	Account exists
Load Account	Name, password		Account exists	
Choose Book	Title	ISBN, Cost	Book in stock	
Add to Order	ISBN, Cost	Order number, Amount		Inventory decreased
Select Credit Card	Type	Valid type?		
Debit Credit Card	Credit card no., Amount	OK?		Account debited, Ship books, Transfer ownership

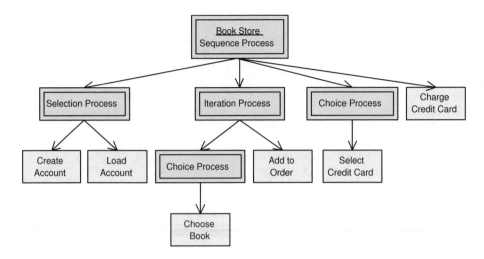

Figure 15.7: A composite service for an online bookstore. The leaf nodes of the graph are simple processes, and the other nodes are composite processes

Selected parts of the OWL-S description of the example in Figure 15.7 are shown in the follow listing. Chapter 8 introduces the constructs used here.

```
<!— DATATYPE —>
<rdf:rdf
 xmlns:service="http://www.daml.org/services/owl-s/0.9/Service.daml#"
 xmlns:process="http://www.daml.org/services/owl-s/0.9/Process.daml#"
 xmlns:proc="http://www.daml.org/services/owl-s/0.9/Process.owl#">

<owl:Class rdf:ID="CreditCardType">
  <owl:oneOf rdf:parseType="owl:collection">
    <CreditCardType rdf:Id="MasterCard"/>
    <CreditCardType rdf:Id="Visa"/>
    <CreditCardType rdf:Id="AmericanExpress"/>
  </owl:oneOf>
</owl:Class>

<owl:Class rdf:about="#Bookstore">
  <process:composedOf rdf:resource="#Bookstore1"/>
</owl:Class>

<rdfs:Class rdf:ID="#Bookstore1">
  <rdfs:subClassOf rdf:resource="proc:Sequence"/>
  </rdfs:Class>

<owl:Class rdf:about="#Bookstore1">
```

```
    <process:components  rdf:resource="#LoginSelection"/>
    <process:components  rdf:resource="#OrderIteration"/>
    <process:components  rdf:resource="#CreditCardChoice"/>
    <process:components  rdf:resource="#DebitCreditCard"/>
</owl:Class>
```

```
<owl:Class  rdf:ID="Login">
  <rdfs:subClassOf  rdf:resource="proc:CompositeProcess"/>
  <process:composedOf  rdf:resource="#Login1"/>
  <rdfs:subClassOf  rdf:resource="#Login1"/>
  <rdfs:subClassOf>
    <owl:Restriction  owl:cardinality="1">
      <owl:onProperty  rdf:resource="#Name"/>
    </owl:Restriction>
  </rdfs:subClassOf>
  <rdfs:subClassOf>
    <owl:Restriction  owl:cardinality="1">
      <owl:onProperty  rdf:resource="#Password"/>
  </rdfs:subClassOf>
  <rdfs:subClassOf>
    <owl:Restriction  owl:cardinality="1">
      <owl:onProperty  rdf:resource="#Address"/>
    </owl:Restriction>
  </rdfs:subClassOf>
  <rdfs:subClassOf>
    <owl:Restriction  owl:cardinality="1">
      <owl:onProperty  rdf:resource="#ValidAddressPrecondition"/>
    </owl:Restriction>
  </rdfs:subClassOf>
  <rdfs:subClassOf>
    <owl:Restriction  owl:cardinality="1">
      <owl:onProperty  rdf:resource="#AccountExistsEffect"/>
    </owl:Restriction>
  </rdfs:subClassOf>
</owl:Class>
```

```
<rdfs:Class  rdf:ID="#Login1">
  <rdfs:subClassOf  rdf:resource="proc:Choice"/>
</rdfs:Class>

<owl:Class  rdf:about="#Login1">
  <process:components  rdf:resource="#CreateAccount"/>
  <process:components  rdf:resource="#LoadAccount"/>
</owl:Class>
```

Processes may be characterized via their function (and data flow) or action. If the processes are composite, they comprise both control flow and data flow.

The *service grounding* portion of the upper-level model for services describes how to access and use a service. It is generally implementation-specific. It includes descriptions for message formatting, transport mechanisms, protocols, and serializations of types. A service model plus a specification of the grounding provide everything needed for using a service. The OWL-S service grounding model builds upon WSDL. An OWL-S atomic processe corresponds to WSDL operation, and inputs and outputs correspond to WSDL messages.

There are many formalisms for describing a business process, such as BPMI and BPML, and OWL-S was designed to be neutral with respect to a particular formalism. It instead just provides the necessary vocabulary and properties for a process model. In this way, OWL-S enables different process models to be described and compared, much as NIST's PSL (see Section 13.5) does for general processes and activities.

Both the OWL-S service model and BPEL4WS provide a mechanism for describing a business process model. However, they can be contrasted in terms of their expressiveness, representations, semantics, discovery support, execution support, and fault handling, as follows:

Expressiveness. OWL-S augments the input and output specifications of BPEL4WS with preconditions and effects: this enables the side effects of services to be encoded. Applications can then reason about how services may be composed to achieve a desired goal, while determining the changes the services will make to their environment.

Representations. Because OWL-S is written in OWL, which supports inheritance and other reasoning about relationships, constraints, and property ranges, OWL-S classes can draw properties from each other, resulting in more easily generated and concise representations. Moreover, the representations for services can then be searched and reasoned over more efficiently. BPEL4WS describes services using structured XML contained in WSDL port type definitions, which is constrained only by XML schema. XML schemas support only limited inferencing.

Semantics. BPEL4WS is not based on a formal semantics, whereas the interpretation of the OWL-S process model is defined in three ways: (1) by a translation to first-order logic, (2) by a translation to an operational semantics using Petri Nets, and (3) by a comparable translation to a subtype polymorphism. Further, the OWL-S service profile and service model provide sufficient information to enable automated discovery, composition, and execution based on well-defined IOPE descriptions and a process model.

Fault Handling. BPEL4WS defines a mechanism for catching and handling faults that is similar to the mechanism in Java. One may also define a compensation handler for actions that cannot be explicitly undone. OWL-S does not define recovery protocols, but Petri Net translations of OWL-S descriptions may be extended to support them.

Execution Monitoring. Neither BPEL4WS nor OWL-S directly support query mechanisms to expose the state of executing processes, although BPEL4WS lists this as future work. Petri Net translations of OWL-S descriptions may be extended to support execution monitoring.

Transactions. BPEL4WS may be extended with WS-Coordination (see Section 12.3) and WS-Transaction (see Section 12.4) to provide a context for predefined transactional semantics.

15.6   Composition as Planning

The composition of Web services is similar to the planning problem that has been investigated extensively in artificial intelligence. Classical planners, using STRIPS-style operators, are ill-suited for Web service composition, because two of the key assumptions are not satisfied: (1) the world is static except for the actions of the planner, and (2) the planner is omniscient. However, there are several recent AI planners that are appropriate, because they can deal with the situation calculus semantics of the OWL-S process model. Such planners would proceed as follows in constructing a composition of Web services that satisfies a requirement:

1. Start by specifying an initial plan, which combines aspects of state requirements and aspects of activity performance.

2. Using a library of process descriptions in the form of partial plans, standard operating procedures, or descriptions of activities that are considered executable, automatically or with user guidance or control, refine the initial outline plan to get a more detailed plan.

3. Support execution and execution monitoring of the partial plans, making selections of undecided parts at run-time.

4. Recognize variances to the expected and required outcomes of the plan steps and begin repairing or replanning to recover dynamically.

O-Plan [Tate and Dalton, 2003] is an example of a planner that follows this procedure.

The Plan Domain Description Language (PDDL) is a unified language for specifying actions. It is intended to be a basic planning language that specific planners can extend in different ways. PDDL includes the ability to represent preconditions and effects of actions (STRIPS-style), conditional effects, hierarchical actions composed of subactions and sub-goals, universal quantification, and safety constraints. For a given domain it can express the predicates, the actions, the structure of compound actions, and the effects of actions.

An important research problem is how to decide how deeply to plan, that is, how many Web services should be identified during planning and how many should be discovered during plan execution. A solution will likely involve a model for the run-time execution context, which is difficult to determine at plan-time in advance of the plan's use, especially for a large open environment such as the Web.

15.7   Rules

In simple terms, it is helpful to distinguish between two main kinds of knowledge: *knowing that* and *knowing how*. Know-that deals with concepts and is static; know-how deals with processing information and is active. Ontologies are a mechanism for making our systems know that, whereas rules are a mechanism for making our systems know how.

Rules are needed for expressing the individual decision making of interacting parties, as well as the contracts that bind them to each other. Nailing down the desired decisions is not trivial and often involves augmenting a given specification incrementally. Likewise, contracts are frequently partial, especially early in the process of being designed. This is the reason that rules are highly suited to specifying such decision making and contracts. The apparent perspicuity and naturalness of rules is the main reason why they are becoming popular in modern practice. Rules are increasingly being used to specify a variety of situations, such as business needs, conventional behavior, and policies. Consequently, rule engines are becoming a routine component of software architectures.

In particular, rules are desirable because they

- Can be created in a modular manner. You can add rules incrementally. As you do so, the set of rules being developed becomes more complete and potentially yields behavior that is closer to what is desired.

- Are inspectable. Unlike imperative programming languages, rules are declarative. Thus, they can be read and understood in terms of their explicit content. You can modify a rule and expect the modifications to have an obvious effect.

- Are executable. Unlike textual descriptions or even some formal specifications, rules can be directly executed. Thus, behavior specified using rules is attained by executing those rules. There is no additional step of converting specifications into formally executable implementations.

For the above properties to hold, it is natural that rules may conflict with each other. Therefore, an approach that can accommodate conflict handling, especially using priorities, is desirable. Grosof and colleagues [2004] have developed one such approach, known as *courteous logic programming*.

Further, it helps to support *procedural attachment*, which means that the rule engine may invoke procedures external to itself. This is a valuable feature, because it means that rules can be used for specifying the top levels of the desired behavior (decision rules or contracts), while leaving the detailed aspects of the behavior to be specified by existing programs. The procedures invoked based on the rules could be Web services and other transactions that are required by the given business process. Rule engines that support procedural attachments are sometimes called *situated*.

In the Semantic Web vision, rules are a higher-level abstraction than ontologies.

signed on, then the customer is considered authenticated." Thus, when a request is received, the variables are bound and a query is generated so as to determine whether the customer is authenticated.

An interesting and important aspect of ECA rules is in how the events are modeled and described. If the event sublanguage is restricted to individual events, it is still useful, but its usefulness might be quite restricted. The above examples involve simple events. Composite events are needed to ensure that the right subtle occurrences of interest are detected and responded to. However, if the event sublanguage is made complex, then the specified composite events can become difficult to detect.

When the actions performed by one rule yield events that trigger further rules, a theoretical challenge is determining whether the set of rules would ever terminate under certain circumstances.

15.7.2.2   Inference Rules

We can think of an inference rule as an if-then statement whose *antecedent* (the part before then) and *consequent* (the part after then) are treated as logical formulas. Typically, for a rule to have some general import, the antecedent would use one or more free variables. To specify the desired conclusions unambiguously, the consequent would use variables that are introduced in the antecedent. In essence, the antecedent declares the variables, which the consequent then uses.

There are some restrictions on the logical forms of the antecedent and consequent. The antecedent of a rule must be a conjunction of zero or more clauses. That is, if all the antecedent clauses hold, then the consequent can be legitimately inferred. There is no loss in disallowing disjunctions, because they could be captured by having multiple rules. That is, if you wish to capture something like

$$\text{if } a_1 \text{ or } a_2 \text{ then } c$$

you can capture it via a pair of rules

$$\text{if } a_1 \text{ then } c$$

$$\text{if } a_2 \text{ then } c$$

This works because if one of a_1 and a_2 hold, then the corresponding rule would yield the consequent c.

The consequent of a rule is also restricted to be a conjunction of one or more clauses. Disjunctions are disallowed because they would greatly increase the complexity of reasoning: the interpreter would not know which of the consequent clauses may be inferred and thus would need to maintain a set of branching possibilities.

Technically, *facts* are simply inference rules whose antecedent happens to be true. This corresponds to the case where there are zero clauses in the antecedent.

There are two main ways in which to apply inference rules. First, in *forward chaining*, the rules justify what conclusions can be drawn from the given premises. The process begins when some premises are asserted as facts. A rule matches when its antecedent becomes true for some binding of the variables occurring in it. When a rule matches, its consequent is then asserted for the given bindings of any variables in it. The consequent is separated into its conjuncts so each can be asserted. Often the new assertions would cause further rules to be matched and their consequents to be asserted. The computation spreads (forward) in this manner.

Second, in *backward chaining*, the rules justify reasoning about what must be true so that a given conclusion can be drawn. The process begins when some target conclusion is posed as a *query* or *goal*. A rule matches when a binding of the variables occurring in its consequent would make the given query true. This leads to the antecedent of the rule being considered a query in the next step. The antecedent would in effect be separated into subqueries (subgoals) corresponding to its logical structure. Typically, these new subqueries would match with additional rules or with facts. The process bottoms out either when enough of the subqueries have matched with facts so that the original query can be answered as true (with the discovered variable bindings as the key answer) or when the search has exhausted the possibilities and the original query cannot be answered as true.

Let us consider some examples of inference rules from a service perspective.

- If Alice is trusted and Alice has given an endorsement to Bob, then Bob is also trusted. In the forward chaining sense, this rule would fire when the two base facts in the antecedent begin to hold regardless of their order. Assume that Alice is initially not trusted. Now, if Alice endorses Bob, nothing would happen. Later if Alice begins to be trusted, the antecedent would match and Bob would begin to be trusted. In the backward-chaining sense, suppose we wish to find out if Bob can be trusted. The rule suggests that we establish both of the following facts: Alice is trusted and Alice has endorsed Bob. These facts could be established merely because they are given as such or because they may be inferred from further rules whose antecedents can be established.

  In general, however, the above kind of rule would not be expressed with specific names of the people in whom trust might or might not be placed. Instead, it would be expressed using variables. In other words, we would state something along the lines of "if a party is trusted and it endorses another party, then the second party is trusted as well." Now the rule would match with any number of facts and queries and could even be chained with itself. For example, if we know that Alice is trusted, then Alice becomes the first party of the antecedent, and if Alice endorses Bob, then Bob becomes the second party of the antecedent. In the meanwhile, if Bob has endorsed Charlie, then (given that Bob is now trusted) Charlie would be trusted as well. And, of course, the rule could be chained with other instantiations of itself.

- If a transaction has been initiated, it will be completed.

- If a transaction has been initiated, it will be completed. However, if an exception occurs, then it will not be completed (and this rule overrides the previous rule).

- If a transaction has been initiated, it will be completed. However, if an exception occurs, then it will not be completed. But if there is a compensating operation for the exception, then it will be completed (and this rule overrides the previous rule).

15.7.3   Jess

Jess is a fast and lightweight rule engine written in Java. It provides to services and agents an ability to reason declaratively using knowledge expressed in rules and facts. To process rules, Jess uses the Rete algorithm, which is an efficient mechanism for solving the many-to-many matching problem between facts and rules [Forgy, 1982]. Jess supports both forward and backward chaining, working memory queries, and the ability to manipulate and reason about Java objects directly.

The features, capabilities, and usage of Jess are best seen through examples. To construct a Jess system, you should first specify the structure for facts using deftemplate. For example, a system for reasoning about automobiles and their relationships might have facts with the following form:

```
(deftemplate automobile
    ''A general representation of a car.''
    (slot manufacturer)
    (slot model)
    (slot year (type INTEGER))
    (multislot accessories)
    (slot color (default silver))
)
```

This would allow you to define facts like this:

```
(deffacts example-facts ''Specific cars.''
        (automobile (manufacturer Ford)
                    (year 1966)
                    (model Cobra)
                    (accessories radio tachometer CD-player)
                    (color blue))
        (automobile (manufacturer BMW)
                    (model 328)
                    (year 1936))
        (automobile (manufacturer Mercedes-Benz)
                    (model 300 SL Gullwing Coupe)
                    (year 1955))
)
```

which represents a silver (the default color) 1936 BMW 328 Roadster, a blue 1966 Ford Cobra, and a silver 1955 Mercedes-Benz 300 SL Gullwing Coupe. Notice that the slots can be used in any order, or not used at all.

Rules can be used to take actions based on available facts. For example, consider the following definition of a rule:

```
(defrule paint-car ''Make all cars silver.''
  ?fact <- (automobile (year ?y) (color ?c))
  (test (neq ?c silver))
  =>
  (modify ?fact (color silver))
  (printout t ''Repainted a '' ?y ''car silver.'' crlf)
)
```

When Jess is run, this rule would be matched with each of the three facts about automobiles and, for those whose value for the color slot is not already silver (the Ford Cobra), the color would be changed to silver and a message about the change would be printed.

15.7.4 SWRL: Semantic Web Rule Language

The Semantic Web Rule Language is a markup language for expressing rules in a standardized manner and melding them with ontologies. It is best used for communicating rules among different services or agents. SWRL can be mapped to the proprietary formats of various rule engines that different services might internally employ. By incorporating inference and reaction rules, SWRL covers the major rule families of interest.

SWRL extends the set of OWL axioms to include a kind of rules (Horn clauses, described in Section 15.7.5), thus enabling rules to be combined with an OWL knowledge base. SWRL rules are in the form of an implication between an antecedent and a consequent. Atoms in these rules can be of the form $C(x)$, $P(x, y)$, or $sameAs(x, y)$, where C is an OWL description, P is an OWL property, and x, y are either variables, OWL individuals, or OWL data values. The following is an example rule written in the Jess and CLIPS Lisp-like syntax followed by the SWRL XML syntax, which borrows several constructs from RuleML. This example captures the idea that a parent's parent is a grandparent.

```
(defrule grandParentRule-1
  (parent ?k ?p)
  (parent ?p ?g)
  ==>
  (grandParent ?k ?g)
)
```

```
<ruleml:rulebase label="grandParentRule">
  <ruleml:imp>
    <ruleml:_body>
      <swrlx:individualPropertyAtom
            swrlx:property="hasParent">
        <ruleml:var>k</ruleml:var>
```

```
        <ruleml:var>p</ruleml:var>
      </swrlx:individualPropertyAtom>
      <swrlx:individualPropertyAtom
              swrlx:property="hasParent">
        <ruleml:var>p</ruleml:var>
        <ruleml:var>g</ruleml:var>
      </swrlx:individualPropertyAtom>
    </ruleml:_body>
    <ruleml:_head>
      <swrlx:individualPropertyAtom
              swrlx:property="hasGrandparent">
        <ruleml:var>k</ruleml:var>
        <ruleml:var>g</ruleml:var>
      </swrlx:individualPropertyAtom>
    </ruleml:_head>
  </ruleml:imp>
</ruleml:rulebase>
```

The SWRL syntax, like most XML-based syntaxes, is more verbose, but also has some additional expressiveness in making the conjunctions of clauses explicit and marking the variables and constants explicitly. This simplifies transforming it into other syntaxes of interest. In general, SWRL would be used for exchanging rules. However, the rules would be processed via tools such as XSLT to produce the syntax desired by one's rule engine or application.

15.7.5 Complexity and Expressiveness

Rules correspond to the so-called Horn clauses in logic, which are defined as follows. A literal is either an atomic proposition (such as p) or the negation of an atomic proposition (such as $\neg p$). A clause is a literal or a disjunction of literals, e.g., $p \lor q \lor \neg r \lor \neg s$. A Horn clause is a clause in which at most one of the literals is positive, e.g., $p \lor \neg q \lor \neg r \lor \neg s$. In essence, by converting the disjunction into implications, we see that in a Horn clause a conjunction of zero or more literals implies no literal or a single literal (the one that was negative). For the above example, this would be $(q \land r \land s) \rightarrow p$, more commonly written as $p \leftarrow (q \land r \land s)$. Recall that a conjunction of no literals is true. If there is no positive literal, then the conjunction of literals is false.

Programming with rules of the kind discussed here is better known as *logic programming* or LP. Importantly, LP does not support concluding a formula that is a disjunction or is existentially quantified. For example, we cannot express a rule that states that a transaction must either be ongoing or be aborted. By contrast, the predicate calculus enables such conclusions, but is intractable: decision procedures might never terminate for false conclusions. Even if the predicate calculus is restricted to just its propositional fragment, the resulting logic remains intractable for practical purposes.

The *datalog restriction* applies to the syntax and prevents logical functions of nonzero arity. Given a set of Horn rules, if the number of logical variables per rule is bounded by a constant, the complexity of computing the conclusions from the set of rules is polynomial in the number of rules. That is, it is tractable.

However, the predicate calculus has the shortcoming that it is monotonic and thus cannot easily express *defeasible* reasoning. Defeasible reasoning involves what can be informally thought of as jumping to conclusions. The classic example is to express that birds can fly. Knowing that Tweety is a bird you can conclude that Tweety can fly. But this conclusion is defeasible. When you learn that Tweety is a penguin, you might change your expectation about whether it can fly. Similar situations arise in contracts. For example, we might require that a buyer may not return goods for a refund. However, if the goods were received damaged and are returned within three days of receipt, then the buyer would be given a full refund. A rule-based language can capture such requirements perspicuously.

Description logic, on which OWL DL is based, obeys the datalog restriction. It is decidable but intractable. However, as explained above, description logic has some expressive limitations. Consequently, formalisms such as rules are essential.

15.7.6 Negation, Nonmonotonicity, Priorities

It is helpful to distinguish between two kinds of negation:

- Weak negation (termed "not"), which indicates nontruth, but does not indicate falsity. For example, "not raining" means that the truth of "raining" has not been established, but does not mean that it is definitely dry. That is, weak negation roughly corresponds to not knowing that the given proposition is true.

- Strong negation (termed "neg"), which indicates falsity and hence also nontruth. For example, "neg raining" means that it is dry. Thus, "neg raining" entails "not raining," but not the other way around.

Strong negation corresponds to the negation in classical binary or two-valued logic. Not true means false—this is popularly known as the law of the excluded middle. However, weak negation is often essential in practice, because we may simply lack the knowledge to establish that something is definitely false. For complete predicates, weak negation collapses into strong negation.

Weak negation turns out to be especially useful, because it enables us to capture knowledge in a flexible, elaboration-tolerant, manner. Such knowledge representations are defeasible in the sense introduced above. Consider the previous example. Suppose we wish to record that birds fly (by default) and that penguins do not fly (by default). We could simply state that birds fly *unless* we know for a given bird that it cannot fly, and that penguins do not fly unless we know for a given penguin that it can fly.

Rules may conflict. In such cases, there are usually few or no principled reasons for choosing between the rules. Some of the few domain-independent means for prioritizing rules are as follows. One, prefer rules whose antecedents use predicates that are more selective. For example, since penguins are birds, a rule that is stated in terms of penguins would have

15.7. Suppose you are developing a Semantic Web service that can provide a client with the weather forecast for a city, and can also provide a list of tourist attractions for a city. Both of these are described within a single WSDL file.

If you document this using OWL-S, should you create two atomic processes and combine them into a composite process using the choice-element? Or should you create two separate OWL-S descriptions? Or is there a better way to describe your service(s)? Please explain what you think is the best OWL-S representation.

15.8. Imagine that there are two existing and deployed Web services as described in Exercises 4.2 and 4.3. Would it make sense for one of these to make use of the other? If so, describe the composition using OWL-S.

15.9. Imagine that there are two services you would like to compose: service S_1 followed by service S_2. However, there is a mismatch in the data types. S_1 produces results as data type D_1, but S_2 requires an input of type D_2, where $D_1 \neq D_2$. Fortunately, you discover a translation service, S_3, that will convert without loss data from type D_1 to type D_2. Write an OWL-S description of the resultant composed service. You may assume that all services are atomic processes.

15.10. **Situation Calculus:**

In the situation calculus, axioms that describe how the world changes when actions are performed are called *effect axioms*. A *frame axiom* describes how the world stays the same. Consider the "blocks world" domain, whose objects are blocks, tables, and situations. The predicates in this domain are

On(x, y, s), which states that object x is on top of object y in situation s.

ClearTop(x, s), which states that there is nothing on top of object x in situation s.

Block(x), which states that x is a block.

Table(x), which states that x is a table.

There is only one action, PutOn, whose effect axiom is

$$\forall x, y, z, s \; ClearTop(x, s) \wedge ClearTop(y, s) \wedge On(x, z, s)$$
$$\Longrightarrow$$
$$On(x, y, Result(PutOn(x, y), s))$$
$$\wedge [Table(z) \vee ClearTop(z, Result(PutOn(x, y), s))]$$

Write a *frame axiom* that describes the predicate ClearTop, i.e., a block will remain clear if nothing is put on it.

15.11. Consider the examples of agents and their percept, action, goal, and domain descriptions given in the table below. Which type of agent architecture (simple reflex, goal-based, or utility-based) is appropriate for each? Explain your choices.

Application	Percepts	Actuators	Goals	Domain
Medical diagnosis	Symptoms, findings, patient's answers	Questions, tests, treatments	Healthy patient, minimize cost	Patient, hospital
Part-picking robot	Pixels of varying intensity	Pick up parts and sort into bins	Place parts in correct bins	Conveyer belt with parts

15.12. Suppose you are constructing agents that will operate in each of the following applications: (a) an assembly-line welding robot, (b) a triage-nurse agent that classifies incoming patients, (c) a homework-grading agent, (d) a traffic-light controller, (e) a subway train controller, and (f) a gate agent that assigns arriving airplanes to gates at an airport. For each agent, state what would be its (i) goals, (ii) environment in terms of being fully or partially observable, deterministic or stochastic, episodic or sequential, static or dynamic, continuous or discrete, and single agent or multiagent, (iii) perceived inputs, (iv) actions, (v) reasoning ability in terms of being logical, utility-based, or goal-based, and (vi) architecture in terms of being reactive or deliberative.

15.13. Consider the following composed service. A customer (a prospective traveler) contacts a travel agent to book a trip. The traveler specifies a destination, a departure date, and an arrival date. A trip consists of an air ticket to the specified destination, a hotel reservation while the traveler is in the destination city, and a rental car. The travel agent would request an air ticket, a hotel reservation, and a car rental reservation from the corresponding services. Assume each service responds with a success or failure. The air ticket and hotel are both necessary in order to complete the trip. The car rental is desirable but not essential.

Write a set of rules, preferably in SWRL syntax, to capture the above composition from the perspective of the travel agent.

15.14. Augment the problem of Exercise 15.13 so that the air ticket, hotel, and car rental services can carry out conversations. In particular, allow each of these services to send back a cancellation to the travel agent *after* they have sent a success message. The cancellation indicates that the corresponding reservation is no longer valid, presumably because of circumstances beyond their control. A cancellation leads to the remaining reservations being canceled and a cancellation message being sent to the traveler.

Write a set of rules to capture the augmented composition from the perspective of the travel agent.

15.15. Construct a small knowledge base for a rule-based inferencing engine, such as Jess or Algernon. Your knowledge base should contain at least 10 forward rules. At least one

Table 16.1: Dimensions of MAS: Agent

Property	Meaning
Adaptivity	Ability of an agent to learn, from teachable to autodidactic
Autonomy	Independent to controlled
Interactions	Direct or via facilitators or mediators; declarative or procedural
Sociability	Interest in others: autistic, aware, responsible, or a team player
Friendliness	Cooperative to competitive to antagonistic

Table 16.2: Dimensions of MAS: System

Property	Meaning
Dynamism	Changing membership
Scale	Number of agents
Control Structure	Hierarchy to democracy
Coordination	Self interest
Uniqueness	Homogeneous to heterogeneous
Interface Autonomy	Communication: specify vocabulary, language, and protocol
	Intellect: specify goals, beliefs, and ontologies
	Skills: specify procedures and behaviors

16.1 Applicability in Service-Based Systems

As explained in Chapter 1, the World Wide Web is evolving into the Semantic Web. The idea behind the Semantic Web is to enable automation so that Web pages are accessed programatically. This transition from human users to software would enable greater productivity by supporting more precise search for information and enabling more flexible and robust business processes.

Web services are central to the above vision. Agents further advance the vision, because they provide greater flexibility in how services are used and created. There are agents that make use of the Web as it is now. A typical kind of such agent is a shopbot, an agent that visits the on-line catalogs of retailers and returns the prices being charged for an item that a user might want to buy. Shopbots operate by a form of "screen-scraping," in which they download catalog pages and search for the name of the item of interest, and then the nearest set of characters that has a dollar-sign, which presumably is the item's price. The shopbots also might submit the same forms that a human might submit and then parse the returned pages that merchants expect are being viewed by humans.

However, such techniques are not easy to apply for developing and maintaining large interacting systems. The Semantic Web will make the Web more accessible to agents by making use of semantic constructs, such as ontologies, so that agents can understand what is on a page.

Typical agent architectures have many of the same features as service-oriented architectures. Agent architectures provide yellow-page and white-page directories, where agents advertise their distinct functionalities and where other agents search to locate the agents in order to request those functionalities. However, agents extend Web services in several important ways:

- A service knows only about itself, but not about its users, clients, or customers. Agents are often self-aware at a metalevel, and through learning and model building gain awareness of other agents and their capabilities as interactions among the agents occur. This is important, because without such awareness a Web service would be unable to take advantage of new capabilities in its environment, and could not customize its service to a client, such as by providing improved services to repeat customers.

- Services, unlike agents, are not designed to use and reconcile ontologies. If a client and the provider of a service happen to use different ontologies, then the result of invoking the Web service would be incomprehensible to the client. Agents can mediate such differences.

- Agents are inherently communicative, whereas services are passive until invoked. As new information becomes available, agents can provide alerts and updates. Current standards and protocols make no provision for different interaction patterns, such as *subscribing* to a service to receive periodic updates.

- For services to apply naturally in open environments, they should be modeled as being autonomous. Autonomy is a natural characteristic of agents, and it is also a characteristic of many envisioned Internet-based services. Among agents, autonomy generally refers to social autonomy, where an agent is aware of its colleagues and is sociable, but nevertheless exercises its independence in certain circumstances. Autonomy is in natural tension with coordination or with the higher-level notion of a commitment. To be coordinated with other agents or to keep its commitments, an agent must relinquish some of its autonomy. However, an agent that is sociable and responsible can still be autonomous. It would attempt to coordinate with others where appropriate and to keep its commitments as much as possible, but it would exercise its autonomy in entering into those commitments in the first place.

- Agents are cooperative, and by forming teams and coalitions can provide higher-level and more comprehensive services. Current standards for Web services provide limited support for composing functionalities.

16.2   Multiagent Architecture

As natural loci of autonomy and decision, agents promise to address these challenges. They perceive, reason about, and affect their environment. They can be designed to be adaptive and communicative. Agents in an information environment can play a number of distinct roles.

The roles of greatest interest to a workflow setting are agents that represent users, agents that represent resources, and brokers that keep track of users and resources and help them find and interact with each other.

When a workflow is constituted in terms of distinct roles that agents can instantiate, the agents can be set up to respect the constraints of their users and resources. Being aware of their local situation enables agents to adapt to a workflow. User agents negotiate with one another and with resource agents to ensure that global constraints are not violated and that global efficiencies can be achieved. Agents can include functionality to identify different kinds of exception conditions and react appropriately, possibly by negotiating a special sequence of actions. More importantly, agents can learn from repeated instances of the same kinds of exceptions. With this learning ability, agents can process the updated set of constraints that emerge when system requirements change.

Relaxed transaction processing refers to modeling and flexible execution of transactions for which the ACID properties are suitably relaxed. Workflow agents can implement a form of relaxed transaction processing. Relaxed or extended transactions are activities consisting of several tasks, or operations, that do not satisfy one or more of the ACID properties. Implementing ACID transactions makes stringent demands that cannot be met in an open environment, such as the Internet. For example, if the workflow in Figures 13.2–13.4 were modeled as an ACID transaction, we would have to ensure that the user could not be told the order was received until after it had been processed—or worse, that an order was received only if it was completed. Of course, these are not reasonable behaviors. Moreover, they are impractical, because they require delaying one task until another task, which might not occur until much later, catches up.

So without transactions how can we ensure consistency? Resource agents working in conjunction with user agents can contribute to a solution. By keeping track of their interactions and how stored data is being accessed and updated, these agents can help maintain overall system consistency. They do not do this in the lock-step manner of an ACID transaction, but they can ensure consistency at intervals sufficient for the particular workflow. By describing at a high level how different components of a workflow ought to be treated, relaxed transactions serve as the basis for designing the behavior of such agents. However, additional functionalities, such as negotiation, become necessary.

16.3   Agent Types

To support an architecture in which heterogeneous components can interoperate, negotiate, and achieve periodic consistency, a variety of agent roles are needed. Each agent role corresponds to the different software components. Figure 16.1 shows a multiagent system architecture in which each agent has a specialized function. The agents communicate using an agent communication language such as FIPA ACL (introduced in Section 18.1), whose sentences wrap sentences from a content language such as SQL. Such an architecture could provide a user with the appearance of homogeneity among heterogeneous resources, and act as a cooperative partner in finding and managing information.

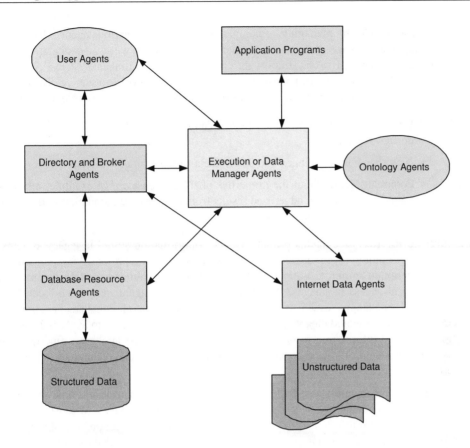

Figure 16.1: Agent-based system architecture showing the *de facto* standard agent types

User agent. User agents act as an intermediary between users and information systems, providing access to such information resources as data analysis tools, workflows, and concept-learning tools. They support a variety of interchangeable user interfaces (for example, query forms, graphical query tools, menu-driven query builders, and query languages), result browsers, and visualization tools.

User agents maintain models of the other agents in their environment, in order to be able to interact with them more effectively. For example, a user agent might contain a mechanism to select an ontology from an ontology agent. The ontology would enable the user agent to present a customized interface that contains terminology familiar to the end user.

Broker agent. Broker agents implement directory services for locating appropriate agents with appropriate capabilities. They manage a namespace service and may store and forward messages and locate message recipients. Brokers might also function as communication aides by managing communications among the various agents, databases, and application programs in an environment. A broker agent works in cooperation with a directory service. Brokers simplify the configuration of multiagent systems. An agent requests the broker to recruit one or more agents who can provide a service. The broker then uses knowledge about the requirements and capabilities of registered agents to:

- Determine the appropriate agents that can be requested to act as providers for a desired service.

- Negotiate with these agents to determine a suitable set of service providers.

- Potentially learn about the properties of the responses. For example, a broker might determine that advertised results from an agent are incomplete and so seek out a substitute for this agent.

Resource agent. Resource agents provide access to information stored in legacy systems. The three common types are classified by the resource they represent. *Wrappers* implement common communication protocols and translate commands and results into and from local access languages. For example, a wrapper agent may use a local data-manipulation language such as SQL to communicate with a relational database or OQL for an object-oriented database. *Database agents* (e.g., those able to apply SQL) manage specific information resources, and *data-analysis agents* apply machine learning techniques to form logical concepts from data or use statistical techniques to perform data mining.

Resource agents apply the mappings that relate each information resource to a common context for purposes of translating messages meaningfully. At most n sets of mappings and n resource agents are needed for interoperation among n resources and applications, as opposed to $n(n-1)$ mappings that would be needed for direct pairwise interactions among n resources without agents.

Workflow agent. Workflow agents are a kind of resource agent that apply to different workflows. The idea of treating a workflow as a resource was discussed in Section 13.3.4. Workflow agents can coordinate the workflows they manage and thereby provide for larger, possibly enterprise-wide, workflows.

Execution agent. Execution agents supervise query execution, operate as script-based agents to support scenario-based analyses, or monitor and execute workflows. This third function can extend over the Web and be expressed in a format such as the one specified by the Workflow Management Coalition (WfMC). Such agents might be implemented as rule-based knowledge systems.

Mediator agent. Mediator agents are specialized execution agents. Mediators work with brokers to determine which resources might have relevant information. They also decompose queries to be handled by multiple agents, combine the partial responses obtained from multiple resources, and translate between ontologies.

Security agent. Security agents provide system-wide authentication and authorization, and can be used to enforce appropriate usage policies for information resources.

Ontology agent. Ontology agents manage the distributed evolution and growth of ontologies. (Chapter 6 introduces ontologies and their uses in information systems.) They provide a common context as a semantic grounding, which agents can use to relate their individual terminologies. A third function of ontology agents is providing remote access to multiple ontologies.

16.4 Life Cycle Management for Agents and Multiagent Systems

An *agent management system*, as shown in Figure 16.2, handles agent creation, registration, location, communication, migration, and retirement. It provides the following services:

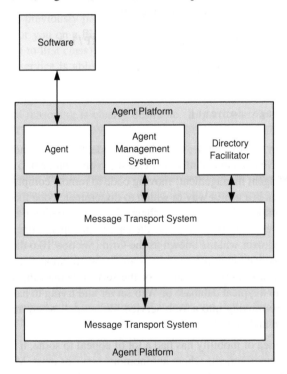

Figure 16.2: The components of an agent management system

White pages. These include support for agent location, naming, name resolution services, and access control services. Agent names are represented by a flexible and extensi-

a knowledge base, it is convenient to place it in the knowledge base apart from any other domain-specific problem-solving knowledge. A reasoner, typically a rule-based system of the sort described in Section 15.7.2.2, would work on the knowledge base, drawing and retracting conclusions based on changing percepts. The conclusions drawn and the justifications for them would be recorded by the knowledge base; as appropriate, even the conclusions that are no longer valid may be recorded in the knowledge base (suitably annotated so they are not confused with the valid conclusions).

16.5.1   Truth Maintenance Concepts

There are two main views of logical consistency. One is *well-foundedness*, which states that all beliefs, except premises, should be justified by other beliefs, and these justifications should be acyclic. The other view is *coherence*, which states that the beliefs should hold together as a coherent body, even if they lack external justification. Under well-foundedness, an agent cannot hold unsupported beliefs, but under coherence it can. It is recognized that human behavior is often closer to coherence than well-foundedness, although traditional logic favors the latter. Both approaches are used successfully for agents through a kind of tool called a truth maintenance system (TMS). The well-founded view is implemented in a *justification-based TMS* and the coherent view in an *assumption-based TMS*. This book considers only justification-based TMSs.

The above motivations leads us to an architecture of a TMS-based agent as shown in Figure 16.3. Here, the problem solver represents domain knowledge, e.g., in the form of rules and procedures, and chooses what to focus on next. The TMS keeps track of the current state of the search for a solution. It uses constraint satisfaction to maintain consistency in the inferences made by the problem solver.

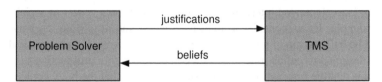

Figure 16.3: The architecture of a TMS-based agent

A truth maintenance system performs a form of propositional deduction. Importantly, it does so in an incremental manner meaning that the beliefs are updated as the data (premises or rules) are added or removed. Because a TMS maintains justifications for the conclusions it supports (and for some that it does not support: typically those that a client had queried for), it can often update the beliefs in an efficient manner. This is because, with each change, only the logically connected parts of the knowledge base are affected. As a further benefit of the data structures, a TMS can also explain its reasoning. For a supported conclusion, it can tell us why, i.e., what premises and rules support that conclusion.

TMSs support a form of atomic update to a knowledge base. When an update is applied to a knowledge base, its TMS will propagate the changes throughout the knowledge base before it returns control. At that point, the knowledge base will be consistent (unless an inconsistency is detected and declared). Likewise, a few updates may be applied as a group. When the effects of all have been propagated, then the new state of the knowledge base is exposed to other clients.

Table 16.3 shows how the integrity of knowledge should be maintained by an agent.

Table 16.3: Knowledge integrity

Property	Meaning
Stability	Believe everything justified validly disbelieve everything justified invalidly
Well-Foundedness	Beliefs are not circular, meaning the justifications bottom out
Consistency	No contradictions
Completeness	Find a consistent state, if any

Table 16.4 illustrates some varieties in which inconsistency may be manifested.

Table 16.4: Knowledge inconsistency

Form of Inconsistency	Example
Both a fact and its negation are believed	Believe the goods have been delivered and believe the goods have not been delivered
A fact is both believed and disbelieved	Believe the goods have been delivered and not believe the goods have been delivered
An object is believed to be of two incompatible types	Believe PO-99 is a purchase order and believe PO-99 is a request for a quote
Two different objects are believed to be the same	Believe PO-99 and PO-98 are the same resource when they are not
The cardinality constraints of relationships are violated, e.g., by giving multiple values to a single-valued relationship	Believe C's shipping address is A_1 and believe C's shipping address is A_2 and believe that $A_1 \neq A_2$ and believe that shipping addresses are unique

16.5.2   Multiagent Truth Maintenance

Single-agent TMSs meet all the requirements of Table 16.3. However, additional problems arise when knowledge is distributed, and different agents must achieve consistency.

In light of the above, different degrees of logical consistency in a multiagent system may be defined. Table 16.5 shows the main degrees of inconsistency that may arise in a multiagent system.

Table 16.5: Degrees of logical consistency

Degree	Meaning
Inconsistency	One or more agents are inconsistent
Local Consistency	Agents are locally consistent
Local-and-Shared Consistency	Agents are locally consistent and all agents are consistent about any data they share
Global Consistency	Agents are globally consistent

Local consistency leaves open the possibility that the different agents may be in serious disagreement. However, global consistency is typically not tractable or even essential. In many cases, it is enough that the agents be in agreement about the data that they share. For this reason, the distributed JTMS (DTMS) maintains local-and-shared consistency and well foundedness [Huhns and Bridgeland, 1991b]. In this approach, each agent has a justification-based TMS, but the justifications can be external, i.e., based on what another agent said. Agents keep track of what they told to whom, so they can suggest updates when their original assertions are no longer supported.

16.5.3   Consistency Maintenance for a Long-Lived Service

Figures 16.4–16.6 show an example of the use and operation of multiagent truth maintenance as two agents interact.

Figure 16.4 shows the initial state of the knowledge bases for the two agents, an investor and a stockbroker. First, the investor asks the stockbroker to recommend a stock. The stockbroker recommends XCorp, which causes the investor to believe that he should buy that stock, as shown in Figure 16.5. However, the stockbroker then learns (not shown) that the basis for his recommendation, namely, that XCorp is cash-rich, is no longer valid. He revises his beliefs and notifies the investor that he has retracted his recommendation for XCorp. The final knowledge bases for the agents are shown in Figure 16.6.

16.5.4   Conflicts among Agents

It is interesting to note that the agents maintain different beliefs about whether or not they can afford to buy XCorp stock. Allowing the difference in belief—this global inconsistency—is useful here. It allows different viewpoints to be represented, it simplifies the representation of knowledge, in that the predicate afford really should have an additional argument indicating *which* agent can or cannot afford the stock, and it eliminates the interactions that would be

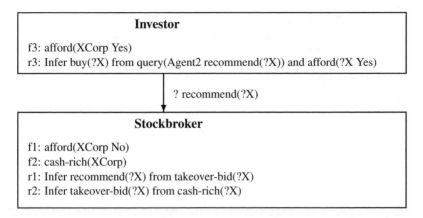

Figure 16.4: Initial knowledge bases of two interacting TMS-based agents, before the investor queries the stockbroker for a recommendation

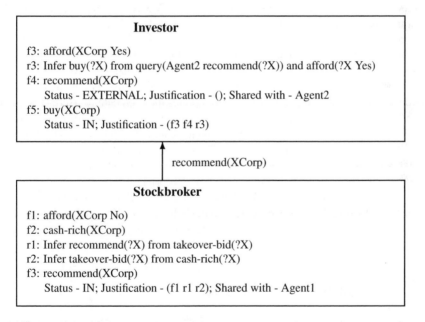

Figure 16.5: Knowledge bases of two interacting TMS-based agents after the stockbroker has replied to the investor's query

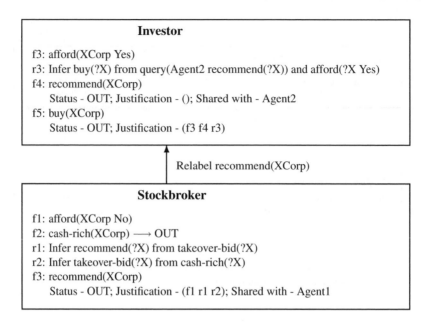

Figure 16.6: Final knowledge bases of two interacting TMS-based agents after the stockbroker has notified the investor that a fact must be relabeled

needed to resolve the difference. Of course, the multiagent TMS would detect, and subsequently correct, the difference if the agents ever share or discuss this predicate.

The above is a case where the predicate used by each agent is the same, but its interpretation is clearly different, since each agent applies the predicate to itself. The same steps could be performed if the agents had differing beliefs about the same fully instantiated predicate. For example, the agents may differ in their treatment of the cash-rich observation for the same company. As a possible scenario, say the investor deals with two stockbrokers, one who believes that XCorp is cash-rich and the other who believes that it is not. Such a situation may arise if each stockbroker has some proprietary data and rules for concluding whether a company is cash-rich. The above formulation simply would not work in such a case. Instead, it would be necessary to record the beliefs explicitly, so that our investor could capture the facts that stockbroker Charlie believes XCorp to be cash-rich, while stockbroker Diane believes the opposite.

16.6 Modeling Other Agents

Software agents being deployed to implement Web services are typically designed to operate independently and are only minimally aware of their own environment and capabilities. They fail to take advantage of each other's abilities or results.

For example, a shopping agent might periodically access several online Web services to find the best price for a music CD and then purchase it if the price falls below its user's threshold. Other agents might be tracking prices for the same CD, duplicating each other's work.

To be more effective, agents must be aware of each other; therefore they must acquire models of each other. One way to do this is by exchanging messages. ("Hi, agent 87 knows about CD prices.") A second form of awareness involves the state of the agent's own environment, including characteristics of the computer on which it is executing and its network connection. ("How many bytes can I send in one second?") A third involves self-awareness— knowing its own name, age, provenance, ontology, goals, areas of expertise and ignorance, and reasoning abilities.

A model of an agent can have many possible forms and amounts of detail, and it can be used in many ways—for example, to plan a partial order of actions that minimizes contentions for resources with other agents or to decide on a course of action once another agent is actually encountered. If an environment has too many agents to be modeled individually, then an agent might act in accord with *social conventions*, such as that agents working on more important tasks have higher priority for using resources.

Another possibility, if the other agents are organized, is for an agent to learn about that organization. This could simplify its modeling task because it would need to model only a subset of the agents—those with whom it interacts and those that are representative of a particular agent role. Moreover, it would know *a priori* something about the agents on the basis of their role in the organization. For example, it would expect subordinate agents to try to satisfy its requests and commands, and it would expect to receive commands from managers or superiors.

How should one agent represent another, and how should it acquire the information it needs to construct a model in that representation? This question has a simple and elegant answer: the agent should presume that unknown agents are like itself, and it should choose to represent them as it does itself. As it learns more about them, it has only to encode any differences it discovers. This can make the resultant representation concise and efficient. Here are some other advantages:

- An agent has a head start in constructing a model for an unknown or just-encountered agent.

- An agent has to manage only one kind of model and one kind of representation.

- The same inference mechanisms used to reason about its own behavior can reason about the behavior of other agents. An agent trying to predict what another will do has only to imagine what it would do itself in a similar situation.

One ramification of such a representation is that an agent constructed with a belief-desire-intention, i.e., a BDI architecture, would attribute intentions to other agents even if they lacked a BDI architecture or any explicit intentions at all. This is consistent with the *intentional stance*, which is introduced in the next section.

- Zeus, http://zeus.enhydra.org/

- CoABS Grid, http://coabs.globalinfotek.com/

Of these, JADE is the most actively developed and maintained.

16.10 Exercises

16.1. What are some of the advantages and disadvantages of synchronous versus asynchronous communication among agents?

16.2. Multiagent Truth Maintenance. A single agent who knows A and $A \Rightarrow B$ would have its knowledge labeled as follows:

```
fact1:   A
status:            (IN)
shared with:       (NIL)
justification:     (PREMISE)
rule1:   A => B
status:            (IN)
shared with:       (NIL)
justification:     (PREMISE)
fact2:   B
status:            (IN)
shared with:       (NIL)
justification:     (<fact1 , rule1>)
```

If the agent shares fact1 with another agent, fact1's shared with label records the name of the agent, and the agent receiving the knowledge labels its new fact as having status EXTERNAL. Now consider the following situation in which the knowledge is initially local to each agent:

Agent 1	Agent 2	Agent 3
f1: P	r1: $P \Rightarrow Q$	f1: R
r1: $S \Rightarrow T$	r2: $R \Rightarrow Q$	
	r3: $R \Rightarrow S$	
	r4: $Q \Rightarrow W$	

- Suppose that Agent 1 shares f1 with Agent 2, who uses forward chaining to make all possible conclusions from its knowledge. Show the effect of Agent 1 sharing f1 on the *status*, *shared with*, and *justification* fields for all data in each agent.

- Now suppose Agent 3 shares f1 with Agent 2. Show the effect of sharing this knowledge on the *status*, *shared with*, and *justification* fields for all data in each agent.

- Now suppose that Agent 1 retracts f1 by making f1 have status OUT. Show the changes on the *status, shared with,* and *justification* fields for all data in each agent.

16.3. In an agent-based two-phase commit protocol, a coordinating agent sends a Prepare message to each resource agent, which is managing a database. Each resource agent responds to the coordinator with either a Ready message or a Don't Commit message. The first phase of the protocol ends when each agent has responded. During the second phase, the coordinator sends either a Commit message or an Abort message to each of the resource agents.

Consider a simple scenario where Agent 0 is the coordinator and Agents 1, 2, and 3 are the resource agents. Use the notation (M, i, j) to mean that message M is sent from Agent i to Agent j, where M can be P (prepare), R (ready), D (don't commit), C (commit), or A (abort).

(a) Show a possible sequence of messages that could occur if Agent 2 wants to abort and Agent 1 and Agent 3 want to commit.

(b) How many possible sequences of messages are there if the overall transaction commits successfully?

(c) If Agents 1 and 3 want to commit, but Agent 2 does not, how many sequences of messages are there, assuming no failures occur?

(d) If Agents 1 and 3 want to commit, but Agent 2 is busy and does not respond to messages, how many sequences are there?

16.4. In a corporate department, database DB_1 stores information about hardware expenditures and database DB_2 stores information about software expenditures. Your job is to construct an agent that can deal with these two systems on behalf of a user.

DB_1: Hardware Expenses			
Item	**Price**	**Unit**	**Date**
ID_1	27.50	$	Oct
ID_2	13.25	Euro	Dec
ID_3	8.20	Yen	Dec
ID_4	86.45	$	Jan

DB_2: Software Expenses			
ID	**Cost($)**	**User**	**Month**
ID_1	18.50	Smith	Nov
ID_2	11.20	Jones	Dec
ID_3	68.99	Huhns	Apr
ID_4	47.30	Singh	Dec

The user, who wants to find the total expenditures for December, sends the following query to YourAgent:

SELECT Sum(Cost) FROM Expenditures WHERE Month='Dec';

(a) What FIPA message should YourAgent send to DB_1?

(b) What FIPA message should YourAgent send to DB_2?

effective communication among agents. This is where the long lifetime of an organization can pay off the most.

There are subtleties in human organizational theory, but this book concentrates on the simple ideas that fit in well from a computational perspective. It mainly uses organizations as a source of metaphors for how to structure systems of collaborative services, each modeled as an agent, into a coherent whole.

Coherence is how well a system of agent-based services behaves as a unit. It depends on agents that act in accord with the roles they assume, the authorizations of those roles, and the commitments among them. The commitments of the agents lend coherence to their interactions over time. Organizations serve the purpose of packaging these commitments into well-defined bundles. Further, they provide a locus for resolving inconsistencies.

17.1   Contracts

Traditional computing approaches to actions focus on their causes and effects. But with agents, we also need to distinguish between right and wrong, and legal and illegal. We need to make this distinction as agent developers, and we might also want the agents to make the distinction about themselves. This is so they can be trusted to act according to a set of proscribed ethics and laws, thereby properly representing humans in contractual settings. With emerging applications in mind, we review some of the essential concepts of agent jurisprudence.

17.1.1   Legal Concepts

The law involves the interactions of entities with one another. Thus legal concepts are inherently multiagent in orientation. Arguably, much of corporate law is about the creation and manipulation of contracts among legal entities (people, corporations, and governmental agencies). Contracts, e.g., service agreements, are about behavior and are crucial in open environments.

A contract represents a legal relation among specified parties. Contracts can also exist among agents, as representatives of human actors. For simplicity, we consider contracts that involve no more than two agents at a time, although some recent work treats more general settings.

Interesting legal ideas for agents originate from the work of the American jurist, Wesley Newcomb Hohfeld (1879-1918) [Hohfeld and Cook, 2001]. Hohfeld observed that terms such as "right" are used ambiguously in the vernacular and proposed a uniform terminology to distinguish the various situations. (Notice that this discussion is about legal rights as opposed to the right (ethical) thing to do.)

Each of Hohfeld's terms has a paired *correlate* term, which applies when the same relation is viewed from the perspective of the other agent. Some correlate pairs are claim and duty, privilege and exposure, power and liability, and immunity and disability.

Claim. This corresponds to what one agent can demand from another. A claim is the most common kind of right. For example, an agent Alice who has rendered services to an agent Bob has a claim to be paid by Bob.

Correlatively, Bob has a *duty* to pay Alice.

Privilege. This arises when one agent is free from the claims of another. In other words, it is the absence of a duty to refrain from a given act. You may or may not exercise your privilege. For example, Alice has a privilege to read Bob's files if Alice has no duty not to do so. Alice cannot be forced to read Bob's file, but if she wishes to read them, she cannot be prevented (without taking the privilege away).

Correlatively, Bob has an *exposure* to Alice's reading his files.

Power. This is the ability of an agent to force (if it so desires) the alteration of a legal relation between itself and another agent. For example, Alice's privilege to read Bob's files may have arisen because of an explicit access control assignment by Bob. That is, if Bob owns the files, he has the power to grant anyone a privilege to read them, but may or may not have the additional power to take away that privilege once granted.

The correlate of power is *liability*. Let's assume Bob has the power to take away Alice's privilege to read his files. Then Alice is liable to Bob for losing that privilege. Notice that Alice is also liable for gaining a privilege, which only goes to show that the technical meanings of terms need not have all the connotations of their informal meanings.

Immunity. This is a freedom from the power of another agent. For example, if Bob owns some files, then Alice lacks the power to take away his privilege of reading them. Thus, by fact of ownership, Bob has *immunity* from Alice's taking away his privilege.

Correlatively, Alice has a *disability* to take away the privilege from Bob.

Hohfeld argued that the above selection of terms covers the legal concepts related to contracts and the rights and duties of individuals. We can use these concepts to establish the norms of agent societies, where the agents are aware of the different shades of each other's rights.

Hohfeld's concepts can be used wherever the relationship among agents represents a contract. One major arena for applying these concepts is in defining and testing for the compliance requirements of the interactions among different agents. For example, we can say that an agent who offers to buy a product must pay the amount it originally offered unless the seller releases the offering agent from this duty.

17.1.2 Deontic Logic

Deontic logic is the logic of *obligations*, of what conditions ought or ought not to be brought about (or correspondingly what actions ought or ought not to be performed). The natural language descriptions are generally easier in terms of actions, but the logic is more obvious if we deal with conditions or propositions.

The main operator of deontic logic is O, meaning *it is obligatory that*. For example, Op means that it is obligatory that p be performed or brought about. Specifically, we could state that Opay($\$5$), which would mean that it is obligatory that $\$5$ be paid. To identify the parties involved, we might revise the above predicate to take additional arguments and thus formulate Opay($A, B, \$5$), where we understand pay($A, B, \$5$) as meaning that A pays B $\$5$. The full expression, therefore, means that it is obligatory that A pays B $\$5$.

Deontic logic enables us to reason about obligations. For example, we might have the following axioms:

$$O(p \wedge q) \Rightarrow (Op \wedge Oq)$$

$$Op \Rightarrow \neg O \neg p$$

Such axioms can enable us to reason about what is obliged or not.

Although there is value in being able to reason formally, traditional deontic logic proves to be quite limited in terms of its modeling assumptions. It has no notion of agents, so it may be appropriate only for stating general ethical laws. For example, deontic logic could be used to express requirements such as the following: "it is a bad thing if an innocent man is punished" or "it is wrong to tell lies." But what we are looking for is something more specific where the parties involved are explicit. The following are consequences of an absence of agents.

- Deontic logic does not separate the obligations of one party from those of another. In the above example, we don't know who it is that is obliged to ensure that A pays B. Is it A, is it B (who should remind A), or is it some unknown C who is trying to return his loan from B by using A to carry the money.

- Deontic logic does not state to whom the given obligation is directed. In business settings, if there is a promise or a contract, there would be a party that is the beneficiary of the given promise or contract.

17.1.3 Commitments

The first patch to deontic logic is to introduce the notion of *directed obligations*, which are obligations directed from one party to another. This is certainly a useful step. However, as we shall see in discussions of virtual enterprises (Section 17.2.2) and business protocols (Section 18.3), it is generally the case that the obligation of one party to another is bounded by the scope of their ongoing interaction. In other words, obligations derived from the virtual enterprise may last no longer than the virtual enterprise in question. Further, there is always the element of conflict, which means that the parties to a contract may be in the need for some adjudication. These considerations suggest that there is an organizational structure to the obligations, which bounds the scopes of the obligations.

The notion of *commitments* (for historical reasons, sometimes referred to as *social commitments*) takes care of the above considerations. Commitments are a legal abstraction. They subsume directed obligations as well as the Hohfeldian concepts. Importantly, commitments (1) are public, and (2) can thus be used as the basis for compliance (discussed in

Section 18.3.2). Commitments support the following key properties that make them a useful computational abstraction for service-oriented architectures.

Multiagency. Commitments associate one agent or party with another. The party that "owes" the commitment is called the *debtor* and the other party is called the *creditor*. Each commitment is directed from its debtor to its creditor.

The directionality is simply a representational convenience. In practice, commitments would arise in interrelated sets. For example, a typical business contract would commit one party to pay another party and the second party to deliver goods to the first party.

Scope. Commitments arise within a well-defined scope. This scope functions as the *social context* of the commitment. In other words, the scope is itself modeled as a multi-agent system within which the debtor and creditor of the given commitment interact. For example, the parties to a business contract can be understood as forming and acting in a multiagent system in which they create their respective commitments and act on them. The multiagent system may have a short or a long lifetime depending on the requirements of the application. Conceivably, the multiagent system for a one-off interaction would be dissolved immediately, whereas some multiagent systems may even last longer than the specific agents that belong to them.

Manipulability. Commitments can be acted upon and modified. In particular, commitments may be revoked. If we were to prevent modifying or revoking commitments, we would end up ruling out some of the most interesting scenarios where commitments can be applied. For example, irrevocability would be too limiting for the kinds of open applications where service-oriented architectures make sense. Irrevocability would prevent considering errors and exceptions that may occur outside of the administrative domain of the given business partner. For instance, it may simply be impossible for a vendor to deliver the promised goods on time if the vendor's factory burns down or there are difficulties with shipping. However, we must be careful that commitments are not revoked willy-nilly, which would make them worthless. When restrictions (sensitive to a given context) are imposed on the manipulation of commitments, they can support the coherence of computations.

Services, although collaborative, retain their autonomy. They can exercise their local policies for most decisions and can be considered as being constrained only by their commitments.

17.1.3.1 Commitments Formalized

We write commitments using a predicate C. A commitment has the form $C(x, y, p, G)$, where x is its debtor, y its creditor, p the condition the debtor will bring about, and G a multiagent system, which serves as the organizational context for the given commitment. A base-level commitment has a simple form, e.g., $C(b, s, \mathrm{pay}(b, s, \$10), D)$, where b is a buyer, s a seller, and D is the context denoting the business deal between them.

17.1.3.2 Operations on Commitments

It helps to treat commitments as an abstract data type. This data type associates a debtor, a creditor, a condition, and a context. The following operations are then natural for commitments.

- Create: I promise to send you $10.

- Discharge (satisfy): I actually send you $10.

- Delegate (change debtor): now my friend is committed to pay you $10.

- Assign (change creditor): now I am committed to pay $10 to your colleague.

- Cancel: I renege on my promise.

- Release (eliminate): you decide to waive receiving the $10, or the government steps in to say that our agreement is null and void.

Create and discharge are obvious; delegate and assign add some flexibility to commitments and are also obvious. Cancel and release remove a commitment from being in effect. Cancel is essential to reflect the autonomy of an agent; just because it made a commitment does not mean that the commitment is irrevocable. However, if commitments could be wantonly canceled, there would be no point in having them, so cancellations of commitments must be suitably constrained (see below). Release helps capture various subtleties of relationships among business partners. A partner may decide not to insist that another party discharge its commitments. Alternatively, the organizational context within which the parties interact may find that a commitment should be eliminated. For example, ordinarily a buyer is expected to pay for goods and a pharmacist is expected to ship medicines that are paid for. However, if the goods arrive damaged then the buyer is released from paying for them (but must return them instead); if the medicine prescription turns out to be invalid, the pharmacist is released from the commitment to ship the medications.

17.1.3.3 Metacommitments

Commitments arise in terms of specific conditions at run-time, but when collaborations are being designed, they are typically formulated as commitments about commitments. Also, negotiations involve communications such as proposals, which can be interpreted as commitments about commitments. For example, a merchant may send a quote to a prospective customer. The content of the quote would be something like: *if you pay me $10, I will send you a book*. In other words, a commitment (from the merchant to the customer) to send a book would be created only if the customer pays. This overall commitment is already present in the message. Such commitments about commitments are termed *metacommitments*. The following exemplify some metacommitments from a seller s to its customer b regarding deal d. Exercise 17.1 returns to these examples.

- The seller will notify customers of any change to their order, o_i:

  $\mathsf{C}(s, b, \mathrm{change\_order}(b, o_1, o_2) \Rightarrow \mathrm{notify}(s, b, o_2), d)$

- The seller will guarantee its price quotes, i.e., if the goods (g) are ordered and the amount quoted is paid, the seller will deliver g goods:

  $\mathsf{C}(s, b, \mathrm{quote}(s, b, g, q) \Rightarrow (\mathrm{pay}(b, s, q, g) \Rightarrow \mathsf{C}(s, b, \mathrm{ship}(s, b, g))), d)$

- The seller will guarantee the quoted delivery date.

- The seller will accept the customer's requests to update orders that have not yet been shipped.

Notice that the above descriptions grossly simplify the domain-specific predicates. In practical settings, these would be replaced by potentially quite extensive documents describing business deals. Not surprisingly, commitments or any other modeling approach cannot provide a free lunch in terms of modeling business interactions. However, commitments enable us to structure the interactions and to apply generic techniques for enacting them, which simplifies key aspects of the modeling and enactment of interactions among autonomous parties.

Metacommitments apply not only to negotiations but to specifying the rules of encounter more generally. The rules are stated in terms of manipulations of the existing commitments. For example, an agent may wish to, or must, violate a commitment. Or an agent may delegate its commitment to another. Such operations are allowed, but may be constrained via metacommitments.

Metacommitments can be realized through reaction or ECA rules (as discussed in Section 15.7.2.1). Metacommitments help generalize the ideas introduced in Section 16.5 for consistency maintenance and in Section 16.2 for relaxed transaction processing via interacting agents.

17.1.3.4 Contrast with Database Commit

We previously encountered the word "commit" in connection with the commit of a database transaction. When a transaction commits, it enables its results to be made permanent in the database and to be accessed by other transactions. Thus, there is a relationship between database commits and agent commitments. However, there are two key differences. First, transaction commits are durable, but only because the lifetime of a transaction does not extend beyond its commit. Thus a transaction cannot interact with another transaction. Second, the flow of information from the results of a transaction occurs through the database and does not recognize any organizational structure.

17.2 Spheres of Commitment

To perform even simple protocols reliably, we must ensure that the parties to an interaction agree on its current state and where they desire to take it. This requires an element of collaboration among the interacting parties. Collaboration can be quite subtle, but it has some

Now VE can act as the seller in a trade where the other party is the buyer. Figure 17.1 describes an execution scenario involving VE, its customer, and two service providers H and V. Here VE accepts an order from the customer to supply a requested assembly based on goods from H and V. When it turns out that V is unable to fulfill its part of the order, it notifies VE and recommends an alternative. VE recomputes the assembly based on the customer's specifications and determines that H's order must be modified. VE notifies the customer that the order has been changed, but honors the previously quoted price.

17.3 Achieving Collaboration via Conventions

A *convention* [Jennings, 1996] provide a means of managing commitments in changing circumstances. Commitments provide a degree of predictability so that agents can take the future activities of others into consideration when dealing with interagent dependencies, global constraints, or resource utilization conflicts. As situations change, agents must evaluate whether existing commitments are still valid. Conventions constrain the conditions under which commitments should be reassessed and specify the associated actions that should then be undertaken: either retain, rectify, or abandon the commitments.

If its circumstances do not change, an agent will endeavor to honor its commitments. This obligation constrains the agent's subsequent decisions about making new commitments, since it knows that it must reserve sufficient resources to honor its existing commitments. For this reason, an agent's commitments should be both internally consistent and consistent with its beliefs.

Conventions help an agent manage its commitments, but they do not specify how the agent should behave towards others if it alters or modifies its commitments. However, because agents typically have mutually dependent goals, it is essential that each inform the relevant agents of any substantial change that affects them. A convention of this type is a social one. If communication resources are limited, the following convention might be appropriate:

LIMITED–BANDWIDTH SOCIAL CONVENTION

INVOKE WHEN
       Operation performed on commitment

ACTIONS
Rule1: IF commitment satisfied
       THEN inform agents involved in related commitments

Rule2: IF commitments dropped because unattainable
       THEN inform creditor as well as SoCom

When agents decide to pursue a *joint action*, they jointly commit themselves to a common goal, which they expect will bring about the desired state of affairs. The minimum information that a team of cooperating agents should share is (1) the status of their commitment to the shared objective, and (2) the status of their commitment to the given team. If either of these changes, then joint commitments require that all team members be informed. Because joint actions may depend upon the entire team, a change of commitment by one participant can jeopardize the team's efforts. Hence, if an agent comes to believe that a team member is no longer jointly committed, it also needs to reassess its own position with respect to the joint action. These three basic assumptions are encoded in the following convention that represents the minimum requirement for joint commitments:

BASIC JOINT–ACTION CONVENTION

INVOKE WHEN
        Status of commitment to joint action changes
        Status of joint commitment of a team member changes

ACTIONS
Rule1: IF Status of commitment to joint action changes
        OR IF Status of commitment to attaining joint action in
           present team context changes
        THEN inform all other team members of these changes

Rule2: IF Status of joint commitment of a team member changes
        THEN Determine whether joint commitment still viable

Conventions provide a ready computational means to engineer collaboration where the member agents can be trusted. If a designer can ensure that certain conventions will be followed, then collaborative behavior can be achieved by having the agents follow suitable conventions for interactions.

17.4 Policies

To understand fully the results and effects of a business process, it is necessary to consider the context in which it is executed. One part of the context for an organizational process is provided by the policies that are applicable to it. For example, we can express a policy that an employee of NSCU may reserve a coach-class but not a business-class airline seat as follows:

$$\text{has}(X, \text{right}(X, \text{reserveBusinessClass}, \text{employee}(X, \text{NSCU}))) \wedge$$
$$\text{has}(X, \text{prohibition}(X, \text{reserveFirstClass}, \text{employee}(X, \text{NSCU})))$$

The above snippet is in the Rei language, which is based on deontic logic [Kagal et al., 2003]. Besides permissions (rights) and prohibitions, the Rei language allows the specification of obligations, dispensations, and delegations.

How may policies for security, authentication, and bookkeeping be enforced? Clearly, insofar as they are the local policies of autonomous entities, each service should apply them

If the agents value pieces differently, then each would divide the interval differently, as shown in Figure 17.2. In this example, the agent G would be assigned its leftmost piece (from the start to G_1), the agent B would be assigned its rightmost piece (from B_2 to the end), and agent R would be assigned its middle piece (from R_1 to R_2). Each agent would have one of the pieces it thought was fair, and the other agents (in its estimation) would have smaller pieces. No agent would be envious. Amazingly, part of the interval would be left over (from G_1 to R_1 and from R_2 to B_2). This protocol works for any domain that can be linearized and for any number of agents.

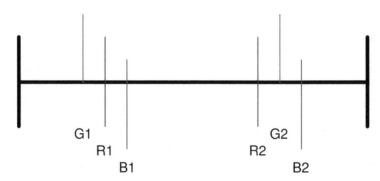

Figure 17.2: Agents R, G, and B divide the interval into what each sees as three fair pieces (R's division is from the left end to R1, from R1 to R2, and from R2 to the right end; similarly for G and B)

17.5.2 Negotiation Fundamentals

As shown above, negotiation is central to adaptive, cooperative behavior. The major features of negotiation are (1) the language used by the participating agents, (2) the protocol followed by the agents as they negotiate, and (3) the decision process that each agent uses to determine its positions, concessions, and criteria for agreement. A variety of approaches have been used to describe negotiation, but they have several features in common. First, it is assumed that there are only a small number of negotiating agents, typically two. Second, the actions relevant to negotiation are propose, counterpropose, support, accept, reject, dismiss, and retract. Third, negotiation requires a common language and a common abstraction of the problem and its solution. The next subsections discuss *environment-centered* or *agent-centered* techniques for negotiation, respectively.

17.5.2.1 Environment-Centered Negotiation

Environment-centered techniques focus on the following problem: "How can the rules of the environment be designed so that the agents in it, regardless of their origin, capabilities, or intentions, will interact productively and fairly?" The resultant negotiation mechanism should ideally have the following attributes:

Efficiency. The agents should not waste resources in coming to an agreement.

Stability. No agent should have an incentive to deviate from agreed-upon strategies.

Simplicity. The negotiation mechanism should impose low computational and bandwidth demands on the agents.

Distribution. The mechanism should not require a central decision maker.

Symmetry. The mechanism should not be biased against any agent for arbitrary or inappropriate reasons.

In particular, three types of environments have been identified: worth-oriented domains, state-oriented domains, and task-oriented domains [Rosenschein and Zlotkin, 1994b]. We consider only task-oriented domains here.

A *task-oriented domain* is one where agents have a set of tasks to achieve, all resources needed to achieve the tasks are available, and the agents can achieve the tasks without help from each other. However, the agents can benefit by sharing some of the tasks. For example, each agent might be given a list of documents that it must access over the Internet. There is a cost associated with downloading, which each agent would like to minimize. If a document is common to several agents, then they can reduce their cost by accessing the document once and then sharing it. What strategy and protocol should they use?

The environment might provide the following simple negotiation mechanism and constraints: (1) each agent declares the documents it wants, (2) documents found to be common to two or more agents are assigned to agents based on the toss of a coin, (3) agents pay for the documents they download, and (4) agents are granted access to the documents they download, as well as any in their common sets. This mechanism is simple, symmetric, distributed, and efficient (no document is downloaded twice). To determine stability, the agents' strategies and protocols must be considered in greater detail.

An optimal strategy is for an agent to declare the true set of documents that it needs, regardless of what strategy the other agents adopt or the documents they need. Because there is no incentive for an agent to diverge from this strategy, it is stable.

With the monotonic concession protocol, the agents successively compromise their positions by agreeing to accept additional tasks. Negotiation ends when neither agrees to compromise further.

A Zeuthen strategy, in which each agent must be aware of how the other values each task, requires the agent having an advantage to make a concession. This process repeats until the agents reach agreement. The result is an optimum task allocation. Notice, however, that it requires the agents to be aware of each other's valuations, which might be a violation of their autonomy. The degree to which each agent might yield some of its autonomy could itself be a subject for negotiation, possibly occurring in advance of the real negotiation.

17.5.2.2   Agent-Centered Negotiation

Agent-centered negotiation mechanisms focus on the following problem: "Given an environment in which my agent must operate, what is the best strategy for it to follow?" Most such negotiation strategies have been developed for specific problems, so few general principles of negotiation have emerged. However, there are two general approaches, each based on an assumption about the particular type of agents involved.

For one approach, message classifiers together with a semantics are used to formalize negotiation protocols and their components. This clarifies the conditions of satisfaction for different kinds of messages. To provide a flavor of this approach, consider the following example how the commitments that an agent might make as part of a negotiation are formalized. (The following Lisp-like snippet shows some predicates that are not defined in this book, merely to give a flavor of what such formalizations seek to capture [Haddadi, 1995].)

(Intend a (Achieves a ϕ)) $\Rightarrow$
(Willing a ϕ) $\wedge$ (Goal a (Achieves a ϕ)) $\wedge$ ($\forall x : x \neq a \Rightarrow \neg$(Commit$_a$ a x (Prevent ϕ)))

This rule states that an agent a may form and maintain an intention to achieve an objective ϕ only if (1) it is willing to achieve ϕ, (2) it has a goal to achieve ϕ individually, and (3) it is not committed to another agent x to prevent ϕ.

The second approach is based on an assumption that the agents exhibit economic rationality. Under a further assumption that there is a small set of agents with a common language and common problem abstraction trying to reach a common solution, Rosenschein and Zlotkin [1994a] developed a *Unified Negotiation Protocol*. Agents that follow this protocol create a *deal*, that is, a joint plan between the agents that would satisfy all of their goals. The *utility* of a deal for an agent is the amount it is willing to pay minus the cost of the deal. Each agent wants to maximize its own utility. The agents discuss a *negotiation set*, which is the set of all deals that have a positive utility for each agent.

In formal terms, a task-oriented domain is a tuple $\langle T, A, c \rangle$, where T is the set of tasks, A is the set of agents, and $c(X)$ is a monotonic function for the cost of executing the tasks X. A deal is a redistribution of tasks. The utility of deal d for agent k is

$$U_k(d) = c(T_k) - c(d_k)$$

The conflict deal D occurs when the agents cannot reach a deal. A deal d is individually rational for agent k if $U_k(d) > U_k(D)$. Deal d is Pareto Optimal if there is no deal d' such that d' is better than d for at least one agent and not any worse than d for any of the other agents. The set of all deals that are individually rational and Pareto Optimal is the negotiation set. There are two possible situations, and a third for state-oriented domains:

Conflict. The negotiation set is empty.

Compromise. (State-oriented domain only) The agents prefer to be alone, but since they are not, they will agree to a negotiated deal. For example, agent Alice needing a document from some site might be able to download it without help, but agent Bob accessing the same document might lock the site, causing a high cost for Alice. Alice should then negotiate a deal with Bob about sharing the document.

Cooperation. All deals in the negotiation set are preferred by both agents over achieving their goals alone.

When there is a conflict, then the agents will not benefit by negotiating—they are better off acting alone. Alternatively, they can "flip a coin" to decide which agent gets to satisfy its goals. Negotiation is the best alternative in the other two cases.

17.5.2.3   Requirements and Properties

Good negotiation protocols should offer more than just fair or envy-free distributions. They should be well-defined and readily available so that they can be easily implemented and used. Protocols are naturally based on communicative acts (see Section 18.1) and incorporated into standard agent frameworks such those for FIPA. Adopting a language with precise semantics will make it easier for agents to use the protocols without misunderstandings.

An efficient algorithm for cake cutting, such as the auctioneer example, requires just $N - 1$ cuts—the minimum possible number of cuts required to divide the resource among N agents. The trimming protocol would require $N(N-1)/2$ cuts, and the successive-pairs protocol would use $N! - 1$ cuts. The divide-and-conquer protocol would require approximately $N\log_2 N$ cuts, the minimum number needed for performing the division without auctioning.

17.5.3   Requirements for a Negotiation Language

A semantic correspondence produced by the process of matchmaking is no guarantee that the client can actually use the service. For instance, the service might not accept the type of credit card that the client plans to use. Making sure that the client can, indeed, use the service (i.e., the service satisfies the functional goals of the client) and that the terms of usage are acceptable to both sides (the qualitative goals) is the purpose of negotiation and contracting.

Negotiation and contracting are enabled by a formal description of the capabilities of a service. (Note that WSDL is not enough for this purpose either, since it is not capable of stating what a service does—only how to invoke it.) A description of service capabilities is similar, but not identical, to an advertisement. Advertisements require different types of ontologies (for instance, the concept of "cheapness" is less likely to be useful in contracting, whereas it is obviously useful in advertising). In addition, contracts may have behavioral, process-like aspects, which would describe what the service will do in response to certain client actions. For instance, a contract might say that the service will collect escrow if the client cancels the contract after a certain date.

The language must support, in an extensible manner, the description of a service so as to support several task aspects of service contracts and negotiation about such contracts. The task aspects include:

- Representation of a wide variety of attributes or aspects of a deal (i.e., contract), particularly pricing and contingent provisions.

  - Frequently-needed contract characteristics besides pricing include, for example, quantity, form and timing of payment, delivery and shipping details including timing, refunds, cancellation, deposits, methods of recourse, performance penalties or bonuses, quality of service, business partner qualifications, reputation or rating information, notifications, roles of different parties to the contract (e.g., buyer, seller, broker, banker, auditor, notary, and escrow), contract phases and renegotiation timepoints, choice of security protocols, and currency units.

- Representation of committed or proposed contractual agreements, i.e., of contracts or contract proposals.

- Representation of business partner qualifications.

- Communication among contracting parties (or relevant other third parties) of proposed or committed deals, including of bids and offers, and requests for proposals.

- Modification of (proposed or committed) deals, especially during negotiation.

- Execution of contract provisions, including drawing inferences or performing procedural business actions, e.g., making authorizations.

- Monitoring of execution of committed deals, including applying contingent provisions, e.g., for exception handling or for notifications during long-running services

- Hypothetical reasoning about the proposed contract by contracting parties (or by third parties, such as adjudicators of disputes), including testing or evaluating a deal during selection, matchmaking, or negotiation. Such hypothetical reasoning should support simulation and verification.

A service contract (committed or proposed and partial or complete) is an artifact that may be the final or an intermediate result of a process of negotiation. Different modes of such negotiation include: simple "take it or leave it"; bilateral bargaining; auction; and other kinds of conversation. A negotiation process may itself be another service (possibly a semantic Web service).

A language for contracting and negotiating about services should

- Complement the development of standard or at least common ontologies for frequently needed contract characteristics (e.g., those listed above).

- Complement other emerging standards efforts relevant to e-contracting. Candidate examples of such other emerging standards efforts include ebXML UBL, UN/CEFACT and ANSI EDI, Oasis Legal XML, and American Bar Association (and their European counterparts') proposals on e-contract law.

- Represent commitments in such a way as to mesh well with methods of dispute resolution and recourse, both legal and reputational.

- Be compatible with, and ideally extend, the contract aspects of existing industry standards for Web services, e.g., via concepts of roles and commitments.

Since negotiating agents are autonomous, they can in principle deceive or mislead each other. Therefore, an interesting research problem is to develop protocols or societies in which the effects of deception and misinformation can be constrained. Another aspect of the research problem is to develop protocols under which it is rational for agents to be honest with each other. The connections of the economic approaches with human-oriented negotiation and argumentation have not yet been fully worked out.

17.6 Exercises

17.1. Formalize the example metacommitments of Section 17.1.3.3 that are not formalized there. State your assumptions about the other predicates used.

17.2. Formalize the following metacommitments (based on those of Section 17.1.3.3). State your assumptions about any other predicates you require.

- The seller will ship the goods for which it quotes a price provided the buyer commits to paying the stated price for the goods (i.e., the buyer need not have paid up when the shipment is made).

- The seller will refund shipping costs if the delivery date to which it commits is not met.

- The seller may delegate its commitment to ship to a shipping company for goods weighing more than 1 ton and, for orders below $10, may assign the buyer's commitment to pay to a payment agency; for orders above $1 000, the buyer may delegate its commitment to pay to its bank.

17.3. Formalize the example metacommitments of Section 17.2.2. State your assumptions about the other predicates.

17.4. Develop an OWL ontology for commitments and metacommitments such as may be used to represent the solution of Exercise 17.3.

17.5. Describe how three agents might negotiate to find a common telephone line for a conference call. Assume that Agent A has telephones lines L_1, L_2, L_3; Agent B has L_1, L_3; and Agent C has L_2, L_3.

The negotiation proceeds pair-wise: two agents at a time. The agents negotiate in order: A, B, C, A, B, C, A, ... Also, alternate lines are chosen in the order specified above for each agent.

For example, the following steps can occur:

communication, and agent-based applications.

Of these four, the first two are straightforward. Message transport merely involves the invocation of a lower-layer functionality. Initially, FIPA was tied to CORBA's IIOP protocol, but now bindings for other protocols, such as HTTP, have been added. Agent management involves the specification of the states that an agent can go through, e.g., to come alive, register with a naming service, execute, suspend, resume, and die.

Agent communication deals with messages exchanged by the agents. The details of the messages would depend heavily on the particular application; these are left to be specified separately, although an XML-like syntax facilitates parsing. An additional level of conceptual support can be provided by ontologies and languages such as RDF and OWL (introduced in Chapters 7 and 8, respectively) that capture more of the declarative content of communications. A lot of the effort on agent communication has sought specialized primitives through which messages can be structured.

18.1.1   Speech Act Theory

Communication has been studied by philosophers of language; this work is a natural source of inspiration for agent communication [Austin, 1962]. The philosophical theory that Austin originated is called *speech act theory*ACL, because it views communication as action.

Agent theorists formulate messages as communicative acts. Specifically, by sending a message, an agent may not only describe the current state of the world, but also change it. The typical change would be through the commitments of the communicating parties that get created, discharged, or modified. For example, by pushing the SUBMIT button on an e-commerce site, you would commit yourself to paying the resulting bill or authorizing your credit card company to pay it on your behalf. Communicative acts prove challenging because they deviate from the truth functional norm of traditional logic: actions are just actions and may be justifiable, rational, authorized, or not, but they are not true or false.

Speech act theory considers three aspects of a message:

Locution. How it is phrased, e.g., "It is hot here" or "Turn on the air conditioner."

Illocution. How it is meant by the sender or understood by the receiver, e.g., a request to turn on the air conditioner or an assertion about the temperature.

Perlocution. How it influences the recipient, e.g., the recipient turns on the air conditioner, opens the window, or ignores the speaker.

Illocution is the core aspect, because it reflects what the given communication itself is while ignoring its phrasing or how it is treated by the recipient.

Speech act theory also suggests that different communicative acts can be constructed by combining a given proposition with different message types or *communicative act types*. For example, consider the proposition "the door is shut." An agent may inform another agent of it, or request it to be made true, or promise to make it true, and so on. In natural language, inferring the message type can be difficult, because it can involve a variety of human psychological and social factors. For Web services, determining the message type is trivial, because it can be explicitly encoded, but determining the agents' beliefs and intentions is impossible, because the internal details of the agents are not known. To emphasize that ACLs need not be like human languages or spoken, the term *communicative act* is used instead of *speech act* in the modern agents literature.

18.1.2 Semantics

The relevant outcome on agent communication languages is that communicative acts provide a principled basis for identifying various patterns of communication. The patterns here describe whether a given message is an assertion (the way the world is claimed to be), a request (what the sender would like done), a prohibitive (what the sender would like to prevent), a permissive (what the sender authorizes), a promise (what the sender will be obliged to do), or a declarative (what the sender brings about by fact of saying it).

Declaratives are the most subtle. They apply in settings where something is named: by saying, "Bob and Alice are now man and wife," someone could be reporting a fact (i.e., an assertion) or marrying them (i.e., a declarative). The latter works only if the context is right, e.g., Alice and Bob have obtained a marriage license, have affirmed their wish to get married, the ceremony is presided over by a duly authorized person, and so on.

Because of the deviation of communications from traditional logic, it has been difficult to give a clearcut semantics to agent communication languages. Some approaches, termed the *mentalist approaches*, consider the beliefs and intentions of the communicating agents. For example, an assertion is considered valid if the sender believes its content and a promise is considered valid if the sender intends to bring it about. These *sincerity conditions* might be useful heuristics in some cases, but cannot be enshrined in the semantics of an ACL. Recall that an ACL is expected to be application independent and used by agents who stand for autonomous services.

The mentalist approaches, including those that underlie FIPA's attempt at a standard semantics, are motivated from traditional artificial intelligence. They are better suited to understanding communication in user interfaces, where the agent conversing with a human effectively belongs to the human or at least has the human's interest at heart. When these approaches are lifted to the open world with a variety of autonomous parties, they simply do not apply. Further, these approaches assume that beliefs and intentions can be unambiguously identified from a piece of code implementing an agent. This is impossible to do in any principled manner. We cannot uniquely ascribe beliefs and intentions to even a trivial agent (even one built as a one-line Java program), and thus cannot characterize its communication.

Consequently, alternative approaches that give precedence to the social aspects of communication have gained popularity. These approaches, termed the *public approaches*, characterize communications based on observable behavior as it affects the agents' commitments. The public nature of this semantics is better suited to open systems; we cannot see inside an agent, but we can potentially observe whether its behavior is compliant.

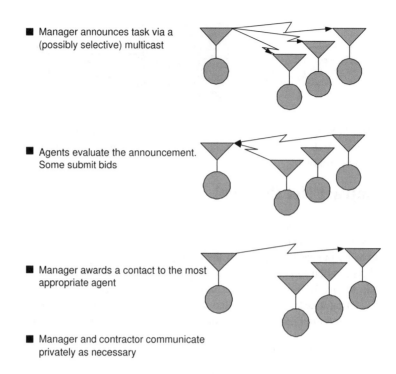

- Manager announces task via a (possibly selective) multicast

- Agents evaluate the announcement. Some submit bids

- Manager awards a contact to the most appropriate agent

- Manager and contractor communicate privately as necessary

Figure 18.3: The Contract Net Protocol is an important generic protocol, which proceeds in much the same way as goods and services are contracted for competitively among human organizations

CNP provides for *directed contracts* to be issued without negotiation. The selected contractor responds with an *acceptance* or *refusal*. This capability can simplify the protocol and improve efficiency for certain tasks.

A task announcement includes descriptions of *addressee, eligibility specification, task abstraction, bid specification*, and *expiration time*. The tasks may be addressed to one or more potential contractors who must meet the criteria of the eligibility specification. The task abstraction, a brief description of the task, is used by contractors to rank tasks from several task announcements. The bid specification tells potential contractors what information must be provided with the bid; a returned bid consists of a brief specification of the agent's capabilities that are relevant to the task, which gives the manager a basis for comparing bids from different potential contractors. The expiration time is a deadline for receiving bids.

CNP offers graceful performance degradation in case of errors. If a contractor is unable to provide a satisfactory solution, the manager can seek other potential contractors for the task. A manager might not receive bids from a potential contractor for several reasons: (1) the contractor is busy with other tasks, (2) it is idle but ranks the proposed task below other tasks under consideration, and (3) it is not capable of working on the given task. To handle these cases, a manager may request immediate response bids to which contractors respond with messages such as *eligible but busy*, *ineligible*, or *uninterested* (the task is ranked too low for the contractor to bid). The manager can then make adjustments in its task plan. For example, the manager can wait until a desired busy contractor is free.

The roles of agents are not fixed in advance in the CNP. Any agent can act as a manager by making task announcements; any agent can act as a contractor by responding to task announcements. This flexibility allows for further task decomposition: a contractor for a task may act as a manager by soliciting the help of other agents in completing that task. The resulting manager-contractor links form a control hierarchy for task sharing and result synthesis.

CNP can be understood at three different levels: (1) it is a high-level communication protocol, (2) it is a way of distributing tasks among self-interested agents, and (3) it is a means of self-organization for a group of autonomous agents. CNP is best used in settings where there is a well-defined hierarchy of tasks involving a coarse-grained decomposition of the given problem so that the parts can be easily assigned to different agents, and where the subtasks minimally interact with each other, but cooperate when they do, so that the agents can carry them out without an excessive coordination overhead.

The CNP does not use models for the internal construction of its participants, thus allowing them to be heterogeneous. Further, their internal reasoning is not modeled either, thus supporting their autonomy. A manager decides whom to invite, prospective contractors decide whether to bid, and the manager decides to whom to award the contract. Accommodating this autonomy also makes the protocol resilient to certain kinds of low-level errors. For example, a network partition between the manager and a contractor in the announce and bid stages is not too significant, as long as some suitable contractor is reachable.

However, the protocol has some key limitations. One, it lacks a formal semantics for the various stages. Two, it assumes a particular order of messages to be exchanged. We can imagine scenarios where another order of messages might be appropriate, e.g., if a contractor decides to advertise its capabilities to catch the attention of potential managers. Such variations cannot be formalized since there is no formal semantics. Three, a task might be awarded to a contractor with limited capability when a better qualified contractor happens to be busy at award time. Four, a manager is under no obligation to inform potential contractors that an award has already been made, possibly causing them to reject alternative contracts during their unrequited wait. Five, because there can be many concurrent managers, a contractor might have to consider several contracts simultaneously. If as a result the contractor misses bidding deadlines, then it might be deemed unresponsive or incapable, neither of which is the case. The protocol does not provide any guidance or support for how the contractor should handle concurrent negotiations. Six, a CNP-compliant agent is allowed to break its commitments unilaterally when it receives a better offer or a better task. The protocol does not specify a recovery mechanism for such unfortunate situations.

- its messages (and which role may send them to which role), e.g., *offer* or *pay*;

- its states, e.g., *shipped* or *paid*;

- its transitions, e.g., the *pay* message causes a transition from *shipped* to *paid.*

Alternatively, the states and transitions need not be explicit and there could be other representations to capture the sequencing constraints on messages. For simplicity, let us confine this discussion to explicit states and transitions.

FSMs capture the states as simple identifiers (all that matters is that a state be distinct from other states), and the transitions as relations in the usual manner. The rule-based approaches express states and messages as formulas and transitions as rules.

But when we have the flexibility of rules, we can go further by making the rules apply on metacommitments (as suggested in Section 17.1.3.3). Doing so gives us the processing ease of rules coupled with the representational power of commitments. When business protocols are modeled using commitments, their creation, satisfaction, and manipulation facilitates understanding exactly what the protocols do, and better captures user requirements. The messages can be given a meaning in terms of the propositions and commitments they affect. Moreover, additional metacommitments characterizing the various roles can be added. For example, a purchase protocol may allow that faulty goods can be returned for a refund. Table 18.1 summarizes three possible approaches for specifying business protocols.

Table 18.1: Specifying business protocols

	Roles	**Messages**	**States**	**Constraints**
FSM	Identifiers	Tokens	Tokens	Reachability of final states; no dead-ends
Rule-Based	Identifiers	Tokens with rules	Formulas	As above
Commitment-Based	Identifiers; constrained via metacommitments	Tokens with rules about commitments	Formulas involving commitments	Metacommitments

18.3.1 Compiling Business Protocols

An advantage of the FSM approaches is that they are easier to interpret than rule-based approaches, which require a rule engine. Therefore, in some cases, it is preferable to use FSM approaches at run-time even though rule-based approaches might be better for the purposes of design.

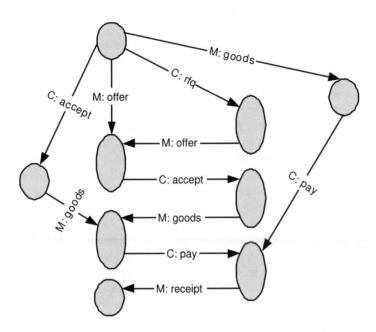

Figure 18.6: FSM representation of an enhanced version of NetBill

Figure 18.6 shows an enhanced version of NetBill that accommodates some of the possible extensions discussed on page 389. The idea is that the extensions that can be anticipated can, in principle, be compiled to yield a larger FSM.

18.3.2   Compliance with Business Protocols

Autonomous parties apply their own policies in deciding how they act. To promote the flexibility of their policies, we would like the protocols to be specified using rules and commitments rather than as finite state machines. But the more flexible we make the protocols the more challenging is the problem of verifying compliance.

At the outset, it is clear that in service-oriented architectures we cannot hope to examine the implementations of the agents who provide the various services. Therefore, compliance verification inherently relies on observing and analyzing the (externally visible) behavior of the communicating parties. Consequently, to determine compliance with a protocol presupposes that the protocol is suitably specified and the parties who wish to determine compliance are able to make distinguishing observations with which they can discern any violations of the stated protocol. Often, such information might not be available to all concerned. For example, how could you determine whether an auction company is complying with its stated rules for declaring winners in auctions? The only way you could do so is if you could see all the data that the company has. For smaller environments, and for specialized protocols, some agents might be able to band together to detect if one of the other parties is not complying with its stated function.

In general, to ensure compliance requires a framework of trust to be in place. Such a framework would be used to find the right parties with whom to collaborate so that the elements forming the basis for determining compliance are themselves trustworthy. Or, a trusted regulatory agency would be used that makes trustworthy assertions about each party's compliance with the given protocol.

To achieve run-time compliance checking, each agent maintains a (local) model of the messages it has sent or received. In this manner, it keeps track of the pending commitments of which it is the debtor or creditor. Consequently, an agent always knows which of its commitments are pending. This can help it plan to discharge those commitments or to resolve the commitments through other means, e.g., by delegating them to other agents. Each agent also knows which commitments it has discharged. At the same time, each agent knows which commitments of others are pending. When those commitments are time bound and their deadline has elapsed, the agent can conclude with certainty that they have been violated.

The basic approach for determining compliance is simple. Just maintain a data structure indicating the state of each separate commitment, and execute some simple algorithms (such as for reachability) on that data structure. There is flexibility as to where the compliance checker is located architecturally. Compliance is intimately related to considerations of monitoring and enforcement.

18.3.2.1   Monitoring

It is conceptually simplest if a special observer monitors the enactment of a protocol. This means that the observer would be involved in each message exchange that takes place as a protocol is enacted—not a reasonable assumption for a service-oriented architecture.

For this reason, it is most natural if each party assumes some of the responsibility for monitoring the enactment of a protocol. Each party can naturally monitor the messages it sends or has received. (The messages sent by the various parties would have to be secure, i.e., unforgeable.) In this manner, the data necessary to determine compliance can be gathered. As remarked above, a participant may not be able to make the requisite observations so as to verify compliance. When a single participant does not make sufficient observations, the observations of some participants can be combined to determine the compliance of the remaining parties.

A technical aspect of monitoring is to ensure that information on commitments is propagated to the right participants. We wish to maintain the constraint that each party knows what commitments it is the creditor or debtor of, and knows who the corresponding debtor or creditor is. Maintaining this constraint enables each party to decide upon its actions to comply with its commitments (as a debtor) and to check if the debtors of the commitments of which it is a creditor are complying. Of the commitment operations only delegate and assign change the debtor or creditor. For this purpose, additional messages are necessary to ensure that information about these operations flows to the right parties. Figure 18.7 illustrates these message patterns.

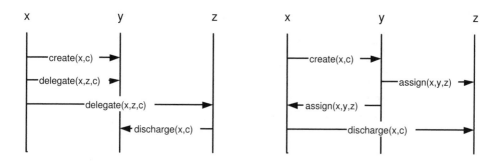

Figure 18.7: Message patterns for operations on commitment

18.3.2.2 Enforcement

Given enough information, a participant or set of participants may verify compliance of some other participants. However, it is another matter to enforce contracts or protocols. Enforcement generally takes us beyond the realm of computing into the real (physical or social) world. The enforcement can be carried out by a computational entity such as a SoCom manager that is suitably instantiated, or potentially by a legal entity that has some purview over the contracts underlying the given protocol.

18.3.2.3 Compliance without a Global Clock

It is worth considering a scenario where the different parties do not share a global clock. An interesting example for this situation is the *fish-market protocol*, which is inspired by a typical Spanish fishmarket. Here an auctioneer attempts to sell a bucket of fish at the highest price he can get. The protocol is simple. The auctioneer announces a price and buyers bid if they like the price. If no buyers bid, then the auctioneer announces a lower price; if more than one buyer bids, the auctioneer announces a higher price; if exactly one buyer bids, the auctioneer hands the bucket to that buyer, who pays the amount bid. Figure 18.8 illustrates a possible enactment of the fish-market protocol (the annotations of the form [1 1 0] are explained below). Traditionally, this protocol is enacted by humans who gather in the physical fishmarket. Thus there is a global clock.

However, let us consider the case where the protocol is enacted in a distributed setting. The essential idea is to employ the notion of *potential causality*, which is a well-known approach for understanding distributed computations [Schwarz and Mattern, 1994]. When we do not assume the existence of a global clock, we must make do with a set of local clocks, one for each participant. The clocks can be thought of as monotonically increasing natural numbers. The local clocks are mutually incomparable, meaning that we cannot directly determine that 10 on the auctioneer's clock is earlier or later than 8 on a bidder's clock. Conceptually, we can think of the system clock as a vector of the local clocks of the participants. Each participant maintains a vector that represents its knowledge of the clocks of the other participants. Each participant has direct knowledge of its own clock and increments it

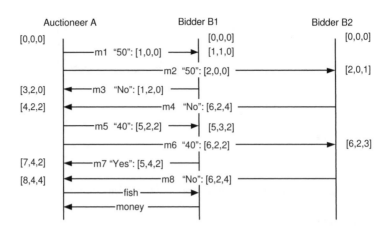

Figure 18.8: An example of the execution of a fish-market auction

whenever it performs some action: an internal action, a message transmission, or a message receipt. This knowledge can be propagated only through messages. For this reason, each message is timestamped with the sender's vector of clocks at that time. When a participant receives a message, it updates its vector of clocks to be the element-wise max of its vector and the message timestamp. Figure 18.8 shows the vector clocks in annotations of the form [1 1 0].

The local observations of each participant correspond to the messages it has sent or received. Figure 18.9 illustrates the local observations of the auctioneer and a bidder based on the enactment of Figure 18.8. The vertices correspond to send and receive events for each message. The message patterns of Figure 18.7 ensure that a message that ultimately resolves a commitment will occur causally after the messages that created the commitment. Thus each agent can search its local model to determine if some commitments remain pending or whether any actions inconsistent with their successful discharge were taken. For example, the auctioneer can verify if a bidder who bid Yes does indeed pay up when he is sent the fish. A bidder can verify that if he gets the fish, it is at the price he bid. However, a bidder cannot verify that the auctioneer is not being deceptive. One way in which the auctioneer can be deceptive is by pretending that other bidders bid Yes for the fish, thus artificially raising the price. Although an individual bidder cannot detect such violations, the bidders as a group, if they pool their observations, can.

18.4   Notes

Some important themes related to agent communication are discussed in Singh [1998].

Potential causality is discussed in texts on distributed computing such as Schwarz and Mattern [1994]; its application to compliance checking was introduced in Venkataraman and Singh [1999].

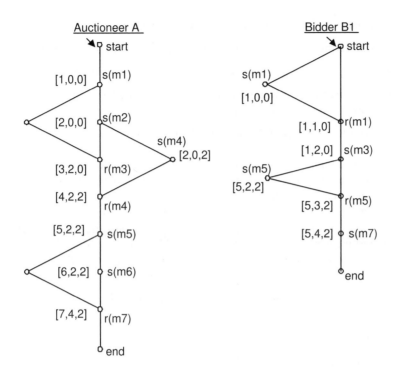

Figure 18.9: An example of the local models of a fish-market auction

18.5 Exercises

18.1. Apply XML to manage communications among agents as follows:

- Write an XML Schema document that describes the FIPA ACL.

- Write an XML-generator for your ACL. That is, write a program in a language of your choice that takes as input your XML Schema document and an ACL expression and outputs the ACL expression in XML.

- Convert the resultant XML-encoded ACL expression into HTML via XSL. That is, write an XSLT transform for FIPA that provides the proper style for rendering ACL expressions.

18.2. WS-Transaction defines coordination types for short-running atomic transactions and long-duration business transactions, such as a saga. Using WS-Coordination, define the coordination type for the CNP.

18.3. Which one of the following is not a benefit derived from using an ontology in agent communications?

e_3, e_4, and e_5 are needed to solve subproblems P_{21}, P_{22}, and P_{23}, respectively; expertise e_6 is needed to solve subproblem P_1. Agent A has expertise e_2 and e_4, Agent B has expertise e_1 and e_3, Agent C has expertise e_5, and Agent D has expertise e_4 and e_6. Show how Agents A, B, C, and D would solve this problem using the Contract Net Protocol.

Part V

Selection

of a cheap long-distance phone service), if expressed in a formal logical language, would be interesting examples of semantic descriptions of service capabilities.

The descriptions can be produced by the service provider or by other parties, e.g., third-party services that generate descriptions for services that have not provided their own descriptions. For example, the Web has many sites about hotels, but most of the sites do not have semantic descriptions of their content or of the Web services they make available. A travel agency service might choose to provide descriptions of these sites that clients of the travel agency can use. The descriptions may be produced in advance or generated on demand for a given service. They can be maintained locally, kept by the third party, sent to a middle-agent matchmaker, or stored at a central registry. To enable more sophisticated matchmaking presupposes that the descriptions are richer, e.g., characterizations of the processes that the services support (as in OWL-S). Richer descriptions enable a more precise selection of services.

On the client side, the goals of the agent must also be described in a formal language. The purpose of matchmaking is to find a "sufficiently good" similarity between the goals of the user agent and the advertised capabilities of the service. Generally, the match is determined by heuristic algorithms aided by domain-specific ontologies defining the terms used in the advertisements as well as descriptions of the agent's goals.

There are two general approaches for semantic matchmaking of services with users, termed explicit and implicit. Imagine stock-profiling services that take as input the New York Stock Exchange symbol for a company, such as "IBM," and return a document charting the performance of the stock over the previous 12 months. Assume that some of these services charge a fee and some are free. An *implicit representation*, as proposed for OWL-S, would describe each service in terms of its IOPEs, where the fee is one of the effects of performing the service. An *explicit representation* would use an ontology to describe each service as being an instance of a suitably defined class, either FreeStockProfiler or FeeBased-StockProfiler.

19.1.1 Applying Ontologies

The greatest challenge for service discovery is dealing with scalability, especially with regard to accommodating complex service descriptions and trust. Capturing the semantics of requests and searches for services, as well as the context of a proposed interaction with the service, requires rich representations of the services and interactions. This is where ontologies are naturally applied. Having an explicit representation enables the principled selection of services, reformulation of requests in a context-sensitive manner, and negotiation about the capabilities of the service providers. Ontologies facilitate specializing and generalizing service needs, as well as facilitating their composition.

Users may formulate descriptions of services they require with sufficient detail to match a class of potentially suitable candidates, transmit these descriptions to other agents (matchmakers or peers), and accept recommendations of possible services meeting the specified criteria. For example, a digital camera might need to locate the nearest printer that can print a color picture with sufficient quality and within a specified time. After it formulates an appropriate description of the printer it needs, the camera can use the description to search a directory or send it to other agents for their suggestions on printers.

Paolucci *et al.* [2002] suggest that a matching engine should:

1. support flexible semantic matching on the basis of shared ontologies;

2. offer some control over the matching to the requesting service;

3. encourage advertisers and requesters to be honest with their service descriptions;

4. be efficient and minimize the number of false negatives and false positives.

Semantic matching algorithms can be based on OWL, thereby enabling a matchmaking agent to recognize semantic similarities between a request and an advertisement despite syntactic differences. Paolucci *et al.*'s algorithm attempts to identify advertisements that can be of use to the requester. This is achieved by comparing the IOPEs specified in the service model with those specified in the request. To accommodate flexible semantic matching, the request is matched on the basis of the subsumption hierarchy provided by a given ontology that includes the concepts being matched, rather than on the basis of syntactical similarity between the request and the advertisement. As such, the ontology provides the context in which the request and advertisement are interpreted. On this basis, a request for cars matches an advertisement for vehicles, since cars is subsumed by vehicles.

A match between an advertisement and a request occurs when all the outputs of the request are matched against the outputs of the advertisement and all the inputs of the advertisement are matched against all the inputs of the request, i.e., when the service is capable of satisfying the needs of the requester and the requester provides all the inputs the matched service needs for its operation. Hence, if even one of the requested outputs is not matched against the outputs of the advertisement, the match fails.

Based on the semantic equivalence of the request and the advertisement, the following categories of matches have been identified:

- *Exact*: when the requested outputs are the same as the advertised outputs. In such cases, the advertised output can be used to satisfy the requested output completely.

- *Plug-In*: when the requested output is subsumed by the advertised output. For example, a service that provides all types of vehicles is a plug-in match for a request for cars. The advertised service can be substituted in place of the requested service. Such a match sacrifices precision, but is capable of satisfying the request.

- *Subsumes*: when the requested output subsumes the advertised output. For example, a service that provides cars is a subsuming match for a request that expects vehicles. Such a match sacrifices recall, because the advertised service would not fully satisfy the request.

An exact match is a special case of the plug-in and subsumes matches.

their interest in participating in the specified abstract SoCom. When they all agree to participate, the broker instantiates the given SoCom. Once a SoCom has been instantiated, it functions in the usual manner and its members carry out their activities as specified. Compliance can be checked as described in Section 18.3.2.

19.3   Exercises

19.1. **Project Idea: Repository for Web Services.** Your task is to build a repository for Web services. Such repositories are typically based on the UDDI standard. (Traditional approaches have some restrictions, which are unnecessary for our present purposes.)

The following is a description of how services are to be recorded and retrieved. You are to design and implement the information system that will enable the required tasks.

The basic problem is to select a service implementation, given some knowledge of a desired service type. In general, there are multiple implementations of each service type. The practical challenge is to choose from among several competing service implementations. In particular, we would like to choose appropriate services programmatically, which is where your database comes into play.

Each service type has a name in an appropriate namespace and is defined in terms of one or more domains (real-estate, retail, books, and so on), as well as the roles it can play in various business protocols. The domains are hierarchically organized in an ontology for the repository.

A number of quality attributes are defined for each domain, although some of the attributes could be common among multiple domains. Examples of quality attributes are price, price-performance ratio, availability, average response time, throughput, customer service, timeliness, neatness of shipped goods, and friendly attitude. Here are some example business protocols, along with the roles they define:

- Three-party-escrow protocol with roles buyer, seller, and bank.
- Advance-payment protocol with roles buyer and seller.
- Five-party-real-estate protocol with roles buyer, seller, buyer's realtor, seller's realtor, and escrow company.
- Catalog-search protocol with roles provider and seeker.

Now we can describe some example service types in terms of the roles they are able to play in one or more business protocols:

- Guaranteed-payment service type: can play the role of a bank in a three-party-escrow protocol.
- Book-selling service type: can play the role of a seller in a three-party-escrow protocol; can play the role of a seller in an advance-payment protocol; can play the role of a catalog provider in a catalog-search protocol.

A service implementation must be the implementation of a service type. An implementation is expected to play all the protocol roles that its declared type is given as playing.

Besides the type information as specified above, there is information about the evaluation of a service by its users. That is, consumers or users of services can rate service implementations. They rate a given implementation along as many or as few of the attributes that are defined for the given type as they care to rate on. For example, one book customer might rate a service implementation in terms of its price, a second customer may rate it in terms of the timeliness of its service, and a third customer may rate it on both attributes.

The above are simply examples. The approach should be generic enough for new business protocols, roles, domains, service types, attributes, and service implementations to be added at run-time. All the relevant type information along with any ratings are stored in a database.

Tasks and Operations. The following are the four major kinds of task that need to be performed by your repository. Each task potentially consists of a number of operations, each of which is something that corresponds to a separate action. For example, inserting a new domain with its attributes is an operation.

- Create new roles, protocols, domains, quality attributes, and service types, and update the above (you can limit the updates with suitable motivation in your report). Declare new service implementations with enough information so that they can be bound and used. Register new service consumers, who would be authorized to publish their ratings of the various service implementations.

- Enable evaluation data (ratings) based on quality attributes; this data would be produced by service consumers based on their experience with a given service.

- Identify service implementations based on type information, such as in terms of service type, roles a service can play, the protocols in which a service may be involved, and the domains of a service.

- Identify service implementations based on the evaluation data. Search for services for which evaluation data exists; among those, choose services that are the best or worst with respect to a given attribute, and best with respect to a given attribute as assigned by a given consumer.

In carrying out this project, you will need to make additional assumptions and design decisions. For example, how would the ratings given by users be represented? How will the ratings be aggregated? How would users and service implementations be identified, and so on?

Chapter 20

Social Service Selection

Let us begin by considering how discovery differs from another major form of information access: retrieval. The difference between retrieval and discovery is the difference between *what* and *who*. Information retrieval is concerned with obtaining information, often from specified sources. Retrieval solves a specific query. Importantly, a notion of correctness is naturally associated with retrieval queries.

Service discovery and location are concerned with finding where to get a given service. The specification of the desired service also corresponds to a query. However, instead of correctness, completeness is the major consideration here. Several services may match the specified requirement and finding the best of them might be difficult.

Retrieval is almost as difficult in closed environments as in open ones, because retrieval involves specific information sources. However, discovery is not as difficult a problem in closed environments as in open ones. In service-oriented architectures, the problem of discovery and selection of services reduces to the discovery and selection of information sources that lead to the desired services.

Inevitably, finding the desired services involves a form of *information navigation* wherein the information sources are organized into a graph. A service is discovered by searching this graph. Traditionally, such graphs are based on edges that can be interpreted as links from one source to another. On the Web, for example, hyperlinks from Web pages provide a basis for this navigation. However, the Web provides no semantics or notion of relevance. Thus, as far as a reader of a page is concerned, all outgoing edges are equal; a page author has no special means to create personalized links.

The DNS is an example of a graph based on referrals from servers to other servers. LDAP is another example of a graph based on referrals. Still another variant occurs through the public key infrastructure, which supports chains of trust. These three examples illustrate increasing flexibility in how the referrals are generated.

When we use a single UDDI registry, it is as if the graph is a simple star, whose center vertex is the registry. In principle, UDDI registries can be federated and thus the graph may have additional structure. The essence of navigation is graph search, although the search may be incremental and distributed across the vertices of the graph.

20.1 Reputation Mechanisms

Effective service discovery and selection require an ability to make recommendations to users about relevant, high quality, and trustworthy services. They also require an ability to evaluate continually the performance of different service providers, which itself involves obtaining evaluations from users, finding evaluations that others may already have given, and aggregating such evaluations in a natural and adaptive manner.

Reputation approaches are commercially applied at popular e-commerce sites, e.g., at OnSale Exchange, eBay, and others. Typically, these sites offer an opportunity to the participants of a transaction to rate each other. The ratings consist of a numeric rating along with some text comments. The ratings are revealed individually and in aggregation to others. OnSale allows its users to rate and submit textual comments about sellers. The overall reputation of a seller is the average of the ratings obtained from his customers. In eBay, sellers receive feedback ($+1$, 0, -1) for their reliability in each auction and their reputation is calculated as the sum of those ratings over the last six months. In OnSale, newcomers have no reputation until someone rates them. On eBay, newcomers start with zero feedback points.

Current approaches for reputation suffer from some shortcomings. One, they tend to sustain the idea of a central authority even where ratings are supplied by different users. The market or e-commerce site where the transactions take place is itself the authority that:

- Authenticates users.

- Records, aggregates, and reveals ratings.

- Provides the conceptual schema for capturing ratings (typically a number and text), specifying their processing, e.g., how to aggregate them and how to decay them over time.

- Owns ratings, meaning that the ratings cannot be used by the participants for purposes unapproved by the market. This point has been the subject of legal action between an e-commerce site and its registered users.

Such authorities can exist only under rigidly constructed and administered computational environments. Two, if multiple authorities (e.g., reputation agencies) may exist, there is no basis for selecting among them. Three, the integrity of the ratings can be compromised because of collusion or retaliation. Four, the users of ratings do not know the parties who provided the ratings. Conversely, a rater's ratings, once given, may be revealed to all.

Some independent sites also deal with reputations, but they tend to deal with general topics and well-known sites, rather than with specific (not widely known) service providers. It would be difficult to store a reputation for providers without a reliable means of identifying them, even if pseudonymously.

20.2   Recommender Techniques

Recommendation comes down to making a prediction of a user's needs or interests. Recommender systems have been widely deployed for product selection. *Content-Based Filtering* is a static approach for selecting among Web sites (or other kinds of information, such as news items) [Dumais et al., 1988]. It involves filtering Web sites or documents in terms of the words that occur in them. This approach could be applied to services by indexing the text descriptions of services based on the words that occur in them. However, this would be a step backward from current Web service standards, which involve formal, structured descriptions of services, and support discovery based on those descriptions.

Another major family of approaches is *social information filtering*. Of these, the most widely used is *collaborative filtering* (CF), e.g., used at well-known e-commerce sites such as amazon.com [Breese et al., 1998]. In CF, a user's ratings for different products are stored centrally—the ratings are often simply captured as the products a given user purchased. A user is given recommendations based on the ratings by other users who are similar to the given user. In simple terms, if Alice and Bob both bought books A, B, C, and D and Alice bought book E, a CF system may recommend that Bob also buy E.

The consumer for whom the recommendation is given is called the *active user*. The idea here is to predict how an active user would rate an item based on ratings by others and the active user's ratings on other products and services. A common approach is to use Pearson correlation to weight the users relative to an active user. An equivalent approach is to use vector similarity between users instead, where the users are modeled in terms of a multidimensional space.

20.2.1   Model-Based Approaches

This class of approaches first builds a model from the given users and then uses the model for making predictions about the active user. After clustering the users ahead of time, the active user is placed in one of the clusters. Alternatively, one might build a Bayesian network representation, e.g., a decision tree.

20.2.2   Memory-Based Approaches

Memory-based approaches consider the ratings of all users directly instead of via an intervening step of building a model. The prediction for the active user's rating is the weighted sum of ratings by others, where the weight corresponds to the similarity between the active user and each of the other users. There are two aspects to making a prediction. One is to figure out the users' averages so that the base-line rating can be captured. The next is to calculate the similarity or dissimilarity between the ratings of different users. From a technical standpoint, dissimilarity is as useful as similarity, because it gives us as sound a basis for making our prediction.

To describe the prediction algorithm formally, let I be the set of items being rated, n be the number of users, I_i be the set of items rated by user i, and v_{ij} be the rating given by user

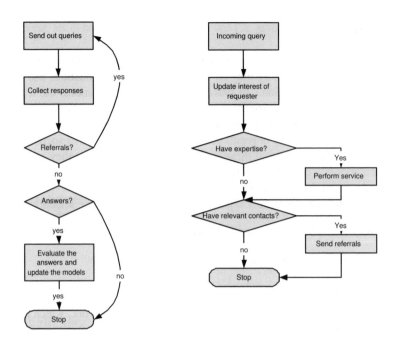

Figure 20.1: Querying (left) and responding (right) for processing referrals

modeled expertise is adjusted) or of the service obtained by following their referrals (their modeled sociability is adjusted).

Some of the acquaintances that are modeled with a high enough combination of expertise and sociability become the given agent's *neighbors*. The agent initially sends its queries to its neighbors. Also, when it issues referrals, the referrals point to some of its neighbors. Each agent selects its neighbors based on their modeled expertise and sociability. Thus the neighborhood relation changes dynamically, reflecting the social network of the agents, subgraphs of which (with appropriate structure) can be thought of as communities.

20.3.2 Advantages of Referrals

Selecting services based on referrals offers some advantages over other approaches. For instance, centralized recommender systems hide the identity of the sources of the recommendations that they aggregate. Consequently, there is no one to trust (or to blame). However, the opinions of those whom you know and trust should be more valuable. In referral systems, the participants reveal their ratings to those whom they trust, so the ratings would be more likely to be honest. Conversely, the ratings that a principal obtains originate with its trustworthy peers. When the identities of the participants are mutually known, they can also understand each other's context of usage and, where they have similar needs, can help make better judgments.

20.3.3   Evaluation

The effectiveness of referral-based approaches has been evaluated through simulations. Work by Yu and Singh [2002] indicates how the effectiveness of a referral network gradually improves and how it adapts around problems or "attacks," such as when a service provider begins to provide poor quality of service. Yolum and Singh [2003] discuss the emergent properties associated with different strategies and assumptions regarding referral networks. The details are left as an exercise.

20.4   Social Mechanism for Trust

The idea behind a social mechanism for trust is to enable information about the trustworthiness of service providers to be shared by the various participants. Intuitively, a service consumer A would rate another consumer or a provider B based on its direct observations of B or the ratings of B given by B's witnesses. There are some important requirements for this approach to work effectively. One, the ratings of the witnesses must themselves be considered. Two, the mechanism should prevent the double-counting of evidence. In other words, a risk to be avoided is that one participant conveys the ratings it received from another, leading to rumors about the trustworthiness of the other participants. Three, it should be possible to introduce new participants without making it possible for a participant to exploit others, change its identity, and then come back to exploit them some more.

20.4.1   Empirical Basis

As explained above, judgments of trustworthiness must be on empirically observable criteria. In principle, the criteria could be quite subtle, although for computational settings, they generally tend to be quite straightforward.

To make an empirical approach work, the first challenge is to account for evidence favoring or opposing the opinion that the given provider is trustworthy. Reputation systems, discussed above, pool all the ratings in a central system. Let us now consider how, in a distributed setting, an individual agent may use its local evidence to form its opinion about the trustworthiness of a given provider.

The local reasoning could involve heuristics about summarizing data, but it helps to have a principled basis behind the summarization. In simple terms, an agent A considers the hypothesis that a given provider B is trustworthy. To evaluate this hypothesis, it helps to consider a theory of evidence. The main such theories are *Bayesian* [Pearl, 1988] and *Dempster-Shafer* [Shafer, 1976].

The Bayesian approach considers evidence as either supporting or refuting a hypothesis. In other words, it does not distinguish between A's lack of belief and A's disbelief in the given hypothesis. Specifically, if A does not believe that B is trustworthy, then A believes that B is not trustworthy. Lack of belief must be modeled indirectly through equiprobable prior probability distributions. By contrast, in the Dempster-Shafer approach, lack of belief does not imply disbelief: it might just be that not enough evidence is available. Lack of belief in any particular hypothesis implies belief in the set of all hypotheses, which is referred to as the state of uncertainty. Initially, the agent A may have some uncertainty regarding the trustworthiness of B. As A accumulates evidence regarding the trustworthiness of B, A can develop beliefs for or against that hypothesis.

Let us consider a proposition T, which indicates the trustworthiness of provider B. The *frame of discernment*, i.e., the set of possibilities, is $\Theta = \{T, \neg T\}$. Dempster-Shafer theory applied to trust considers three sets of hypotheses: $\{T\}$ (is trustworthy), $\{\neg T\}$ (is not trustworthy), and $\{T, \neg T\}$ (could be either). The last indicates lack of belief in the trustworthiness. The above are all the nonempty subsets of $\{T, \neg T\}$. A *basic probability assignment* (BPA) m is a function that yields a probability for each of the above. Thus the value of m falls in the real interval $[0, 1]$ and the sum of the m values for the different sets is 1. Notice that $m(\{T\}) + m(\{\neg T\})$ may be less than 1, which is where this approach differs from the Bayesian approach.

20.4.2   Local Belief Ratings

The first kind of evidence used by A to evaluate the trustworthiness of B is the ratings of services provided by B. Intuitively, these ratings are obtained from users. The rating scales would be specific to applications. For concreteness, let us assume that the ratings are normalized to be in the range $[0, 1]$. For convenience, an implementation may discretize the ratings, e.g., to lie in the set $\{0.0, 0.1, \ldots 1.0\}$.

It is a good idea to bias the evaluation toward recent observations so that a provider's current behavior can be given greater importance than prior behavior. This is easily accomplished by recording just the last H service episodes, i.e., the set of ratings being considered is $\{s_0, \ldots s_{H-1}\}$. The list can be primed with some default value, possibly based on nonempirical factors.

It is also natural to establish separate thresholds for trustworthiness and nontrustworthiness [Marsh, 1994]. Ratings above Ω contribute to trustworthiness; ratings below ω to nontrustworthiness; those in the middle to uncertainty about trustworthiness. Thus it is natural to compute A's BPA toward B based on the last H ratings of B's services as follows (with j ranging from 0 to $H - 1$):

$$m(\{T\}) = \frac{\sum_{s_j \geq \Omega} 1}{H}$$

$$m(\{\neg T\}) = \frac{\sum_{s_j \leq \omega} 1}{H}$$

$$m(\{T, \neg T\}) = 1 - m(\{T\}) - m(\{\neg T\})$$

20.4.3 Combining Evidence

The above discusses a local empirical basis for estimating trustworthiness. Another important basis is in the opinions that are gathered from others. This is especially important when the agent does not have sufficient local evidence. The Dempster-Shafer theory provides us with a rigorous basis for combining the evidence from different sources. It only requires that the underlying bodies of evidence be nonoverlapping, so there is no double counting.

For a subset $\hat{A}$ of Θ, Dempster's rule defines $m = m_1 \oplus m_2(\hat{A})$ to be the sum of all products of the form $m_1(X)m_2(Y)$, where X and Y range over all subsets whose intersection is $\hat{A}$. The commutativity of multiplication ensures that this rule yields the same value regardless of the order in which the functions are combined. In our case, this yields the following formulas. Here c indicates the extent of conflict between m_1 and m_2; this captures the evidence from the two sources that is ignored. If the conflict is total, i.e., $c = 1$, then the combination is not defined.

$$c = m_1(\{T\})m_2(\{\neg T\}) + m_1(\{\neg T\})m_2(\{T\})$$

$$m(\{T\}) = \frac{m_1(\{T\})m_2(\{T\}) + m_1(\{T\})m_2(\{T, \neg T\}) + m_1(\{T, \neg T\})m_2(\{T\})}{1 - c}$$

$$m(\{\neg T\}) = \frac{m_1(\{\neg T\})m_2(\{\neg T\}) + m_1(\{\neg T\})m_2(\{T, \neg T\}) + m_1(\{T, \neg T\})m_2(\{\neg T\})}{1 - c}$$

$$m(\{T, \neg T\}) = \frac{m_1(\{T, \neg T\})m_2(\{T, \neg T\})}{1 - c}$$

You can confirm that $m(\{T\}) + m(\{\neg T\}) + m(\{T, \neg T\}) = 1$. Let us now consider examples of how the above is applied. Suppose

$$m_1(\{T\}) = 0.8, m_1(\{\neg T\}) = 0, m_1(\{T, \neg T\}) = 0.2$$

$$m_2(\{T\}) = 0.9, m_2(\{\neg T\}) = 0, m_2(\{T, \neg T\}) = 0.1$$

That is, m_1 and m_2 agree. Then $c = 0$, and m is obtained as follows:

$$m(\{T\}) = 0.72 + 0.18 + 0.08 = 0.98$$

$$m(\{\neg T\}) = 0$$

$$m(\{T, \neg T\}) = 0.02$$

Alternatively, if the two bodies of evidence conflict, the computation is slightly more complex. Suppose

$$m_1(\{T\}) = 0.8, m_1(\{\neg T\}) = 0, m_1(\{T, \neg T\}) = 0.2$$

$$m_2(\{T\}) = 0, m_2(\{\neg T\}) = 0.9, m_2(\{T, \neg T\}) = 0.1$$

Here $c = 0.72$, $1 - c = 0.28$, and m is obtained as follows:

$$m(\{T\}) = 0.08/0.28 = 0.29$$

$$m(\{\neg T\}) = 0.18/0.28 = 0.64$$

$$m(\{T, \neg T\}) = 0.02/0.28 = 0.07$$

20.4.4   Gathering Opinions

The above approach for combining opinions has the nice effect that each agent can locally summarize its evidence for or against the trustworthiness of a given provider. The agents need only share their summarizations.

In principle, the agents' opinions could be gathered centrally or gathered by whichever agent has an interest in estimating the trustworthiness of a given provider. For central gathering, we would use a traditional reputation system; for distributed gathering, a natural idea is to use the framework of referrals. A key requirement is to ensure that the opinions that are combined are based on independent evidence. A risk is that agents may merely relay the opinions they heard from others, thus causing cyclic beliefs. A simple approach to avoid this is to require that agents convey only the beliefs derived from their local observations, not the beliefs derived from information received from other agents. In other words, only the direct witnesses may report their observations. Thus, if a referral process is used, it is restricted to finding the witnesses; the evidence collected from the witnesses is combined through the above technique.

20.5   Identity

Approaches for service selection based on reputation (or any kind of knowledge of service quality stored by agents) face a common challenge: in the on-line world, users and service providers can change their identities more easily than in the physical world. This is called the problem of *cheap pseudonyms* [Friedman and Resnick, 2001]. Thus reputations and recommendations might be bogus. One way to mitigate this threat is to set the default levels of trust to be low and let each party work its way up. Then no one can exploit others merely by inventing a new identity. However, such a strategy has a side-effect that new participants are treated suspiciously and thus would find it harder to break into the social network. Section 25.4 returns to this point.

20.6   Exercises

20.1. Propose a distributed system architecture and algorithm for resolving URNs (described in Appendix B) that uses a central authority (e.g., to begin its search process). The

URNs not only might identify abstract entities such as namespaces, but also could be names of services. The location to which a service-naming URN would be resolved would be a service instance.

20.2. Propose a distributed system architecture and algorithm for resolving URNs that does *not* use a central authority. Each party must begin its search through it neighbors. Base your approach on the treatment of discovery via referrals as in Section 20.3. How does your approach compare to the approach for Exercise 20.1? In particular, can the resolution process be context sensitive? That is, could the location (service instance) to which a name (identifying a service interface) resolves depend upon who initiates the resolution process? How would new service interfaces be added? How would new service instances be published, i.e., associated with the service interfaces?

21.1   Market Environments

A computational market can constitute an environment in which agents can operate. As described in Section 15.2, agent-based systems involve the design of both agents and their environment. Market environments are used to control the allocation of resources among agents. They can also be used to control the actions of their participating agents. Markets can be used by services to decide which users to serve, and by users to decide with which services to contract. Criteria for the decisions can be based on price systems or barter systems. Prices simplify a mechanism for service selection, because the many factors contributing to the value of a service are compressed into a single number.

For example, imagine there are four agents that would like to make use of a Web service that performs financial portfolio analysis. As shown in Figure 21.1, each agent would like to reserve the service for a specific length of time and before a specific deadline. The solution shown satisfies three of the agents, which end up with positive utility (their resultant values are greater than their costs), and the service receives a positive utility for each hour of its availability (the income exceeds the fixed cost, i.e., the reserve price). There might be several factors characterizing both the needs of the agents and the capabilities of the service, but all of these are represented by the prices shown.

Market-oriented programming is an approach to distributed computation based on the market-price mechanisms of buying and selling. Its focus is on characterizing and designing environments in which agents interact, rather than the agents themselves. It can be effective for coordinating the activities of many agents with minimal communication among them, and it is most appropriate for solving problems of distributed resource allocation. It does this by representing resources to be allocated as goods in a computational economy. At equilibrium, the market has computed the allocation of resources and dictates the activities and consumptions of the agents. Market-oriented programming has the following features:

- The state of the world is described completely by current prices, and communications are offers to exchange goods at various prices.

- Agents do not need to consider the preferences or abilities of others.

- Under certain conditions, a simultaneous equilibrium of supply and demand across all of the goods is guaranteed to exist, to be reachable via distributed bidding, and to be Pareto optimal.

There are two types of agent in the market: (1) consumer agents, which exchange goods, and (2) producer agents, which transform some goods into other goods. Both types of agents bid so as to maximize their profits (or utility). That is, they are *self-interested*. It is assumed that there are enough agents for the impact of an individual agent on a market to be negligible.

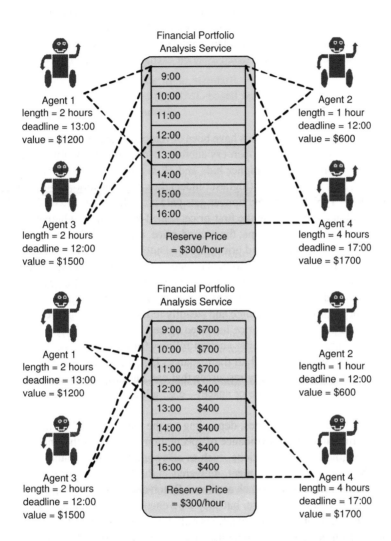

Figure 21.1: An example allocation of a Web service for financial portfolio analysis among four potential user agents. The agents' requirements (e.g., Agent1 is willing to pay $1200 for two hours of time before 13:00), the initial state of the service (top), and a solution (bottom) that maximizes the income of the service are shown

21.2.2.3   Application: Markets for Information

As another application for auctions, consider the University of Michigan Digital Library (UMDL), which implements an open market where information providers, such as publishers, can sell their goods to individual consumers. UMDL is composed of collection interface agents (CIAs), user interface agents (UIAs), mediators, and facilitators. CIAs provide an interface between the publishers' databases and UMDL. UIAs provide an interface between a user, or a user's browser, and UMDL. Mediators match buyers with sellers, while facilitators handle the economic infrastructure.

Almost all UMDL transactions have an associated dollar value and are facilitated by auction agents. For example, when a user issues a query, his UIA first finds an auction agent that is advertising the correct type of mediator agents—those that can answer the query. The UIA bids in this auction for mediator service. Once the auction clears, the query is sent to the winning mediator for processing. The mediator in turn might also use an auction to find the services it needs. Once suitable documents are identified in particular collections, the CIAs responsible for them are paid, using some third-party payment mechanism, and the documents are sent to the UIA, which forwards them to the user.

UMDL implements, in its own microcosm, a working model of the envisioned worldwide electronic commerce infrastructure. It showcases both the power and the drawbacks of using automated auctions. Research results show that these auctions serve as effective load balancers. For example, when one agent crashes or bogs down from a heavy load, the flow of queries is automatically directed toward underemployed agents, because they can offer lower prices. On the other hand, the use of auctions adds an extra layer of overhead to the system, which sometimes increases response time.

21.2.3   Agent Economies

If we are going to have online auctions with agents buying and selling items automatically, we should first consider such a system's dynamics. Classical economic theory tells us that in a market populated by rational agents, prices of goods will approach their marginal costs. However, this is based on goods of a given type being mutually indistinguishable, the agents being utility maximizers, and there being a large number of agents so that the influence of any one of them on the market is minimal.

In an economy of agents, some of the necessary conditions for a competitive market may not be met, and prices will not reach the equilibrium point. For example, a buyer agent may be unable to assess the quality of a purchased service reliably, as in the UMDL. Sellers can take advantage of this scenario by providing lower quality services. More complex buyers might then retaliate by modeling the sellers' behaviors. Sellers might in turn build models of the buyers' models of the sellers' models, and so on. This situation leads to an escalation of modeling levels and to price dynamics not seen in human auctions.

Another possibility is building agents with myopic decision functions that maximize only the agents' immediate payoff. Kephart *et al.* [2000] studied such a scenario and found that these systems could lead to continual price wars. Rather than settling on an equilibrium price, the agents would underbid each other until the price became too low, at which point they would start the price war again, but at a higher price.

These examples warn us that the price dynamics of market systems composed of software agents will not always be the same as those of market systems composed of human agents. We must be careful when choosing the auction types, protocols, and agents that will populate the future online auction landscape.

21.2.3.1 Understanding Agent-Based Economies

At first glance, it seems that market mechanisms can be easily applied to agent-based information economies. However, information economies involve human as well as software agents and, by comparison, software agents are immeasurably less sophisticated, less flexible, less able to learn, and notoriously lacking in common sense. The science of agent-based economies is not yet fully understood and, given these differences, it is entirely possible that agent-based economies will behave in unexpected ways.

This is exactly what has been found in a simulation of an agent-based economy [Kephart et al., 2000]. In the simulation, it is assumed that there is a mixture of buyer agents, some searching for the lowest price and the rest buying from the first seller who meets the price they have in mind. It is also assumed that sellers operate according to one of three different pricing strategies:

- *game-theoretic*, based on a mixed-strategy Nash equilibrium, in which each seller sets its price before observing the prices set by others.

- *myopically optimal*, by which prices are set according to the known mix of buyers and other sellers' current prices.

- *derivative*, whereby each seller sets prices according to its own profit trend—increasing its prices whenever profits are rising and decreasing prices whenever profits are falling.

The first two strategies require perfect knowledge of buyer statistics and the number of sellers, whereas the third requires only local knowledge.

When all sellers follow the same strategy, the derivative strategy yields the highest profits for all. However, with a mixture of sellers following different strategies, those following the myopically optimal strategy fare best. Interestingly, if sellers are allowed to change their strategies, they end up with the myopically optimal strategy, which leads not only to repeated price wars, but also to lower profits for all.

21.2.3.2 Brand Names as a Component of Quality

One potential outcome of a mass market of interacting services is the loss of brand-name identity. In our present economy, sellers rely heavily on brand-name influence, and buyers learn to depend on known brands. First-generation services focus only on price and are oblivious to brand names: to them, a mom-and-pop garage operation is indistinguishable from a major manufacturer, and there is no such thing as brand loyalty.

Part VI

Engineering

the behavior of the services than the use of small methods to get and set variables, as is done for typical data type definitions.

Arm's length interactions. There is a tendency (among programmers accustomed to conventional closed systems) to think in terms of detailed interactions among services and their users. Arm's length interactions avoid such unnecessary coupling, making it easier to evolve one of the interacting parties without affecting the others.

Another reason for maintaining arm's length relationships is to improve the performance of the overall system. This is because arm's length relationships mean that there are fewer events and less data that the parties need to share with each other. This leads to reduced communication latency and more independent, asynchronous computations by the various components.

Reconfigurability. To facilitate deployment, services should be able to be relocated. Service registries provide a mechanism to ensure location transparency. Another consideration is allocating computational resources to a service implementation. Grid computing, or more generally utility computing, permits the late binding of service instances to resources, thus enabling high-demand services to be allocated more resources, thereby improving overall system behavior.

Caution. Since services apply in open environments, they are subject to greater threats than components in conventional closed settings. Therefore, services must deal with other parties cautiously. This entails the elementary kinds of software engineering considerations, such as checking inputs before initiating processing. Further, there is a significant need for the usual kinds of security considerations, such as authentication and authorization. It also leads to the question of trust. Each party in an interaction must be able to trust the others. Next there must be a means to determine compliance of the various parties with respect to the stated protocols.

Narrow interfaces. Interfaces should be kept as narrow as possible. It is almost always a bad idea to expose the details of implementations or even software frameworks. For this purpose, an implementation detail is anything that is not explicitly encoded in a contract.

What is the proper granularity for Web services? If they are too small, then there is too much overhead in describing them, locating them, composing them, and invoking them. If they are too large, then they might not fit an application's requirements precisely. It can be attractive to treat Web services simply as distributed objects that can be invoked in a manner that is independent of programming languages. This view proves quite problematic in practice. Objects provide the wrong set of abstractions for dealing with autonomous, heterogeneous components. Further, Web services are not designed for the fine-grained invocations of local objects. Conversely, Web services should not be based directly on legacy systems, which may be unduly coarse. If the system design can be refactored into a small number of reusable services, then greater value can be extracted from it.

22.2   Quality of Service

Current standards for Web services, such as WSDL, support descriptions of the functional aspects of a service, but not the nonfunctional aspects. Examples of nonfunctional aspects are: who designed the Web service, who implemented it, when it was implemented, and where its documentation is located. Many of the nonfunctional aspects are related to quality of service [Mani and Nagarajan, 2002; Maximilien and Singh, 2002], which is described by the following parameters:

Availability. Availability is a measure of whether a service is present and ready for immediate use, or likely to be ready when it is needed. It can be specified as the probability that the service is available or specified in terms of the particular times it is usable or, alternatively, the times when it is under repair. For example, a bank might balance its accounts every night between midnight and 1:00 a.m., and a cash withdrawal service at an ATM might be unavailable during this time.

Accessibility. Accessibility is a quality measure representing the likelihood that a service can satisfy a request at a given point in time. A service might be available but inaccessible due to network traffic or because it is busy. For example, a Web service selling tickets for a popular sporting event might be inaccessible because of a high demand from client applications trying to purchase tickets. High accessibility can be achieved by building scalable systems that can serve requests consistently despite variations in volume.

Integrity. Integrity measures how well a service maintains the correctness of its interactions with respect to a client, as well as its own state. If a single invocation of a service is viewed as a transaction, then integrity refers to the proper execution of the transaction, including obeying all interaction protocols, operating according to its advertised specification, and satisfying all applicable constraints, so that the transaction will either complete successfully or abort entirely.

Performance. Performance is of two main types: throughput and latency. Throughput is the number of service requests served during a given time period. Latency is the round-trip time between sending a request and receiving the response.

Reliability. Reliability measures how often a service meets or exceeds its advertised capabilities, typically specified as inversely proportional to the service's number of failures per time interval.

Compliance. This quality reflects how well a service complies with the appropriate rules, laws, standards, and any established service-level agreements. Strict adherence to correct versions of standards (e.g., SOAP 1.2) by services is necessary for their proper invocation.

Security. Security specifies how well a service provides confidentiality and nonrepudiation by authenticating the parties involved, encrypting messages, and maintaining access

There is no correct unique or canonical ontology for a given domain. How people conceptualize a domain differs and how they express their conceptualizations in a formal representation differs too. This means that we can evaluate ontologies only by judging how good they are at supporting the applications of interest. The lack of a canonical form is also why merging related ontologies is difficult: similar ideas might have been captured differently in the given ontologies.

Since ontology construction is about creating a conceptualization, it is inherently an engineering art rather than an exact science. Like other forms of software engineering or requirements capture, it is a task that requires experience and good judgment. Some ideas for methodologies for ontology development can be borrowed from the world of conceptual modeling of databases and information systems, although such efforts were often of a more local scope than ontologies that would be used to support service-oriented computing.

The following steps in creating an ontology are based on the discussion in Noy and McGuinness [2001].

- Consider the following key questions to organize your thoughts.

    - What applications will your ontology support? These can be captured as the kinds of *competency* questions that you expect your ontology to answer.

    - How will the ontology be used and maintained? Is it going to be standard or merely one person's take on the world?

    - What existing ontologies could you reuse and enhance? It is best not to build an ontology from scratch if you can avoid doing so.

    - What are the key concepts in the ontology? What attributes do they have? What relationships do they have with other concepts?

- Formulate a hierarchy of the main concepts. Established techniques are bottom up, top down, or inside out. When a number of applications exist, the bottom-up approach of identifying their commonalities makes sense. When a domain is extremely well understood, a top-down strategy can capture its essence quickly. For domains that are novel and thus imperfectly understood, the inside out strategy is a desirable one. In this strategy, you first place the main concepts that you wish the ontology to cover. Then you elaborate them and enhance the structure underlying and overlying them.

- Formulate a hierarchy of the main relationships and define their properties. These include their domain and range, as well as further restrictions. Define the attributes of the concepts and properties, such as value type and cardinality.

The most important point to realize is that ontology development is not a one-shot effort. Except for domains that are so well-understood as to be of limited practical value (and boring to boot), you would not begin with a clear picture of the conceptualization you are building. It is, therefore, wise to proceed in an iterative manner, where the first tentative design decisions can be revised without large effort.

22.3.2   Ontology Guidelines and Conventions

The following guidelines are based on the authors' experience with ontologies in several domains:

- Classes should be named distinctly. They correspond roughly to nouns, and should be expressed in the singular form. Properties correspond to verbs.

- Classes should correspond to the *intrinsic* attributes of the objects of interest. For example, we may classify shoes according to their size, but not according to whether they are currently clean or not. A shoe will typically not change its size, but will alternate between clean and dirty throughout its life. This might affect the value of one of its properties, say the value of cleanliness for a given shoe, but not the class to which the shoe belonged. That is, ontologies should concern schemas and not specific data.

- There is often a choice between creating a subclass or creating a property and allowing it to take a variety of values. For example, we could create subclasses for shoes of different colors, or we could create a property called color and use it to capture the various colors. We should prefer explicit subclasses for colors if we plan to capture different properties for them, such as how formal they are.

- A class with a single subclass suggests that the class and its subclass be merged.

- A class with a large number of subclasses may indicate that either some of the subclasses should be coalesced or an intermediate layer of subclasses should be introduced to make the structure manageable.

- Siblings should be at about the same level of generality. This restriction makes the ontology easier for people to understand.

- As many constraints as possible should be captured explicitly. Doing so enables some potential errors to be detected automatically.

- Properties can have default values. When used, such values serve as a convenience. In general, it is not too good to rely on such defaults, because they can give the impression of a class being fully fleshed out while not meriting any assurance of having correct values.

- Different classes should generally have different properties and should participate in different relationships. This holds even when one of the classes is the subclass of another. The subclass and superclass should be different enough for it to be worth keeping both in the ontology.

- Ontologies in the narrow sense have classes rather than instances. Instances are the province of databases and knowledge bases, whose schemas can be specified via ontologies. Keep in mind that there are no subinstances.

22.3.2.1 Guidelines for RDF and RDF Schema

RDF is mostly a straightforward notation and has only a few quirks.

- The XML syntax of RDF often proves quite cumbersome. It might be preferable in some cases to use an alternative, more compact, notation such as N-Triples for development. This should be translated into XML for publishing.

- rdf:ID applies within the given namespace. Thus an ID resolves to a URI given by the base URI of the scope where it occurs, followed by a # followed by the value of the ID. If the base URI changes—as may happen when the document is moved—the resolved URI for the entity with the rdf:ID would also change. For things that may move, rdf:about with an absolute URI is better than rdf:ID, because the absolute URI would not change. Another approach is to use the xml:base attribute in the rdf:RDF element, which sets the base URI regardless of where the element is placed.

- Use an explicit rdf:label. It is not safe to assume that the last "word" in a URI would be interpreted as a label.

- Use rdfs:isDefinedBy, which points to an authoritative description of the resource. Its superproperty rdfs:seeAlso, which points more generally to relevant descriptions, can also be useful. Use rdfs:comment for additional human-readable information.

- Follow the standard naming convention wherein classes are identified by names that begin with an uppercase letter, whereas properties are identified by names that begin with a lowercase letter.

22.3.2.2 Guidelines for OWL

The OWL syntax allows malformed restrictions and multiple ways to capture the same meaning.

- A restriction that lacks an owl:onProperty element would have no semantic effect, although interpreting the formulation purely as RDF may have some effect, which is lost. It is best to avoid such formulations.

- A restriction with more than one owl:onProperty element could have surprising consequences, such as equating the extensions of two well-formed restrictions. The behavior would be quite unintuitive and should be avoided.

- Constructs such as sameAs and equivalentClass or equivalentProperty or sameIndividualAs differ ever so slightly. For classes, properties, or individuals, it is best to use equivalentClass, equivalentProperty, and sameIndividualAs, respectively.

- OWL makes it easy to scatter the meaning of a class. To maximize readability, it is best to place the constructors in the same owl:Class element unless there are excellent reasons for placing constraints in different elements.

- Recall that the domain and range of a property are global restrictions, since they apply to a property by itself independent of the class whose instances to which it is being applied. Further, multiple domains and multiple ranges are implicitly intersected. For this reason, these should be used with care. However, the necessary requirements should be captured.

- It is best to keep instance and class data in separate documents. An ontology should be mainly about classes and only refer to key individuals, such as might be needed to define other classes.

22.4   How to Create a Process Model

Processes are created by one of the following four general strategies:

- By searching forward from a start state towards a goal state to enumerate successively the possible sequences of intermediate states.

- By searching backward from a goal to a start state.

- By successively refining an abstract process description into a concrete description, commonly termed task decomposition.

- By successively defining the operations needed to transform an initial state into a final or goal state.

The strategy used most often is task decomposition in which, given an overall task or goal, the system's first step is to decompose it into smaller, more manageable pieces. This is typically done using a divide-and-conquer approach. This reduces complexity: smaller subtasks require less capable components (agents or services) and fewer resources. Task decomposition must consider the resources and capabilities of the components, and potential conflicts among tasks and components. These four strategies correspond to the various styles of artificial intelligence planning, which is beginning to be used for planning service compositions.

22.5   How to Design Agent-Based Systems

Agent-based systems are unique types of service-oriented computing systems that deal with the knowledge, intentions, and responsibilities of their components. Several agent-oriented software engineering (AOSE) methodologies have been proposed. One approach, followed by AUML (which extends UML) [Bauer et al., 2001], is to extend methodologies intended for conventional software systems. Another approach, followed by Tropos [Kolp et al., 2002] and Gaia [Wooldridge et al., 2000], is to support the particular characteristics of agents that mandate the use of agents in the system being developed. Below is a longer discussion of Gaia.

Gaia supports the design of both individual agents and systems or societies of agents. Multiagent systems are viewed as being composed of a number of autonomous interactive agents that operate in an organized society in which each agent plays one or more specific roles. Gaia specifies a multiagent system in terms of the roles that agents play and the interaction protocols among the roles.

Roles have four attributes: *responsibilities*, *permissions*, *activities*, and *protocols*. There are two types of responsibilities: *liveness properties* that try to ensure that the role exhibits positive behavior in completing its assigned tasks, and *safety properties* that try to prevent the role from exhibiting negative behavior or to ensure the role maintains acceptable conditions during task execution. Permissions specify what the role is allowed to do and which information resources it is allowed to access. Activities are the tasks that an agent performs. Protocols are the specific interaction patterns that the roles must support.

There are three steps in the Gaia design methodology. The first step is to map roles into agent types and instantiate the agents of each type. The mapping in general is N-to-1, i.e., each agent can fill one or more roles. The second step is to create the service model. Services, similar to OWL-S, are specified in terms of their inputs, outputs, preconditions, and postconditions (effects). Each service is a function that an agent performs, and is derived from the protocols, activities, responsibilities, and liveness properties of a role the agent is filling. The third step is to create the acquaintance model that defines the communication paths among the agent types.

22.5.1   Engineering Cooperation

Choosing the roles that an agent will play is often done by considering the goals of a system and the tasks that are needed for its domain. In an environment with limited resources, agents must coordinate their activities with each other to further their own interests or satisfy system or group goals. The actions of multiple agents need to be coordinated, because there are dependencies among agents' actions, there is a need to meet global constraints, and no one agent has sufficient competence, resources, or information to achieve the goals. Examples of coordination include supplying timely information to other agents, ensuring that actions of agents are synchronized, and avoiding redundant problem solving.

Cooperative services require techniques for distributing both control and data. Distributed control means that agents have a degree of autonomy in generating new actions and in deciding which goals to pursue next. The disadvantage of distributing control and data is that knowledge of the system's overall state is dispersed throughout the system and each agent has only a partial and imprecise perspective. There is an increased degree of uncertainty about each agent's actions, so it is more difficult to attain coherent global behavior.

The actions of agents in solving goals can be expressed as search through a classical AND-OR graph or goal graph. The goal graph includes a representation of the dependencies among the goals and the resources needed to solve the primitive goals (leaf nodes of the graph). Indirect dependencies can exist among goals through shared resources.

Formulating a multiagent system in this manner allows the activities requiring coordination to be clearly identified. Such activities include: (1) defining the goal graph, including identification and classification of dependencies, (2) assigning particular regions of the graph to appropriate agents, (3) controlling decisions about which areas of the graph to explore, (4) traversing the graph, and (5) ensuring that successful traversal is reported. Some of the activities may be collaborative, while others may be carried out by an agent acting in isolation. Determining the approach for each of the phases is a matter of system design.

An intuitive strategy shared by many approaches for developing cooperating multiagent systems is to decompose and then distribute tasks. Such a divide-and-conquer approach can reduce the complexity of a task: smaller subtasks require less capable agents and fewer resources. However, the system must decide among alternative decompositions, if available, and the decomposition process must consider the resources and capabilities of the agents. Also, there might be interactions among the subtasks and conflicts among the agents.

Task decomposition might be done (1) by the system designer, whereby decomposition is programmed during implementation, (2) by the agents using hierarchical planning, (3) inherently according to the representation of the problem, as in an AND-OR graph, (4) spatially, based on the layout of information sources or decision points, or (5) functionally, according to the expertise of available agents. Once tasks are decomposed, they can be distributed according to any of the following mechanisms:

Market mechanisms. Tasks are matched to agents by generalized agreement or mutual selection (analogous to pricing commodities).

Negotiation. Task assignments are mutually negotiated among agents, e.g., via the Contract Net Protocol's announce, bid, and award cycles (see Section 18.2).

Multiagent planning. Planning agents assign tasks to others.

Organizational structure. Agents play roles that have responsibilities for particular tasks.

Figure 22.1 illustrates two of the methods of decomposing and distributing tasks.

22.5.2   Diversity versus Complexity

Among the many reasons why agents are attractive, the following two are of interest here. One, agents enable us to construct modular systems from heterogeneous components, potentially created by any number of vendors. Two, the agents themselves embody diverse knowledge, reasoning approaches, and perspectives. This diversity is sometimes essential, because the agents represent people or business interests that have different goals and motivations. Diversity can sometimes be added in by design: it can make an agent system more robust by enabling a variety of viewpoints to be represented and exploited.

However, agents can be complex pieces of software, so the question arises whether a set of agents that are different from each other would unnecessarily add to a system's complexity. The more kinds of agents there are, the harder it is to build and maintain them.

Fortunately, this turns out to be a false dilemma. The agents have to be diverse in content, e.g., knowledge, reasoning techniques, and interaction protocols, but not in the form in which

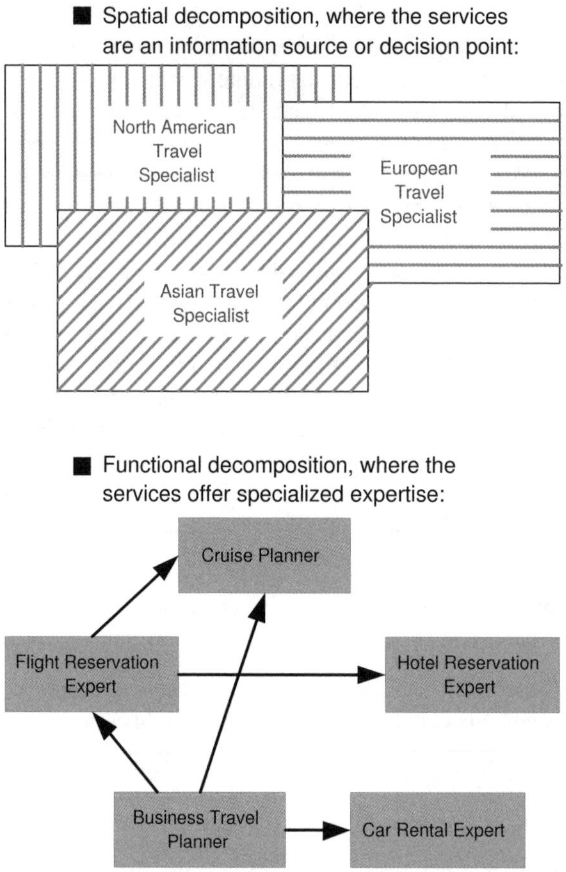
Spatial decomposition, where the services are an information source or decision point:
North American Travel Specialist
European Travel Specialist
Asian Travel Specialist
Functional decomposition, where the services offer specialized expertise:
Cruise Planner
Flight Reservation Expert
Hotel Reservation Expert
Business Travel Planner
Car Rental Expert

Figure 22.1: Examples of task decomposition based on spatial or functional criteria

that content is realized, e.g., the language or toolkit with which they are constructed. Problems arise through unnecessary heterogeneity in construction; the cost of necessary heterogeneity in content is more than recovered through the flexibility it offers.

There are three practical ways you can limit the heterogeneity and its pernicious effects. One, construct agents using a toolkit and, preferably, a common toolkit (or as few as possible, because the choice is often based on past practice or local politics). Two, apply agents in the, by now, conventional roles outlined above. You will be much happier if you keep your broker conceptually separate from your user agent, for example. You could upgrade each agent independently or, if you like, plug in someone's improved version for yours. Three, use standards wherever appropriate. Public standards can make it easier to construct composite systems from heterogeneous and independently developed parts. The more that you and your collaborators can agree on in advance, the fewer problems you will have when you hook up your agents to form the desired multiagent system.

Once an agent-based system is designed using an AOSE approach and taking into account the goal and task requirements of the system, it can be implemented and deployed using an agent development environment, as described in the next section.

22.6   How to Construct Agent-Based Systems

The easiest way to construct an agent or a multiagent system is to use a development environment. There are several such environments available, including Jade, Zeus, FIPA-OS, and CoABS. Each has its strengths compared to the others, but they all provide skeleton agents and implementations of standard FIPA interaction protocols.

Jade, the most popular of the above, is a FIPA-compliant platform, which implements FIPA's agent Management System, Directory Facilitator, and Agent Communication Channel. It supports interagent communication in the form of a message transfer protocol (MTP) over HTTP, CORBA's IIOP, and a Java ORB. Agents can be deployed and execute on a variety of platforms, including PDAs. Messages are represented using the FIPA Agent Communication Language (see Section 18.1). Jade also supports FIPA's standard interaction protocols, such as the Contract Net Protocol. For such protocols, Jade distinguishes initiator and responder roles, based on whether an agent starts or participates in a conversation, respectively. AchieveRE classes are made available to applications, and are used to implement FIPA's *achieve rational effect*. A finite-state machine for the AchieveRE initiator role of the Contract Net Protocol is shown in Figure 22.2.

Each Jade platform has its own Java Virtual Machine, in which each agent is a single thread. Within its thread, each agent is controlled by a scheduler with a nonpreemptive round-robin policy for selecting, executing, and managing behaviors. Constructing an agent involves giving it one or more behaviors, and then customizing the behaviors for the particular responsibilities of the agent. Each behavior implements an intention or performs an atomic task, and the behaviors can be composed to realize complex patterns. For example, three simple *one-shot* behaviors might be used to prepare, commit, and compensate a purchasing operation, respectively, and a *finite-state-machine* behavior would combine the one-shot behaviors into an ACID transaction.

22.7   How to Engineer Composed Services

Constructing an application by composing services first requires that existing services, with the functionalities they provide, be identified. Where essential services are missing, they must be constructed by the application developer or their construction out-sourced. The next step is to select, plan, or specify the desired combination of services. Finally, the composition of services is executed and monitored for success or faults.

Current approaches take a procedural view of service composition by formulating workflow graphs that can be stepped through in a simple manner. Because of this, the main engi-

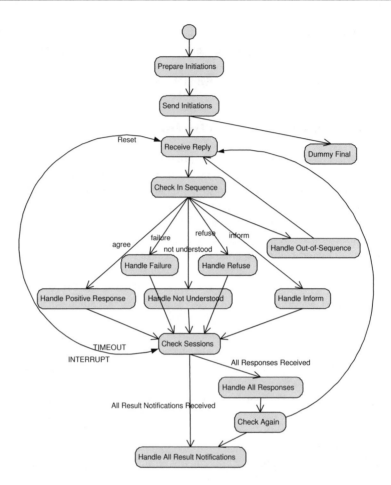

Figure 22.2: A finite-state model in Jade for the initiator role of the Contract Net Protocol

neering challenges that arise concern standardizing on the data, e.g., through syntax or the semantics. However, Web services have attributes that set them apart from traditional closed applications: they are autonomous, heterogeneous, long-lived, and interact in subtle ways to cooperate or compete. Engineering a composition of such services requires abstractions and tools that bring these essential attributes to the fore. When the requirements are expressive, they highlight the potential violations, e.g., the failure modes of a composition.

Engineering composed services thus requires capturing patterns of semantic and pragmatic constraints on how the services may participate in different compositions. It also requires tools to help reject unsuitable compositions so that only acceptable systems are built.

A key aspect in the design of a composed service or a multiagent system is to maintain global coherence, often without explicit global control. This calls for a means to pool knowledge and evidence, determine shared goals, determine common tasks across services, and avoid unnecessary conflicts. The result is a *collaboration*.

Several challenges regarding transport, messaging, and security constraints must be handled for each collaboration. In general, business collaborations are becoming increasingly complex, and most systems will be dealing with multiple collaborations at the same time. If the transactions are legally bound or even otherwise, a nonrepudiation condition may have to be satisfied. Lastly, as usual, there is always a possibility of exceptions.

For the reasons described in several places in this book, the above challenges make automating business transactions using ontologies and agents an appealing approach. Steps toward such technologies are already underway in industry. For example, OAG, OASIS, bizTalk, and RosettaNet are standardizing the syntax and semantics of the *documents* that are exchanged during a B2B transaction. What is also needed is a basis for standardizing (and automating) the *behaviors* that are expected of the participants in a B2B transaction. Further, *misbehaviors* must be handled. For example, a precise specification of a purchase order does not say what to do in the following case: if a company does not receive a response, should it assume that the recipient was not interested or that the PO was lost?

Whether an architecture for an application is specified in terms of a workflow, causal model, process model, goal-subgoal graph, or some other modeling formalism, it must be realized by compositions of available services. These services might be found in a local repository or located across the Internet. Engineering a service-oriented computing system is thus a process of discovering and matching the appropriate services, planning their interactions, and composing them at run-time. This process is illustrated in Figure 22.3. Also note that when deciding on tasks, or decomposing goals into subgoals, it is important to end up with tasks and goals that match services available in a repository or ones that can be constructed easily. An unresolved problem is that the services within the repository are typically not organized.

22.8 Exception Handling

Because service-oriented computing systems have diverse, distributed, autonomous components, it is highly possible that exceptions will occur. It would be good if the exceptions were anticipated, rather than coming as a surprise. It would be even better if a scheme for handling them were ready in advance.

Exceptions have two main connotations: they are things that go wrong and they are deviations from the normal pattern of activity of interaction. Some exceptions can be easy to detect and correct, others not so. To understand exceptions better, it helps to classify them as follows:

Programming. A computation may encounter an exception because of some internal violation of integrity that can be thought of as arising from the programming language engine. Examples of these are divide by zero, array bounds check failure, erroneous typecast attempted, attempt to use uninitialized object, or corruption of program state.

- After the fact: detect and resolve exceptions. These lazy exception handlers are commonly used in practice, sometimes due to a lack of modeling of the possible exceptions. When we do not know what exceptions might occur, we clearly cannot anticipate them with any precision. Even detecting such exceptions after the fact is nontrivial; the only feasible way currently is by making the user responsible for detecting exceptions. The approach of Section 18.3.2 shows how violations of commitments can be detected by each interested party.

In general, there is no easy solution to exception handling. Richer representations of processes identify the exceptions more clearly in the models. Flexible executions of processes via agents help choose suitable behaviors in the face of exceptions. High-level contracts and communications among agents help them interact to enact processes in a manner that can side-step certain exceptions and maintain overall coherence. Whereas all of these help a lot, ultimately there is still a need to model the exceptions that are worth worrying about. To this end, taxonomies of exceptions and exception handling can be invaluable. The MIT Process Handbook [Malone et al., 2003] describes different process types. Klein *et al.* [2002] propose a taxonomy of exceptions that enhances the handbook.

22.9   Knowledge Management Applications

The foregoing has discussed the engineering of SOC systems in general. Let us now turn our attention to some major application classes, the first of which is knowledge management. If you want to find out how to increase the bandwidth into your office, you ask your network administrator, but if you want a new chair, you might not know whom to ask. Your secretary, however, will know, so ask him (or her). In this way we all depend on each other, because few people know everything they need to know to get their jobs done. Studies indicate that people who are more successful have faster networks of more capable experts, and they access this expertise in one-on-one interactions. So *whom* you know is still as important as *what* you know.

In today's large or extended enterprises, where frequent personnel changes make it difficult to conduct business in such a direct way, success requires an ability to exploit the cumulative knowledge of a widely distributed and diverse workforce. Moreover, the basic problem of knowing whom to ask for help has another side: you too have expertise in some area, and you need to find the people who can benefit from it.

Many companies are trying to facilitate such connections with some combination of collaboration software (groupware), such as Lotus Notes or Groove, and an intranet portal navigated by a search engine. If each employee has a home page on the intranet that clearly spells out that person's responsibilities and capabilities, then a search engine can compute an index for the home pages and store it in an online database. For example, along with 7 000 internal Web sites, Hewlett-Packard's intranet has a database of company-wide expertise. A properly conducted search can match the right person to the right task. In some companies, Lotus Notes is then used to keep track of incomplete tasks, and properly motivated employees can browse the list of tasks to find some they can work on.

Unfortunately, a combination of an intranet portal and Lotus notes is insufficient in practice, because it does not support the way most of us work. We do not randomly browse directories to find someone who might be able to help us; instead, we are accustomed to finding the right people through our personal offline networks. Such networks of expertise tend to have only a few levels, and they are not fully connected. In fact, to be useful, they need paths that are short and fast. As described below, there are only a few degrees of separation between most people in social networks, and we can assume this pertains to business as well. One-level referrals are the basis for some successful search engines, such as Direct Hit.

22.9.1   Agent-Based Knowledge Network

In the long run we need a more comprehensive solution. The necessary capabilities are:

- Categorizing: the ability to classify unstructured content automatically.

- Hyperlinking: the ability to add to each item of information appropriate pointers to other relevant items of information.

- Alerting: the automatic notification of users and agents to new information that might be of interest to them.

- Profiling: the construction of models of users and agents to describe their interests and expertise.

This last capability is the most important for a knowledge network, because it involves integrating statements of work, contracts, plans, and corporate strategies with structured data to characterize an enterprise's objectives and work.

The system architecture for a knowledge network must include:

- brokers that manage the metadata relating applications, agents, systems, and people;

- search engines;

- ontology servers to reconcile the semantics of the different components that make up the intranet;

- knowledge bases for each of the active participants in the system;

- agents (of course!) to provide the proactive behavior needed to make the knowledge network an active collaborative service.

How close are we to achieving such a solution? It appears that all of the individual pieces are available, but they are just not integrated with agents into a complete system. The following sections review the current state of portals, groupware, and corporate knowledge management.

trial segments. For example, Ariba has developed a marketplace based on procurement portals and dynamic exchanges for horizontal marketplaces, which includes dynamic trade mechanisms, such as auctions, reverse auctions, and bid/ask exchanges and negotiations. SAP Services Marketplace, an Internet portal for the SAP community, provides online services such as catalog browsing, matchmaking, and ordering from SAP and its partners. PaperExchange provides a vertical marketplace, enabling customers and suppliers to negotiate pricing and transact directly with one another. PaperExchange offers supporting services for logistics, industry-specific job listings, industry events, and a resource directory. VerticalNet also has built a general marketplace as a set of several Web-based marketplaces for specific industrial segments, such as financial services, healthcare, and energy. Each Web site forms a community of vendors and customers in a specific area. Vertical trade communities are introduced in segments with a substantial number of customers and suppliers, a high degree of fragmentation on both the supply and demand sides, and significant on-line access.

Many software vendors are developing Internet-based commerce platforms. Examples are IBM CommercePOINT, Microsoft Site Server Commerce Edition, Oracle Internet Commerce Server, Intershop Communications' Unified Commerce Management, and the Java Electronic Commerce Framework (JECF) from Sun Microsystems. These proprietary tools focus on providing infrastructure services such as security, payment, directories, and catalogs to be integrated with existing systems and the Web.

22.10.1   Business Models for eBusiness Applications

Surveys of small and large companies have shown that one of the most frequently mentioned barriers to eBusiness projects is the lack of an appropriate business model. A business model can be used with marketing strategies as an effective tool to assess the commercial viability of an eBusiness application. A business model is "an architecture for the product, services and information flows, including a description of the various business actors and their roles; a description of the potential benefits for the various business actors; and a description of the sources of revenues" [Timmers, 1999]. In an eBusiness framework, a business model can be viewed in terms of four principal components [Bartelt and Lamersdorf, 2001]: (1) the products and services offered by the business entity, (2) the customer relationships that the business entity creates and maintains in order to generate revenues, (3) the financial aspects of the business, such as cost and revenue structures, and (4) the infrastructure and the network of partners that are necessary in order to create value and to maintain good customer relationships. Possible architectures for business models are constructed by combining interaction patterns with value-chain integration for the possible creation of marketplaces. The following are the most widely realized models [Timmers, 1999].

A basic model of eBusiness is the eShop model. It is based on providing a self-service storefront to a customer by displaying the company catalogs and product offers on a Web site. The business objective is to lower the sales cost. A major concern in this model is the assumption that the customer should be responsible for comparison-shopping between products of different suppliers.

While an eShop model is based on the selling aspect of the business, an eProcurement model focuses on the buying aspect of the business. Typically it consists of a browser-based self-service front end to a corporate purchasing system or ERP. The supplier catalogs are presented to end-users through a single unified catalog, thereby facilitating a corporate-wide standard procurement process. In addition, eProcurement might support calls for tender through the Web, which might be accompanied by an electronic submission of bids. Nonetheless, an eProcurement model does not support dynamic trading. Its business objective is cost savings on purchasing operations. Recently, eAuction models have received much attention for automating dynamic trading. In fact, an eAuction model can be applied in any situation where there is fluctuation of demand or supply. The prime business objective is to increase efficiency, reduce waste, and minimize overall cost.

Alternative models are based on creating value-chain businesses. One model describes service provisioning of specific functions for the value chain, such as electronic payments or logistics. New approaches are also emerging in production and stock management, where new intermediary service providers are formed to provide specialized expertise to analyze and fine-tune production. The business objective is to generate revenue based on a fee-based or percentage-based schema. Another model is based on integration of multiple steps of the value chain in a way that exploits the information flow between the steps for further added value. In particular, this approach may include customized advice to buyers and manufacturers. The business objective is formulated as revenues from consultant or transaction fees.

Clearly these models still encounter the challenge of how to create, populate, and leverage significant growth in services and supply operations in a way that seamlessly integrates customers, suppliers, partners, and competitors in a trading community. An important and promising business model is the eMarketplace. This model supports value chain integration and provisioning in its (1) structure, (2) services, such as advertisement-branding, (3) one-to-one marketing logistics, (4) pre- and post-financing, (5) risk management, and (6) product bundling. Ultimately, the eMarketplace relieves participant business entities of much of the burden to participate effectively in the eBusiness domain. The business objective of the eMarketplace model is based on a combination of subscription fees, transaction fees, and service fees. This approach is important, as it is particularly suited for B2B volume trading.

22.10.2   eMarketplace Architectural Requirements

The feasibility of implementing a business model depends upon the state-of-the-art of the technology, whether for realizing individual functions, for supporting interaction patterns, or for integrating components. A successful engineering foundation for an eMarketplace must accommodate the needs of the eBusiness participants and allow them to extend advanced services to the trading community. As eMarketplaces mature, they must support a broader base of services, ranging from baseline interaction and directory services to specialty services such as online payment, logistics, and dynamic trade. In addition, an eMarketplace should facilitate the many-to-many relationship between suppliers and customers, enabling both to leverage economies of scale in their trading relationships and produce a more liquid marketplace. This in turn allows the use of dynamic pricing models, such as auctions and exchanges, which improve the economic efficiency of the market. To provide smooth and effective integration at the business level, an eMarketplace model should accommodate and support interfaces to the existing business models of the participant entities for procurement processes, customer-supplier interactions, business rules, workflow, and relationships.

Another key factor for the foundation of an eMarketplace is its ability to be integrated into an open environment, because a customer's needs might go beyond the specialist capabilities of any single eMarketplace. The ability of eMarketplaces to interact extends the idea of liquidity and network effect, but should not sacrifice the ability of an eMarketplace to be highly specific to the supply-chain node or target customer group it serves. This model combines the advantages of the sell-side, the buy-side, and the value-chain models.

Because an eMarketplace requires complex interactions among the systems deployed by the participating business units of an enterprise and their customers, there must be substantial amounts of business knowledge within the eMarketplace transactions, activities, and service definitions. Figure 22.4 shows an example supply chain for automobile manufacturing, indicating several tiers of suppliers, which must begin to interact when they begin serving the same customer.

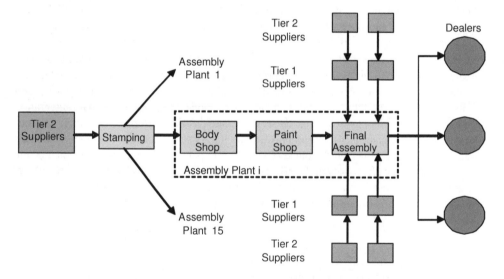

Figure 22.4: Supply chains and the automotive industry

As this example indicates, there must be a large degree of communication, coordination, and cooperation within and among business entities and their systems in the eMarketplace. This, in turn, facilitates consistent behavior among the participants. In other words, the eMarketplace architecture should represent an integrated body of people, systems, information, processes, services, and products. By integrated, we mean the structural, behavioral, and informational integration of the participating business entities, including their legacy applications. Business-process reengineering addresses the challenge of structural integration by reorganizing enterprises along critical business processes, such as the supply-chain and the product life cycle [Hammer and Champy, 1993; Davenport, 1993]. Achieving integration requires the development of an information infrastructure that supports the communication of information and knowledge, decision-making, and the coordination of actions.

22.11 Application to Supply-Chain Automation

The above requirements can be met by a multiagent system approach. Recent research, originated by Parunak [1996] and then successively extended by Singh [2000] and Huhns *et al.* [2002], can partially automate the development and deployment of the agents needed to implement a supply chain. The approach begins with a UML sequence diagram (see Figure 22.5) showing the interactions among a number of independent organizations in a B2B or supply-chain scenario.

The interactions in Figure 22.5 consist of the exchange of structured documents, which the OAG and RosettaNet call business object documents (BODs). For B2B interactions, a ProcessPO BOD is a directive that carries the composite semantics of *request* and *inform*; that is, the sender requests that the recipient evaluate the purchase order and inform the sender of the results. The informal semantics is that ProcessPO will be followed by a response from the recipient and that the response will be either an AckPO or a DeclinePO.

Next, using the formal semantics, a tool converts the messages in the sequence diagram into collaboration diagrams, shown in Figure 22.6. The diagram for each role can be converted directly into a state-machine description for the agent's behavior, enabling automatic agent generation. Each business collaboration is handled by an autonomous software entity, i.e., an agent.

On forming a collaboration, we can generate ebXML-compliant agents automatically, as shown in Figure 22.7. After being installed at each company, the agents manage the B2B supply-chain process. Figure 22.8 shows several of the state-machine behavioral descriptions. At execution time, the software agents exchange ebXML messages to perform the specified collaboration.

22.12 Exercises

22.1. Review your solution to Exercise 8.5 with respect to the SOC design criteria introduced in this chapter.

22.2. Using a tool such as Protégé, develop an ontology that emphasizes meronomy, that is, the part-whole relationship.

(a) Specifically, choose a complex object or concept that has at least three levels of sub-parts. For example, an engine has a part cylinder, which has a part intake valve, which

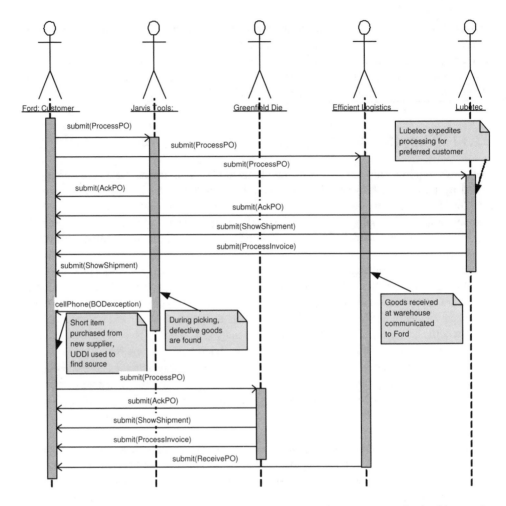

Figure 22.5: A sequence diagram representing a B2B or supply-chain scenario (in this case for Ford and its suppliers) is the first step in automating an MAS implementation of the scenario

has a part valve seat. Use the guidelines presented in this chapter when constructing your ontology.

(b) Include in your ontology some concepts that are objects, such as a bolt, and some concepts that are substances, such as steel.

(c) Include both intrinsic and extrinsic properties for the objects in your ontology. An *intrinsic* property is one that holds for all portions of an object, such as its density. An *extrinsic* property is one that does not holds for all portions of an object, such as its weight. Identify which of the properties in your ontology are intrinsic and which are

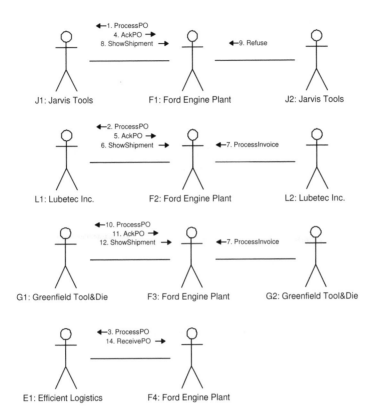

Figure 22.6: A collaboration diagram showing the participant roles for a supply-chain scenario, constructed automatically from a sequence diagram by an analysis of conversations based on Dooley graphs

extrinsic. Are there any that are neither?

(d) Fill in the following table for the reasoning that would be valid, that is, the kinds of inferences you could make, for each combination of an object type and a property type. Hints: Would the density of an object apply to its parts transitively? Would the weight of an object apply to its parts transitively? Would the weight of an object impose any constraints on its parts? Would the weight of a part impose any constraints on the weight of its whole?

	Intrinsic Property	Extrinsic Property
Individual Object		
Substance		

Chapter 23

Service Management

As services proliferate and applications based on them become more prevalent, enterprises will increasingly rely upon them for critical business functions. This introduces a risk, because some of the services might be outside of the enterprise and beyond its control. How can an enterprise mitigate this risk?

One approach is to extend common industry-approved practices to the deployment and control of Web services. A second approach is to employ Web service management techniques for assessing and maintaining quality of service. A third approach is to make use of recently proposed techniques for agent-based redundancy. This chapter describes all three of these approaches.

23.1   Enterprise Resource Planning

Enterprise Resource Planning (ERP) is a collection of tools that management uses to operate a business on a daily basis. ERP usually comprises several modules, such as a financial module, a distribution module, or a production module. Each module uses information that is housed within the database structures underlying the ERP system. ERP helps coordinate different departments within a business, e.g., by alerting management that a sales department is selling 25% more goods than the manufacturing department can produce, and that the shipping department is underutilized. If a designer needs to specify a bolt for part of a design, it makes sense to specify a bolt the company already uses. A salesman visiting a customer needs to know what dealings the business has already had with that customer: what did the customer buy, were there any complaints, were invoices paid on time and, if not, was that because there were warranty problems? With an ERP system, all departments have access to the up-to-date information that is needed for the enterprise to operate smoothly.

The software consulting industry has developed several competing views of enterprises and competing approaches for helping enterprises to automate their information systems and become more efficient. They are based on the economic concept of a value chain: an organized collection of activities that an enterprise performs to design, produce, market, deliver and support its product. An enterprise's value chain and the way it performs those activities are a reflection of its history, its business strategy, its approach to implementing its strategy, and the underlying economics of the activities themselves [Porter, 1985].

One approach for representing and automating a value chain, shown in Figure 23.1, is termed "Global Best Practices." The additional activities involved in supporting this approach to a value chain are (1) develop and manage human resources, (2) manage information resources and technology, (3) manage financial and physical resources, (4) manage environmental, health, and safety issues, (5) manage external relationships, and (6) manage improvement and change.

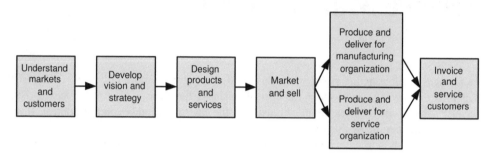

Figure 23.1: The "Global Best Practices" approach to automating value chains for enterprises

An alternative view, shown in Figure 23.2, is termed "Multi-Industry Process Technology." Besides the activities shown, the additional activities needed to implement and use this approach for a value chain are (1) perform business improvement, (2) manage environmental concerns, (3) manage external relationships, (4) manage corporate services and facilities, (5) manage financials, (6) manage human resources, (7) provide legal services, (8) perform planning and management, (9) perform procurement, and (10) develop and maintain systems and technology.

Figure 23.2: The "Multi-Industry Process Technology" approach to automating value chains for enterprises

The two models attempt to achieve the same overall goal of process management and continuous improvement, but the information systems that support them and each of their activities are incompatible, in spite of their overlapping concerns. Section 6.3 describes a methodology for how the semantics and behaviors of different approaches can be reconciled.

Integrated suites of ERP modules are available from several vendors, including SAP, Oracle Corporation, PeopleSoft, and Baan International. SAP's approach, labeled "mySAP ERP," is typical of these, and includes the following functionalities:

- *mySAP Analytics* provides strategic enterprise management and business analytics for forecasting, planning, and asset exploitation.

- *mySAP Financials* performs supply chain management and financial and managerial accounting.

- *mySAP Human Resources* is used for recruiting employees and managing their employment lifecycles.

- *mySAP Operations* provides the management of purchase orders, inventory, production, maintenance and quality, delivery, and sales orders.

- *mySAP Corporate Services* manages real estate, incentives and commissions, and travel.

23.2 WSMF: Web Services Management Framework

Imagine that an electrical parts manufacturer would like to provide an on-line catalog of its parts, with the prices available in any currency that a user would like, such as dollar, rupee, yen, or euro. This might be accomplished by a combination of two Web services: one to look up parts and the other to perform a currency conversion. The catalog is a type of computing resource, and managing it would require the management of both of its Web service components. Also required would be a management interface for the components that enables a manager (either a human or a software application) to monitor and control the components. For example, the components might have to be stopped, reset, and restarted if they are misbehaving.

The Web Services Management Framework is a logical architecture that uses Web services to manage computing resources, such as the on-line parts catalog described above. WSMF defines a standard extensible interface for communicating management information, using a standardized protocol and description language. In the case of the on-line catalog, each component Web service would require its own Web services for stopping and restarting it. That is, an object or resource must provide a management interface before it can be managed, and the management interface is described using WSDL.

23.3 WSDM: Web Services Distributed Management

The purpose of the Web Services Distributed Management framework is to provide a means for managing a set of Web services deployed within an enterprise [Labbe, 2002]. WSDM uses basic Web service standards, such as SOAP and WSDL, to describe and implement its management capabilities. Enterprises use WSDM to document:

- the services that are invoked;

- the applications that invoke services;

- the performance of each service;

- comparisons of the performance of a service with its past performance or with the performance of competing services;

- comparisons of measured QoS with the Service Level Agreements (SLAs) that are in force, i.e., ensuring that each business customer is receiving the service for which it contracted;

- the costs for setting up, using, and maintaining services;

- version numbers or dates of the currently used services and the numbers or dates of the newest versions available.

The WSDM framework defines how the resources in an adaptive enterprise can expose management information about themselves and how they can be managed. Rather than relying on each service to provide management information in an individual form, WSDM provides a common means for the information to be represented. As described in Section 2.3, a Web service is a kind of resource. It is specified as an aggregation of endpoints, each of which offers the service at an address and makes it accessible in accordance with a binding. Manageability is a possible quality of a service, and it can similarly be specified by an endpoint. A manageability endpoint exposes a set of management interfaces through which the underlying service could be managed. A management interface communicates information about changes in business processes and IT infrastructure whenever process and infrastructure events occur. This makes it possible for an enterprise to engage in real-time service monitoring using tools based on the Simple Network Management Protocol (SNMP).

To accomplish this, there must be a definition for a management event. An event is expressed as a 3-tuple in XML syntax consisting of (1) reporterComponentId, the identification of the component that is reporting the situation, (2) sourceComponentId, the identification of the component that is experiencing the situation (which might be the same as the component that is reporting the situation), and (3) situationInformation, the situation itself. reporterComponentId and sourceComponentId are instances of the ComponentIdentification type. The situation information includes a required set of properties that are common across product groups and platforms, yet adaptable to product-specific requirements.

23.3.1 Contingency Plans for Service Failures

When service failures occur, contingency plans must be instantiated and enacted automatically. For example, a Web service provider might change the endpoint of the service by deploying a new version or changing its hosting environment. Upon receiving a service access error, a requester could execute a contingency plan to check the description of the service in a UDDI repository to find the service's new endpoint. Unfortunately, formulating such contingency plans within each application can cause inconsistent approaches to recovery. WSDM recommends that contingency plans be executed centrally. Further, WSDM also recommends that service contingency and monitoring procedures be implemented together, so that service access problems can be fixed before an application's service request fails.

23.3.2   Security and Authentication

Although login and authentication information can be placed in the SOAP header of a Web service invocation, WSDM guidelines recommend that such information be removed from end-user applications and stored at a central proxy layer. Once authenticated, an application would place its secured information into the SOAP request that it sends to a service.

23.3.3   Features and Benefits of WSDM Centralization

Centralizing the management of services enables a number of other benefits to be realized. First, if all SOAP messages requesting services for an enterprise pass through a central proxy layer, then requests to the same service can be batched and connections can be pooled, thereby improving response times. Second, response times can be improved by caching service results. Whether or not this is worthwhile depends on how frequently results are needed and how frequently they change. For example, it would be reasonable to cache stock quotes after an exchange closes for the day, but not during the trading day. Third, a centralized WSDM strategy enables XML transformations to be performed enterprise-wide, rather than made the responsibility of individual developers. Overall, centralized WSDM practices have the potential to yield improved utility and functionality for enterprises.

23.4   Metadata Protocols

The new Web Services Metadata Exchange (WS-MetadataExchange) specification builds on the existing architecture for Web services. It defines how Web services retrieve specific types of metadata used for interactions with other Web services.

The specification defines three request-response message pairs: one retrieves the WS-Policy associated with the receiving endpoint or a given target namespace, a second retrieves the WSDL data, and a third retrieves the XML schema.

23.5   Scalability

Fundamentally, scalability is the ratio between performance and resources. We can think about the scalability of a service-oriented system in terms of its characteristics when applied to a domain that changes in size, or in terms of how the system achieves scalability.

A service-oriented system can cope with a growing application domain by increasing the number of services, the capability of each service, the computational resources available to each service, or the infrastructure resources needed by the services to make them more productive.

Alternatively, a service-oriented system can exhibit scalability one of three ways:

- As the amount of resources available to a fixed system of services increases, system performance should increase.

- As the number of services in a system increases to match increases in the number of entities in a domain, such as patients in a hospital information system or packages being tracked for delivery, the system should continue to function as designed.

- A scalable system should perform better by taking advantage of the additional capabilities offered by the increased number of services.

Scalability can also be dynamic or static. Static systems must be recompiled and restarted when the number of services or the resources available to them change. Dynamic systems can accommodate changes in services and resources during runtime. Obviously, dynamic systems are preferable.

23.5.1   Scalability in Practice

Scalability is not a problem for reactive services, because they do not use any system resources until they receive a SOAP message. Increasing the number of reactive services in a system simply causes a storage problem for the services and a possible communication bottleneck for the messages they exchange.

Nonreactive services behave like proactive, deliberative, agents, which consume resources as long as they exist. They evaluate their current circumstances and then plan their actions to achieve both immediate and long-term goals. They are continuously active, where even deciding to do nothing requires active deliberation and, thus, resources.

Physically, you can achieve scaling as follows:

- Distribute system components, namely, agents and services, across multiple physical machines, using distributed computing technologies such as Microsoft DCOM, CORBA, Java RMI, .NET, and Jini. Unfortunately, this approach can introduce communication latencies.

- Replicate the components on multiple physical machines, using distributed computing technologies similar to those in the distributive approach. Unfortunately, this tactic introduces consistency problems among the multiple copies, which limits its use to services that are mostly stateless and applications that are mostly static (a lot of communication is required to restore consistency when systems change).

- Schedule the components intelligently to execute only when and for as long as necessary to optimize the use of available resources. You can also arrange the components into hierarchies or other organizational structures to make their interactions more efficient.

Each of these techniques has been used successfully to deploy large systems of agents.

23.5.2   Scaling Infrastructure Services for Agents

The services agents require—services provided by the infrastructure and by agents to each other—must also scale. Agent services include name services, location services, directories, facilitators, and brokers. Infrastructure services include message transports, human interfaces, and CPU cycles.

The Internet, though DNS, already has an established means for scalable name services, which agent-based systems can use. DNS essentially scales through replication.

Scaling a directory service is more problematic. A directory service, such as LDAP, consists of attribute-value pairs that an agent can search for matches to its requirements, much as a person searches through a yellow pages. In general, an agent might need to search an entire directory exhaustively for each lookup. An index can shorten the search time, but such indexes are difficult to maintain in a distributed setting.

23.5.3   Scalability Experiments

In investigating the effect of communication on scalability, researchers developing the ZEUS multiagent framework discovered that the maximum communication load grows, at worst, linearly with the number of agents.

A research team at the University of Saskatchewan used the DICE framework to investigate the computational load of creating and executing 1 000 simple agents on a set of 10 remote hosts. The results demonstrated the feasibility of moving agents to less busy hosts for load balancing and also that response times remain reasonable—a few seconds at most—for agents that need to respond to people. Other results with complex rule-based agents (which incorporated the Jess reasoning engine) showed that 400 agents could execute acceptably within the same computational framework.

The DIET framework uses lightweight threads and thread-management techniques to enable more than 100 000 simple agents to execute on a single host machine.

In a different kind of experiment, a team at the University of South Carolina is investigating the scalability of a system of medium-complexity, heterogeneous agents. The agents form geometric shapes on a two-dimensional grid by communicating with nearby agents. Although only 60 agents are involved, they are heterogeneous—60 different programmers constructed them. For online reputation assessment experiments in (human) social networks, the team is scaling the system to more than 500 agents.

In a similar effort for scaling heterogeneous agents in a distributed and dynamic world, the DARPA Control of Agent-Based Systems (CoABS) program has developed an infrastructure called the Grid. The Grid has integrated agents and components from more than 20 independent projects and has operated successfully in a series of naval fleet battle exercises and other applications from information retrieval to military command and control. Built using Sun's Jini services, the Grid can integrate agent-based systems, object-based applications, Web services, and legacy systems. Agents in the Grid communicate point-to-point, so communications scalability, up to the limit imposed by network bandwidth, is not a problem. The Grid relies on a lookup service for registration and discovery that is centralized, which is a potential bottleneck. However, recent experiments with up to 10 000 agents show that registration and discovery are essentially independent of the number of agents registered.

23.5.4 Long-Lived Adaptable Agents

Scalability applies not only to the number of services and their interactions, but also to their lifetimes and the duration of their interactions. Most agents in use today are designed for short lives in relatively static online worlds. For example, an agent might be programmed to access the Web pages of five online stores and find the best price for a given music CD. While this is underway, the sites are presumed to be static and, when finished, the agent dies.

In contrast, future agents—especially those who represent users in their dealings with a ubiquitous computing world—might live for many years. Such agents will learn and adapt as they and their users encounter new situations, making it impractical for them to be recreated from scratch. Their needed infrastructure services must also be designed to exist for many years. They will also need new kinds of services: social services to help them cooperate in solving larger tasks, and legal services to help them meet their obligations and ensure their rights.

23.6 Robust Services via Agent-Based Redundancy

Software problems are typically characterized in terms of bugs and errors, which may be either transient or omnipresent. The general approaches for dealing with them are: (1) prediction and estimation, (2) prevention, (3) discovery, (4) repair, and (5) tolerance or exploitation. Bug estimation uses statistical techniques to predict how many flaws might be in a system and how severe their effects might be. Bug prevention is dependent on good software engineering techniques and processes. Good development and run-time tools can aid in bug discovery, whereas repair and tolerance depend on redundancy.

Indeed, redundancy is the basis for most forms of robustness. It can be provided by replication of hardware, software, or information, e.g., by repetition of communication messages. Redundant code cannot be added arbitrarily to a software system, just as steel cannot be added arbitrarily to a bridge. A bridge is made stronger by adding beams that are not identical to ones already there, but that have equivalent functionality. This turns out to be the basis for robustness in service-oriented systems as well: there must be services with equivalent functionality, so that if one fails to perform properly, another can provide what is needed. The challenge is to design service-oriented systems so that they can accommodate the additional services and take advantage of their redundant functionality.

The authors hypothesize that agents are a convenient level of granularity at which to add redundancy and that the software environment that takes advantage of them is akin to a society of such agents, where there can be multiple agents filling each societal role. Agents by design know how to deal with other agents, so they can accommodate additional or alternative agents naturally.

Fundamentally, the amount of redundancy required is well specified by information theory. If we want a system to provide n functionalities robustly, we must introduce $m \times n$ agents, so that there will be m ways of producing each functionality. Each group of m agents must understand how to detect and correct inconsistencies in each other's behavior, without a fixed leader or centralized controller. If we consider an agent's behavior to be either correct or incorrect (binary), then, based on a notion of Hamming distance for error-correcting codes, $4 \times m$ agents can detect $m - 1$ errors in their behavior and can correct $(m - 1)/2$ errors.

Redundancy must also be balanced with complexity, which is determined by the number and size of the components chosen for building a system. That is, adding more components increases redundancy, but also increases the complexity of the system.

An agent-based system can cope with a growing application domain by increasing the number of agents, each agent's capability, or the computational and infrastructure resources that make the agents more productive. That is, either the agents or their interactions can be enhanced, but to maintain the same redundancy m, they would have to be enhanced by a factor of m.

N-version programming, also called *dissimilar software*, is a technique for achieving robustness first considered in the 1970s. It consists of N separately developed implementations of the same functionality. Although it has been used to produce several robust systems, it has had limited applicability, because (1) N independent implementations have N times the cost, (2) N implementations based on the same flawed specification might still result in a flawed system, and (3) each change to the specification will have to be made in all N implementations.

Database systems have exploited the idea of transactions: an atomic processing unit that moves a database from one consistent state to another. Consistent transactions are achievable for databases because the types of processing done are regular and limited. Applying this idea to software execution requires that the state of a software system be saved periodically (a checkpoint) so the system can return to that state if an error occurs.

23.6.1 Architecture and Process

Suppose there are a number of services, each with strengths, weaknesses, and possibly errors. How can the services be combined so that the strengths are exploited and the weaknesses or flaws are compensated or covered?

Three general approaches are evident in Figure 23.3. First, a preprocessor could choose the best services to perform a task, based on published characteristics of each service. Second, a postprocessor could choose the best result out of several executing services. Third, the services could decide as a group which ones should perform the task.

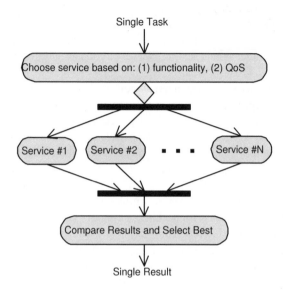

Figure 23.3: Approach for combining N versions of a service into a single, more robust system

The difficulties with the first two approaches are (1) the preprocessor might be flawed, (2) it is difficult to maintain the preprocessor as services are added or changed, and (3) the postprocessor wastes resources, because several services work on the data and their results have to be compared.

The third approach requires distributed decision-making, which is not an ability of conventional Web services. What generic ability could be added to a service to enable it to participate in a distributed decision? The generic capability has the characteristics of an agent, so distributing the centralized functions into the different modules creates a multiagent system. Each agent would have to know its role as well as (1) something about its own service, such as its time and space complexity, and input and output data structures; (2) the complexity and reliability of other agents; and (3) how to communicate, negotiate, compare results, and manage reputations and trust.

23.6.2   Experimental Results

Huhns and colleagues collected one set of 25 algorithms for reversing a doubly linked list and another set for sorting a list. Different novice programmers wrote each algorithm. For sorting, no specifications were given to the programmers (beyond that the problem was sorting), so the algorithms all have different data and performance characteristics. For list reversing, the class structure (i.e., method signatures) was specified, so the differences among the algorithms are in performance and correctness.

Each algorithm was converted into an agent, composed of the algorithm written in Java and a wrapper written in Jade. The wrapper knows only about the signature of its algorithm, and nothing about its inner workings.

Our experiments verified that the same wrapper can be used for both the sorting and list-reversing domains. We also verified our hypothesis that more algorithms give better results than any one alone. Further, we investigated both a distributed preprocessor and a centralized postprocessor for combining the agents' functionality, and found that the postprocessor is generally better, but performs worse for large data sets or selected algorithms with long execution times.

The eventual outcome for application development is that service developers will spend more time on functionality development and less on debugging, because different services will likely have errors in different places and can cover for each other.

23.7 Exercises

23.1. WSDM is a centralized strategy for Web service management—true or false?

23.2. List and provide a short description of three techniques for achieving robustness in Web services.

23.3. Describe the differences between WSMF and WSDM in terms of what things are intended to be managed and what entities perform the management.

Chapter 24

Security

Service-oriented architectures not only make it easier for legitimate entities to access systems from outside an enterprise boundary, but also expose new opportunities for unauthorized and illegal entities to misuse and exploit the services. An SOA is subject to more than the usual threats to security, because it inherently involves interactions among autonomous entities. A simple presupposition underlying contracts and negotiation, for example, is that you can ensure that the various entities are accountable, which is not always the case when the entities are autonomous. To some extent, nothing other than a social or economic notion of trust would apply, but even such techniques must be supported through the appropriate distributed system infrastructure. Further, traditional approaches for ensuring security do not apply across administrative domains or when there are high transaction rates. Consequently, SOAs provide a hot testbed for the elaboration of security techniques. This chapter briefly reviews some of the key techniques with reference to the general Web infrastructure.

24.1 Securing Web Services

Security is fundamental to the successful adoption of Web Services for business applications. First-generation Web services have been largely unencrypted and unsecured, and this has hindered their wider adoption. The most basic security technique for simple applications is to use traditional transport layer security, for example, a secure sockets layer (SSL) connection between two points. For more complex applications involving multiple parties and services, messages among the end points might be encrypted and signed to protect their confidentiality and integrity. Tokens may also be added to messages to assert claims, e.g., about identity checks that have been carried out by a trusted authority.

SSL and its successor, the transport layer security (TLS) protocol, operate between the HTTP and TCP network layers. They provide public-key and private-key encryption and support the digital certificates issued by a certification authority to authenticate parties. TLS works well for browsers, but is unsuitable for the high transaction rates that Web services

- Attribute assertions attest to the values of one or more attributes of the subject, for example, a credit limit for e-commerce.

- Authorization assertions attest to which services the subject is or is not entitled.

As well as XML document types for expressing these assertions, SAML also defines simple request-response message pairs for obtaining these assertions from messages.

WS-Security defines how such assertions and other security tokens can be attached to the headers of SOAP messages. It is an enhancement to SOAP messaging that provides protection quality through message integrity, message confidentiality, and single message authentication. These three mechanisms can accommodate a wide variety of security models and encryption technologies, such as PKI, SSL, X.509 certificates, and Kerberos tickets.

WS-Security also provides a mechanism for associating security tokens with messages. The mechanism is extensible and supports multiple security token formats. For example, a client might provide both proof of identity and proof that they have a Better Business Bureau certification. Security metadata defined by the XML Encryption and XML Signature specifications can also be attached.

A security token, defining the name of the requester for a service, could be added to the header of a SOAP message by the following code (note that wsse denotes the namespace for WS-Security):

```
<wsse:UsernameToken Id="MyID">
    <wsse:Username>Munindar</wsse:Username>
</wsse:UsernameToken>
```

A digital signature, using the XML Signature specification, could also be added to the SOAP header by the following code:

```
<ds:Signature>
  <ds:SignedInfo>
    <ds:CanonicalizationMethod
       Algorithm="http://www.w3.org/2001/10/xml-exc-c14n#"/>
    <ds:SignatureMethod
       Algorithm="http://www.w3.org/2000/09/xmldsig#hmac-sha1"/>
    <ds:Reference URI="#MsgBody">
       <ds:DigestMethod
          Algorithm="http://www.w3.org/2000/09/xmldsig#sha1"/>
       <ds:DigestValue>LyLsF0Pi4wPU ... </ds:DigestValue>
    </ds:Reference>
  </ds:SignedInfo>
  <ds:SignatureValue>DJbchm5gK ... </ds:SignatureValue>
  <ds:KeyInfo>
    <wsse:SecurityTokenReference>
       <wsse:Reference URI="#MyID"/>
    </wsse:SecurityTokenReference>
  </ds:KeyInfo>
```

```
</ds:Signature>
```

The Reference field in the above listing indicates that the body of the message, not the header, is encrypted.

WS-Security is designed to prevent unauthorized entities from reading (confidentiality) or modifying (integrity) a SOAP message and unauthorized entities from obtaining services to which they are not entitled by sending a SOAP message without a valid token.

24.3 WS-Trust

In order to secure a communication between two parties, the two parties must exchange security credentials (either directly or indirectly), using a specification such as WS-Security. However, each party needs to determine if they can "trust" the asserted credentials of the other party. The Web Services Trust (WS-Trust) language uses the secure messaging mechanisms of WS-Security to define additional primitives and extensions for issuing, exchanging, and validating security tokens. That is, *trust is represented through the exchange and brokering of security tokens*. WS-Trust also provides for issuing and disseminating credentials within different trust domains. Using these extensions, applications can engage in secure communications while still using the basic SOAP+WSDL+UDDI+HTTP framework for Web services.

A service first specifies its required claims using WS-Policy. When a request containing security tokens arrives at a service as a SOAP message, WS-Trust requires that the message prove a set of claims, e.g., name, key, and permission. Otherwise, the service will ignore the message or deny the request. To validate a request, the service must have a *trust engine* that (1) verifies that the claims match the requirements of the service's policy, (2) verifies that the signature proves the attributes of the claimant, and (3) verifies that the security tokens are trusted to issue their claims. As an alternative, a trusted intermediary may verify the requester by simply asserting the requester's identity.

24.4 XACML

The eXtensible Access Control Markup Language, is an OASIS specification for expressing access-control policies in XML for information access over the Internet. Once configured, it expresses and communicates the rules and policies that an access-control mechanism can use to decide about access to objects and attributes. XACML postulates two architectural components: a Policy Enforcement Point (PEP), which decides whether to allow or disallow a particular request; and a Policy Decision Point (PDP), which determines whether a policy request should be allowed. A PEP consults a PDP. A PDP consults its repository of XACML policies. Figure 24.1 shows the steps in a typical use of XACML for access control.

XACML specifies both an access control policy language (which allows developers to declare who can do what and when), and a language for expressing requests and responses. This language expresses queries about whether a particular access should be allowed and need to perform their task or achieve their goal. It also advocates separating the above three aspects to enable a finer-grained control of permissions.

24.7 Exercises

24.1. The Extremely Reliable Operating System (EROS) [Shapiro et al., 1999] enforces POLA as an inherent part of its architecture. Discuss how EROS could be used for a distributed system of Web services.

24.2. *Secure Database Queries via E-Mail*

- Construct a database that contains at least two tables, with at least four columns in each table and at least five tuples in each table. The database can be for any domain of your choice.

- Construct an email front-end to your database. Your overall system should be able to receive an email message, parse it to find a query from the sender, apply that query to your database, retrieve the result from the database, and email the result back to the sender. Turn in all source code, examples of the email messages, and screen-shots of the system's operation.

- Construct a public-key encryption system for your email responder. That is, the result of a query should be encrypted by your email front-end using the public-key of the email sender, and then decrypted by the recipient using his private key. Show both the encrypted and decrypted data, as well as both of the keys that you chose.

24.3. SAML can be used in combination with which of the following specifications for security?

- WS-Security.
- WS-Trust.
- XACML.
- WS-Policy.
- XML Encryption.
- XML Signature.

24.4. A Web service for a corporate personnel system has the following two operable policies:

"Permission to use the service must be granted explicitly by the database administrator (DBA) in charge of the system."

"The CEO has the right to overrule the DBA."

Using Rei or another deontic logic, formulate the policies. Then, assuming that the DBA has not granted permission to the CEO to use the service, describe using WS-Security and WS-Trust how a request from Smith, claiming to be the CEO, would be validated so that Smith could use the service.

Part VII

Directions

Chapter 25

Challenges and Extensions

Advances in techniques such as those discussed in the foregoing chapters are fulfilling the promise of personalized, friendly, Web services. The improvements come at a cost, however: greater implementation complexity. Thus, as we come to rely more on the improved services for e-business or information retrieval, we understand less about how they operate.

Abstraction is a classical approach for dealing with complexity. This book discusses several computational abstractions motivated from how people organize their knowledge, their activities, or their contracts. Additional computational abstractions can be based on subtle concepts related to human interactions.

25.1 Trust

Web services routinely put us in interactions with strangers, both people and corporations, or their digital assistants. Services can be defined for information, entertainment, and commerce. For each there is a need for trust. Trust goes beyond security in that it is about managing interactions at the application level. For example, whereas security is about authenticating another party and authorizing actions, trust is about the given party acting in your best interest and choosing the right actions from among those that are authorized.

Interactions among strangers presuppose some level of assurance that the parties will act appropriately. For example, a shipping service cannot be sure that it will receive payments after delivery and may thus require payments before delivery. Or it may seek assurance from a third party that can guarantee the payment. Security mechanisms yield assurances, e.g., that messages will not be intercepted. Likewise, contracts yield assurances, e.g., that the buyer will pay (or face legal action).

In some respects, trust is complementary to assurance. If you trust the other party more, you need fewer assurances about its good behavior. The existence of assurances can also engender trust. And yet you need a level of trust in an assurance mechanism even to believe the mechanism itself. For example, if you use an escrow agency, you would need to trust it.

If you participate in a contract, you need to trust that the contract reflects the contingencies that are of importance to you.

Several interesting scenarios are being envisaged for service-oriented computing. For example, global supply networks will control the complex movement of goods from raw materials to customers without human intervention. As systems become more complex and longer lived, the situations to which they will be subjected cannot all be anticipated, so the systems cannot be fully verified. We will have to trust them, and there must be a principled basis for our trust.

We can think of trust as placing your plans in the hands of another. The trustworthiness of a party reflects how much trust you can place in it. Thus trustworthiness involves the following elements, which affect whether your plan, if placed in the hands of the given party, will come to fruition. The following are some of the ingredients of trust:

Capability. The party should be competent and aware of material facts.

Sincerity. The party should believe what it asserts and commit to acting according to its promises. It should not be deceptive or misleading, even if correct in a literal sense.

Helpfulness. The party should be well-intentioned, and should apply its capabilities to support the needs of the trusting party.

Duty. The party should be under a duty to help you, because of its ethics, its social or team role, or its prior commitments.

Predictability. The behavior of the party should be as easy to predict as possible.

Understanding. The behavior of the party should be understandable or explainable. Unfortunately, as components become more complex it is harder to understand them.

The above can be supported via the following. One, ethics is key. Rational parties are potentially easier to predict provided you have the knowledge they have. Further, you can constrain the actions of a rational entity by making it in its interest to act as you prefer. Thus rationality enhances trustworthiness. Another basis for trust arises from societal conventions, possibly inspired from ethical considerations. For example, we could have a convention that service requesters whose requests have not been processed successfully be given priority on later requests. The presence of such a convention improves the predictability of the system for all concerned and makes its behavior more understandable.

25.2 Ethics

Ethics is the branch of philosophy concerned with codes and principles of moral behavior. It is important to distinguish between the concepts of right and good:

Right is that which is right in itself.

Good is that which is good or valuable for someone or for some end.

The German philosopher, Immanuel Kant (1724-1804), defined the *categorical impera-tive* as an absolute and universal moral law (of the form "Do this") based entirely on reason (as distinguished from his *hypothetical imperative*, which is based on desire: "Do this if you want that"). We can state the categorical imperative in relation to agent behavior as follows: "Agents should act as if the maxim of their action were to secure through their will a universal law of nature." It provides a source of right action. For example, breaking a promise is not right, because if all agents broke promises, the system they support would not function.

Kant's categorical imperative does not contain a way to resolve conflicts of duty. Less stringent formulations specify *prima facie* duties, which do not bind agents absolutely but instead hold generally: "All other things being equal, keep your promises."

The *deontological theories* of ethics, like Kant's, emphasize right before good. They oppose the idea that the ends can justify the means, and they place the locus of right and wrong in autonomous adherence to moral laws or duties. These theories distinguish inten-tional effects from unforeseen consequences. That is, an action is not wrong unless the agent explicitly intends for it to do wrong. This legitimizes inaction, even when inaction has pre-dictable, but unintended effects. For example, a service manager would not be wrong to shut down a service for diagnostics, even if that might leave some users hanging.

In contrast, *teleological theories* of ethics choose good before right: something is right only if it maximizes the good; in this case, the ends can justify the means. In teleological theories, the correctness of actions is based on how the actions satisfy various goals, not the intrinsic rightness of the actions. Choices of actions can be comparison-based or preference-based. A further distinction is between utilitarianism (action should maximize the universal good of all agents) and egoism (action should maximize self-interest). In teleological theo-ries, good may be interpreted in various ways:

- Pleasure, in which case it is called hedonism.

- Preference satisfaction, called microeconomic rationalism, which assumes each agent knows its preferences.

- Interest satisfaction, called welfare utilitarianism.

- Esthetic ideals, called ideal utilitarianism.

What agents need to decide upon actions are not just universal principles (each can be stretched) and not just consequences, but also a regard for their promises and duties.

Along with keeping promises, agents' *prima facie* duties include helping others and repaying kindness. Just to help, a courier service may offer to deliver some low-priority packages for a parcel service when it sends a courier for a high-priority package. Moreover, there may or may not be a ranking among an agent's duties. For example, if the courier ser-vice's truck has a breakdown, it may commandeer a truck belonging to the parcel service, based on a belief that its promises are more important than the parcel service's promises.

25.2.1   Machine Ethics

Isaac Asimov proposed a moral philosophy for intelligent machines in 1940. His collection of short stories, *I, Robot*, included a Handbook of Robotics that defined three *Laws of Robotics*. These were subsequently augmented in *Foundation and Empire* by the "zeroth law," and the four laws were rewritten as follows:

Law 0. A robot may not injure humanity or, through inaction, allow humanity to come to harm.

Law 1. A robot may not injure a human being or, through inaction, allow a human being to come to harm.

Law 2. A robot must obey the orders given it by human beings except where such orders would conflict with the First Law.

Law 3. A robot must protect its own existence as long as such protection does not conflict with the First or Second Law.

An adaptation of these laws for a collection of services might be:

Principle 1. An agent shall not prevent the success of a service interaction.

Principle 2. Except where it conflicts with Principle 1, an agent shall not interfere with other agents.

Principle 3. Except where it conflicts with the previous principles, an agent shall not risk its own success.

Principle 4. Except where it conflicts with the previous principles, an agent shall make rational progress toward team goals.

Principle 5. Except where it conflicts with the previous principles, an agent shall follow established conventions.

Principle 6. Except where it conflicts with the previous principles, an agent shall make rational progress toward its own goals.

Principle 7. Except where it conflicts with the previous principles, an agent shall operate efficiently.

SOA implementations are susceptible to deadlocks and livelocks. However, if we could establish that the components obey these seven principles, then the susceptibilities would disappear, because deadlock and livelock would violate Principle 6. Conversely, because of the obvious connection with the Halting Problem for Turing machines, this suggests that we cannot develop an automatic procedure to verify that an agent system satisfies the above principles.

25.2.2 Applying Ethics

A philosophical approach to SOAs presupposes that the components are autonomous agents that can negotiate with others and can enter into commitments to collaborate with others. However, the ethical theories described above are theories of justification, not of deliberation. That is, agents can use them to explain why an action should or should not be taken, but not how to find possible actions to be taken. Other means are needed to decide on a course of action. An agent can decide what basic "value system" to use under any approach.

The deontological theories are narrower and ignore practical considerations, but they are only meant as incomplete constraints—that is, the agent can choose any of the right actions to perform. The teleological theories are broader and include practical considerations, but they leave the agent fewer options for choosing the best available alternative.

All of these ethical approaches are single-agent in orientation and encode other agents implicitly. An explicitly multiagent ethics would be an interesting topic for study.

25.2.3 Ethical Violations

Ethics governs how an entity *should* behave, but does not specify what should happen when it misbehaves. In human societies, punishments are enacted to deal with legal violations committed by individuals or organizations; these are discussed in Section 18.3. There can be no punishment for ethical violations except social censure or loss of reputation. What punishments should be meted out in computational systems? The challenges involved include detecting violations, determining responsibility (e.g., intent), and enforcing a punishment.

Punishments are often not cost-effective locally, but they are cost-effective globally. That is, if someone is caught shoplifting a $10 CD (which is more than just an ethical violation), arresting, prosecuting, trying, convicting, and incarcerating the criminal costs several hundred thousand dollars. For just this one case, it obviously makes more economic sense to ignore the theft. However, a prison sentence is expected to be a deterrent that prevents millions of other people from committing the same crime. Globally, then, society saves money.

For ethical violations, the situation can be simpler because no formal punishment may be necessary. However, the parties would need to ensure that the punishment (even if no more than loss of reputation) was fair and had consequences that were not disproportionate to the original violation.

25.3 Coherence

To endow agents with ethical principles, we as developers need an architecture that supports explicit goals, principles, and capabilities (such as how to negotiate), as well as laws and ways to sanction or punish miscreants. Figure 25.1 illustrates such an agent architecture that can support both trust and coherence, where coherence is defined as the absence of wasted effort and progress toward chosen goals.

The service networks can involve disparate administrative domains, which control the flow of services.

These private service networks are exemplified by those of AOL, NTT DoCoMo's iMode, and Microsoft. These networks provide either the so-called walled gardens, in which the administrator must approve the services that are available, or a trusted third party, who can be relied on to complete various transactions.

A telecommunication provider's *intelligent network* is a second form of a private service network. Compare such networks to the Internet. It is trivial to add new services over the Internet, because you do not need anyone's approval to do so. However, the services so added are impoverished in that they lack knowledge of the kinds of things only an intelligent network would know, e.g., a user's location or the current throughput or reliability. The complexity and administrative control of telecom networks make it difficult to introduce new services, e.g., those that address narrow market segments or short-lived fads. Context-based services seek to exploit such elements of context in a manner that retains the flexibility of creating, administering, and maintaining services.

25.6   Managing Privacy

For services to be effective, they must be personalized to users. But personalization means that the service provider potentially knows something about the user. In general, for two parties to carry out well-nuanced collaborations presupposes that they develop extensive models of each other. But for the models to be built means that each party's privacy is compromised to some extent. The potential loss of privacy becomes more significant when context-based services are involved, because they end up knowing aspects of the user's context that might otherwise not be known to them.

Having users opt in for services does not quite solve these problems. A user who opts in to join a mailing list faces a small bit of annoyance if the mailing list is not of interest. But a user who opts in to a location-based service faces a potentially significant risk through violation of privacy, because the information about the user is so much more precise. Traditional privacy policies can help only a little, because the information may still be compromised, even if the service follows its published policy.

The approach involves a fairly straightforward application of agent technology. Each user and provider is assigned an agent who represents the user's interests. The user signs on and off from services through his agent. Consequently, the user agent knows what services its user has signed up for currently. It also knows what its user's preferences are regarding each given service. These preferences would, in particular, apply to the sharing of information (about the user) with the service.

Capturing a user's wishes, however, may not be easy. This is because users may choose to reveal different information to different providers. For example, a user may need to inform an office groupware service where he is presently, but only during work hours. However, the user might not allow revealing his location to a service that might send him advertisements for products in his vicinity. When the preferences are more than simply on/off decisions, maintaining them becomes quite tedious: the rules involved could be cumbersome and the user may not understand their ramifications until an unexpected event occurs.

Agents can help by capturing the constraints or policies that a user would like to enforce with regard to their privileged information. An agent-based interface can suggest the constraints that the user may wish to enforce based on properties of the given service. This would require developing an ontology of services and would benefit from self-describing services. For example, an enterprise service might be given location information during office hours and just informed if the user is entertaining only text messages the rest of the time.

These agents may need some information from the underlying network (e.g., their user's location), but otherwise can function like any other service. In other words, privacy management can itself be modeled as a service, one that is specially trusted by the user and can mediate interactions with other services.

25.7 Key Challenges and Recommendations

The main lesson is to recognize the constraints of service-oriented architectures, while respecting the autonomy and heterogeneity of the participants. Let us briefly discuss some challenges that recur in the development of different technologies and approaches for Web services.

Design rules must be formulated for facilitating the application of the various interesting ideas discussed in this book in a manner that minimizes the cognitive load on a designer.

Security and trust are key because SOAs are inherently open.

Scalability is essential for ensuring that systems of practical interest can be dealt with; scalability applies not only to deployments, but also to design tools.

Quality of service applies not just to traditional aspects of performance, such as reliability and availability, but also to application-specific considerations.

User interfaces are the Achilles heel of complex software systems, but should be given deeper consideration to ensure the success of a practical installation. Services, ultimately, are for the benefit of users.

In light of the above, we need more sophisticated means to describe services. For example, it might be helpful to encode how the result of the service is obtained, how accurate it is, and what constraints there are on the validity and uses of the result. Likewise, we need approaches to handle exceptions. The previous chapters discuss how systems can be organized to most naturally handle exceptions, but require a detailed study of domain-specific considerations. Further, there are low-level concerns, such as that a service might not be able to respond to a particular request because it is off-line, does not have an answer, or its response is proprietary.

Advanced abstractions can help significantly to understand services in the broadest terms, specify and seek services at a high level, design service compositions in a principled manner, and validate composed services. But the abstractions must be conceptually simple and well-grounded in theory, and represent the true status of the system, in reflection of its service-oriented architecture. If you can use an adaptive approach, doing so enables dynamically configuring complex systems, and simplifies your task of implementing them because some decisions can be deferred.

Often, technologists forget that real-world problems are often mundane and that a large fraction of the solution effort will go into simple details. Advanced ideas must be incorporated into existing systems. Let common sense be your guide!

Part VIII

Appendices

Appendix A

XML and XML Schema

As shown in Figure A.1, the foundation for interoperation among enterprises and for the envisioned Semantic Web is the eXtensible Markup Language, XML. This chapter describes XML to a sufficient extent to understand services standards and approaches. You will learn how to write well-formed XML data and documents, how to write the rules that such data and documents must obey, and how to validate the documents against the specified rules.

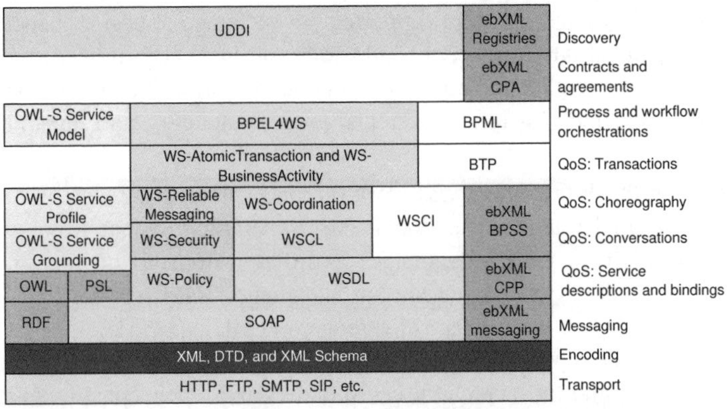

UDDI				ebXML Registries	Discovery
				ebXML CPA	Contracts and agreements
OWL-S Service Model	BPEL4WS			BPML	Process and workflow orchestrations
	WS-AtomicTransaction and WS-BusinessActivity			BTP	QoS: Transactions
OWL-S Service Profile	WS-Reliable Messaging	WS-Coordination	WSCI	ebXML BPSS	QoS: Choreography
OWL-S Service Grounding	WS-Security	WSCL			QoS: Conversations
OWL	PSL	WS-Policy	WSDL	ebXML CPP	QoS: Service descriptions and bindings
RDF	SOAP			ebXML messaging	Messaging
XML, DTD, and XML Schema					Encoding
HTTP, FTP, SMTP, SIP, etc.					Transport

Figure A.1: The relationships among the different proposed standards and methodologies that are the foundation for service-oriented computing. XML and its rules provide the syntax in which all of the Semantic Web and its services and protocols are being described

A.1 Why XML?

It is clear that there is a need to share information among software components. To enable such sharing, the network and data encoding need to be compatible. The use of IP networks

499

throughout, along with standard character encoding schemes, greatly mitigates these lower-level problems. The convergence on HTTP further simplifies the connectivity. Since not everyone uses ASCII, more powerful text encoding schemes were needed and have been invented in the form of the Universal Character Set (UCS) and the UCS Transformation Format (UTF-8). Consequently, the low-level problems are all but eliminated.

But the data needs to be cleanly formatted so that it can be parsed reliably by the recipient to yield the intended structures. In other words, the information that is transmitted must be expressed in some generic manner. In particular, this manner could be independent of the programming language in which the interacting components may have been implemented.

Any format would work, provided the parties to the communication agreed to it. For instance, it could be the famous comma-separated value (CSV) format, which despite its ubiquity remains unstandardized [Python, 2004]. CSV is defined only operationally in terms of how it is interpreted by a given application, leading to various problems. CSV, moreover, is unsuited for representing anything more complex than simple tabular data. In particular, nested object-like structures cannot be represented easily in CSV. More complex structures can be expressed readily in the syntax of the programming language Lisp or in Knowledge Interchange Format (KIF), which also uses a Lisp-like syntax. Another language for expressing complex structures is Abstract Syntax Notation (ASN.1). ASN.1 was standardized by the International Standards Organization (ISO). It is used in some telecommunication protocols, including H.323, and for storing genomic data. Starting as early as the 1960s, Electronic Data Interchange (EDI) standards have supported the exchange of business documents. EDI standards have expanded into a variety of industries and have been implemented, but these implementations were complex and not easy to maintain. Part of the problem was that other key technologies, especially for information and process modeling and semantics, were not well developed.

Also starting in the 1960s, there was work on marking up documents in a manner that would express a formal grammar for documents, separating content from structure, and allowing customization for markup languages in different domains of interest. This work culminated in the Standard Generalized Markup Language (SGML) in the 1970s, which was standardized by the ISO in the 1980s. In essence, HTML and XML are special cases of SGML. SGML is an extremely powerful language. This made it difficult to build robust and usable tools for it, and for its target users to understand it conceptually with high confidence. Consequently, applications of it were mostly limited to large publishing concerns and document management efforts in large companies.

Arguably, any of the above (except perhaps CSV, which is expressively quite weak) could have been the standard interchange language for service-oriented computing. For most purposes, it would not matter what the standard was as long as it was expressive enough and perspicuous enough that robust tools existed for it, and users were comfortable specifying the contents of their communications using it. Beyond that, the choice is a question of historical accident and culture (for example, KIF's association with Lisp may have been a strike against it in the minds of many programmers who were not well-versed in Lisp). As a practical matter, what has happened is that alternative XML-based notations are being proposed for each of the above languages so as to help assimilate their use into XML-based systems.

A.2   XML

The eXtensible Markup Language is an attempt to attach semantics to data through a structured syntax. It turns out that it does not quite attach any semantics, but it streamlines the syntax and provides a basis for introducing semantics.

XML syntax provides a set of rules for describing content, rather than the presentation of that content, and it is used to provide a structure for data. A well-formed XML *document* corresponds to a tree. Elements function as customized opening and closing parentheses. For example, the temperature element in the example below has an opening tag ⟨temperature⟩ and a closing tag ⟨/temperature⟩—the difference is the leading / before the token in the closing tag. The nesting of the document tree is indicated by what is included in an element, i.e., between its opening and closing tags. Elements may include other elements or may include text. Elements may have zero or more *attributes*, each with its associated value. Attribute values must be strings, which can be demarcated via either single (') or double (") quotes. The attributes are placed within the opening tag for the element. This is why the example below shows the opening tag for temperature as ⟨temperature scale="Celsius"⟩. Elements that have no subelements or PCDATA can be written via an *empty tag*, which ends immediately. The element ⟨ambience condition="shade"/⟩ in the example below is an empty tag.

A well-formed XML document consists of a single top-level *element*, which may (and usually does) contain other elements. That is, the document tree is rooted at the top-level element. Two requirements of well-formedness are that each element that begins must end, and no element may have two attributes of the same name. For example, to express the fact that a temperature is 25 ± 2 degrees Celsius, the syntax might be as in Listing A.1.

Listing A.1: XML example

```
<?xml version='1.0' encoding='UTF-8'?>
<temperature scale="Celsius">
  <value>25</value>
  <accuracy>2</accuracy>
  <ambience condition="shade"/>
</temperature>
```

where the first line is a processing instruction (XML boilerplate); value and accuracy are *subelements*; and scale is an *attribute* of the element temperature.

Unfortunately, the specification for XML allows alternative ways of expressing the same information, so that the above fact could also be expressed, equally correctly, as in Listing A.2.

Listing A.2: Alternative XML example

```
<temperature accuracy="2" scale="Celsius">
```

```
   25
   <ambience condition="shade"/>
</temperature>
```

XML, however, is an improvement over HTML, which is intended to express the appearance of data, rather than its meaning. XML enables the separation of the meaning or semantics of the data from the way it is used by an application or rendered on a screen or output device. For example, consider the HTML source code for a table of temperatures in Listing A.3.

Listing A.3: An example of HTML that contains the same data as the previous example of XML

```
<TABLE BORDER=1>
   <TR>
      <TH>Value </TH>
      <TH>Accuracy </TH>
      <TH>Scale </TH>
      <TH>Ambience Condition </TH>
   </TR>
   <TR>
      <TD>25 </TD>
      <TD>2 </TD>
      <TD>Celsius </TD>
      <TD>Shade </TD>
   </TR>
</TABLE>
```

The resultant table would likely have an appearance that resembles Table A.1; the exact appearance would of course depend on the browser being used and the display on which it is rendered.

Table A.1: Rendering the temperature document

Value	Accuracy	Scale	Ambience Condition
25	2	Celsius	Shade

A.2.1   XML and Vocabularies

We can think of the tags used in a markup document as a *vocabulary*. An important difference between Listing A.1 and Listing A.3 is that the XML document uses tags that are specific to the domain, whereas the HTML document uses formatting tags that have nothing to do with the domain of interest. This illustrates the extensibility of XML. XML is a language in which arbitrary domain-specific vocabularies can be constructed. These vocabularies might not be sufficiently expressive to capture meaning (as discussed at length in Chapter 6), but the fact that they are domain-specific enables them to indicate the meaning more perspicuously than the corresponding HTML version. If the parties sharing some information can agree on a vocabulary and its intended meaning, then they can interoperate effectively.

A.2.2   Transforming XML

Table A.1 contains the same data as the previous example of XML, but there is nothing in this example that readily associates "Scale" with "Celsius" and nothing about "temperature." The XML example could be rendered to appear exactly like this table, using the following XSL Transformations (XSLT), but different XSLT documents could be made to appear as plain text for rendering on a cell phone or as the input to a text-to-speech system for speaking the information aloud. Listing A.4 shows a complete XML file that references an XSL stylesheet; Listing A.5 shows the stylesheet itself.

Listing A.4: An XML file that references an XSL stylesheet named TempTable.xsl

```
<?xml version="1.0"?>
<?xml:stylesheet type="text/xsl" href="TempTable.xsl"?>
<TempCollection>
  <temperature scale="Celsius">
    <value>25</value>
    <accuracy>2</accuracy>
  </temperature>
  <temperature scale="Kelvin">
    <value>123</value>
    <accuracy>5</accuracy>
  </temperature>
  <temperature scale="Fahrenheit">
    <value>32</value>
    <accuracy>1</accuracy>
  </temperature>
</TempCollection>
```

Listing A.5: An example of a stylesheet stored in a file named TempTable.xsl that could transform the XML example listed above into the HTML that would appear as a table in a browser

```
<xsl:stylesheet version="1.0"
  xmlns:xsl="http://www.w3.org/1999/XSL/Transform">
  <xsl:template match="/">
    <HTML>
      <BODY>
        <xsl:apply-templates/>
      </BODY>
    </HTML>
  </xsl:template>
  <xsl:template match="TempCollection">
```

```
        <TABLE  BORDER="2">
          <TR>
            <TH>Value</TH>
            <TH>Accuracy</TH>
            <TH>Scale</TH>
          </TR>
          <xsl:apply-templates/>
        </TABLE>
  </xsl:template>
  <xsl:template match="temperature">
    <TR>
        <TD><xsl:value-of select="value"/></TD>
        <TD><xsl:value-of select="accuracy"/></TD>
        <TD><xsl:value-of select="@scale"/></TD>
    </TR>
  </xsl:template>
</xsl:stylesheet>
```

The XML example expresses better semantics than the HTML example, but it still suffers from two major limitations. First, another temperature application might use the tags "Amount," "Precision," and "TempScale" instead of "Value," "Accuracy," and "Scale." XML provides no way to reconcile the different tags. Second, the XML specification (described below) requires that a well-formed XML document be a tree with a single root, but it does not otherwise specify the structure of a valid document, so that many different structures can convey the same information. Section A.3 on XML Schemas addresses this limitation.

A.2.3   Well-Formedness

A well-formed XML document must satisfy the following requirements:

- It must be a tree with a single root. This makes it easier to determine when a document is complete.

- Every start tag must have a matching end tag, e.g.,
  ⟨temperature⟩25⟨/temperature⟩

- All attributes must be quoted, e.g.,
  ⟨temperature scale="Celsius"⟩

- Empty tags must have a trailing slash, e.g.,
  ⟨temperature value="25"/⟩

- All entities (see Section A.2.5) must be declared.

- Child elements must be properly nested, e.g.,
  ⟨temperature⟩ ⟨value⟩25⟨/value⟩ ⟨/temperature⟩

The validity of an XML document can be determined by comparing it to a specified Document Type Definition (DTD) or a specified XML Schema document. A DTD for a collection of data consisting of temperatures like the example at the beginning of this chapter would be as shown in Listing A.6.

Listing A.6: A DTD for a collection of the XML temperature examples

```
<?xml encoding="UTF-8"?>
<!ELEMENT TempCollection (temperature)+>
<!ELEMENT temperature (value, accuracy, ambience)>
<!ELEMENT value (#PCDATA)>
<!ELEMENT accuracy (#PCDATA)>
<!ELEMENT ambience (EMPTY)>
<!ATTLIST ambience condition CDATA "shade">
<!ATTLIST temperature scale CDATA #REQUIRED>
```

The last line specifies that "scale" is a *required* attribute of "temperature" and that the data must be characters, i.e., text. Notice that this DTD does not satisfy the rules for an XML document, which means that two kinds of documents, a DTD and XML, are needed to specify some of the semantics of data. This is corrected by using an XML Schema instead of a DTD, because an XML Schema is itself a well-formed XML document.

A.2.4   Namespaces and Qualified Names

Because XML enables the development of multiple vocabularies, it introduces the possibility of confusion among those vocabularies. In other words, whereas for HTML we can be sure what the ⟨TD⟩ tag means, for XML we cannot quite be sure that ⟨temperature⟩ means for us what it means for another party. That is, although we know we are using a vocabulary, how do we know that we are using the right vocabulary?

Namespaces provide a simple means to keep the different vocabularies separate. By placing the element and attribute names in separate namespaces, we can ensure that we can identify the ones we mean. A namespace is identified via its URI. Typically, XML documents specify the namespaces they are using. The namespace declarations, which relate a local prefix to a namespace URI, are usually placed in the top-level element of the given document. These declarations are made via a reserved attribute and a reserved prefix. For example, Listing A.7 uses xmlns to declare its default namespace, in which the elements weather, temperature, value, and accuracy are declared. This listing uses the xmlns: paradigm to declare places as a namespace prefix to identify elements for describing a place via its latitude and longitude.

Listing A.7: Example of XML namespaces

```
<?xml version="1.0"?>
<weather xmlns ="http://www.temperature.org"
 xmlns:places="http://www.temperature.org/places"
```

```
<temperature scale="Celsius">
 <value>25</value>
 <accuracy>2</accuracy>
 <ambience condition="shade"/>
</temperature>
<places:latitude>33.23</places:latitude>
<places:longitude>155.23</places:longitude>
</weather>
```

Namespaces need not be defined anywhere, but it is good practice to define them as XML Schema documents. They are used via *qualified names*. A *qualified name* consists of a namespace prefix and a local part, which is one of the names declared within the defined namespace.

A.2.5 Using Entities

The !ENTITY construct in XML is a convenient means for creating abbreviations. It is analogous to symbolic constants in programming languages, which you can declare in the preamble of a program and use throughout the program as a means of making your program more readable. This book uses the !ENTITY construct in several places as do many standard documents, so it is a good idea to be familiar with it. Listing A.8 shows a simple example of using the !ENTITY construct.

Listing A.8: An example of !ENTITY in XML

```
<!ENTITY MNH 'Michael'>
<!ENTITY MPS 'Munindar'>
<doc>
<book>
 <title>Readings in Agents</title>
 <editor name='&MNH;'/>
 <editor name='&MPS;'/>
</book>

<goodguys>&MNH; and &MPS;</goodguys>
</doc>
```

In this example, MNH and MPS are first declared as abbreviations. These abbreviations are then used in the remainder of the document via &MNH; and &MPS;, respectively. The notation simply substitutes the definition string for the terms being defined. It can be used for attribute values and for PC Data, but not for the names of elements. Thus the above document is equivalent to Listing A.9.

Listing A.9: An equivalent example to Listing A.8 without !ENTITY

```
<doc>
```

```
<book>
  <title >Readings  in  Agents </title >
  <editor  name='Michael'/>
  <editor  name='Munindar'/>
</book>
<goodguys>Michael  and  Munindar</goodguys>
</doc>
```

Entities can be used as certain kinds of security attacks. For example, recursive entities (not allowed by XML) can cause unbounded computation during expansion if not detected by a processor. For this reason, products such as Apache Axis do not accept entities.

A.2.6 XML Extensions

XML is not just a language, but a family of technologies. Its main specification, e.g., *XML 1.0*, defines the syntax for tags and attributes and the requirements for a well-formed XML document. Other specifications extend the usefulness of XML. For example,

XSL. The eXtensible Stylesheet Language specifies rules for formatting an XML document, including presentation formats such as font color and size. The XSL Transformations (XSLT) is a standardized specification for XSL processors. Implementations of XSLT can convert an XML document into HTML, or one XML document into another, or even an XML document into an unstructured document.

Xlink. Xlink specifies how to add hyperlinks to an XML document. The links can be uni-directional like HTML, as well as bidirectional, multidirectional, and typed.

XPointer. XPointer specifies how to refer to individual items of data inside an XML document.

XPath. XPath is used by XPointer to describe location paths. A location path consists of steps, which in turn consist of an *axis*, a *node test*, and a *predicate*. An individual step allows navigation among documents or within a document to reach a desired location.

XML Schema. XML Schema, described in Section A.3, specifies the structure of a set of similar documents, along with the elements that can appear in a document, their attributes, and their default values. In addition, it can enforce data typing.

A.3 XML Schema

An XML Schema is used to provide the rules that conforming XML data and documents must obey. Consider again the example of XML data at the beginning of the previous section. The example obeys all of the rules for correct XML, but how would we know if this data is structured properly and the data values are valid? For example, would a "Huhns" scale make sense for describing a temperature? For the example data to be deemed valid, we might decide that the following business rules must be satisfied:

- A temperature element must comprise a value subelement and an accuracy subelement in that order.

- The data in a value element must be an integer.

- The data in an accuracy element must be an integer that is ≥ 0.

- A scale must be one of the strings: Celsius, Fahrenheit, or Kelvin.

All of the above business rules can be expressed using an XML Schema. An XML Schema is itself a document, written in XML, that constrains the allowable structure and values for valid XML data. You can think of it as the metadata for the document. It is a formal specification of the rules of an XML document, including a definition for the data types that can appear in the document. For example, an XML Schema that corresponds to the DTD above and expresses the above business rules is shown in Listing A.10.

Listing A.10: An example of an XML Schema for temperatures

```xml
<?xml version="1.0"?>
<xsd:schema xmlns:xsd="http://www.w3.org/2001/XMLSchema"
            targetNamespace="http://www.temperature.org"
            xmlns="http://www.temperature.org"
            elementFormDefault="qualified">
  <xsd:simpleType name="scaleType">
      <xsd:restriction base="xsd:string">
          <xsd:pattern value="Celsius"/>
          <xsd:pattern value="Fahrenheit"/>
          <xsd:pattern value="Kelvin"/>
      </xsd:restriction>
  </xsd:simpleType>
  <xsd:element name="TempCollection">
    <xsd:complexType>
      <xsd:sequence>
        <xsd:element ref="temperature"
                     minOccurs="1" maxOccurs="unbounded"/>
      </xsd:sequence>
    </xsd:complexType>
  </xsd:element>
  <xsd:element name="temperature">
    <xsd:complexType>
      <xsd:sequence>
        <xsd:element ref="value" type="xsd:integer"
                     minOccurs="1" maxOccurs="1"/>
        <xsd:element ref="accuracy"
```

```
                        minOccurs="1" maxOccurs="1"/>
        </xsd:sequence>
        <xsd:attributeGroup ref="TempAttributes"/>
      </xsd:complexType>
   </xsd:element>
   <xsd:attributeGroup name="TempAttributes">
     <xsd:attribute name="scale" use="required">
       <xsd:simpleType>
         <xsd:restriction base="xsd:string">
           <xsd:pattern value="Celsius"/>
           <xsd:pattern value="Fahrenheit"/>
           <xsd:pattern value="Kelvin"/>
         </xsd:restriction>
       </xsd:simpleType>
     </xsd:attribute>
   </xsd:attributeGroup>
   <xsd:element name="accuracy"
                type="xsd:nonNegativeInteger"/>
</xsd:schema>
```

This schema is typical in most respects. It uses the targetNamespace attribute to specify the namespace that it defines.

A *validator* is a program that checks XML data or documents to determine if they conform to a given XML Schema, as indicated in Figure A.2. Listing A.11 is a valid document that references the above XML Schema.

Listing A.11: An XML document that conforms to the XML Schema named TemperatureTable.xsd and found at http://www.temperature.org

```
<?xml version="1.0"?>
<TempCollection xmlns ="http://www.temperature.org"
 xmlns:xsi="http://www.w3.org/2001/XMLSchema-instance"
 xsi:schemaLocation=
  "http://www.temperature.org/TemperatureTable.xsd">
 <temperature scale="Celsius">
  <value>25</value>
  <accuracy>2</accuracy>
 </temperature>
 <temperature scale="Kelvin">
  <value>123</value>
  <accuracy>5</accuracy>
 </temperature>
  ...
</TempCollection>
```

In general, for the simpler cases, types defined in XML Schema can be mapped to the types in established programming languages such as Java, and vice versa. However, it is

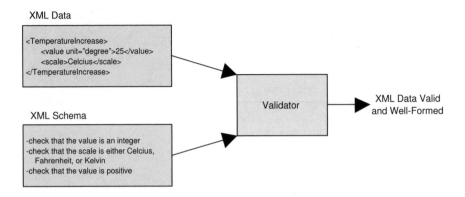

Figure A.2: XML data and documents can be validated by comparing their structure and content to the rules for structure and content expressed in an XML Schema

possible to specify complex types in XML Schema through the use of restrictions that cannot be mapped easily to programming languages. And programming language types, especially those involving references, would not map easily to XML Schema.

A.4   Limitations

XML, in its current form, has the disadvantages that it is best used for transferring text, not binary data. In its typical encoding it is highly inefficient, because the tags are a significant proportion of most messages and have a lot of redundancy. However, binary and compressed formats do exist, e.g., Wireless Binary XML (WBXML) [Martin and Jano, 1999], although these formats are used predominantly in specialized settings, such as those involving wireless devices. Also, WBXML does not handle XML namespaces and requires shared-token dictionaries, which work best for predefined element types. The SOAP standard, discussed in Chapter 2, is able to accommodate binary data through the mechanism of attachments, but the main content remains text based.

An XML document is validated by reconciling it with either a DTD (Document Type Definition) or an XML Schema. Either of these can provide the link among the tags in an XML document, which might be idiosyncratic and local, to definitions of tag structures and syntaxes that could be industry-wide or standard.

A.5   Notes

Xerces, available at http://www.apache.org/xerces-j/index.html, is a well-known open-source XML tool. An XML checker is included in Internet Explorer. The XML Schema Validator (XSV) is available from http://www.ltg.ed.ac.uk/ ht/xsv-status.html. All of these tools check that an XML document is well formed. The validators also check that a document is valid according to a specified DTD or XML Schema. The Saxon tool includes a powerful XSLT engine.

A.6 Exercises

A.1. A new XML namespace is declared using which one of the following:

- the xmlns attribute;
- the DOCTYPE element;
- the namespace attribute;
- the xsi attribute;
- the xsd attribute.

A.2. In the following XML snippet:

```
<choice ret:answer="1"><code>Pick me</code></choice>
<choice xmlns:rdf="http://www.rdf.org">
  <rdf:code>I am wrong</rdf:code>
</choice>
<choice xsi:type="string">
  <code>I am almost correct</code>
</choice>
<choice namespace="http://example.com/x.xsd">
  <code>I am boring</code>
</choice>
<choice><code>Who? me?</code></choice>
```

Which one of the following is a namespace prefix?

- ret
- namespace
- xmlns
- choice
- http://www.rdf.org

A.3. Which of the following is true about XML Schema and DTDs?

- XML Schema is written in XML.
- A DTD can be more specific than XML Schema about the type of subelements an element can have.
- XML Schema is easier to validate.

- XML Schema implements namespaces.
- A DTD cannot be used to define the set of elements a document should have.

A.4. The following XML Schema definition:

```
<xsd:sequence>
  <xsd:element name="part" type="xsd:string"/>
  <xsd:element name="quantity">
    <xsd:simpleType>
      <xsd:restriction base="xsd:positiveInteger">
        <xsd:maxExclusive value="20"/>
      </xsd:restriction>
    </xsd:simpleType>
  </xsd:element>
</xsd:sequence>
```

would validate which of the following XML examples?

- `<part>valve</part>`
  `<quantity>12</quantity>`

- `<quantity>12</quantity>`
  `<part>valve</part>`

- `<part>valve</part>`
  `<quantity>24</quantity>`

- `<partName>valve</partName>`
  `<quantity>20</quantity>`

- `<part>piston</part>`
  `<quantity base="xsd:positiveInteger">4</quantity>`

A.5. In the following XML snippet

```
<alpha delta="gamma">
   beta
</alpha>
```

identify the element, attribute, attribute value, and PC data.

A.6. Namespaces were a necessary addition to XML because of which one of the following?

- The need to use multiple schemas in one document.
- The need to build really large XML documents.

- The proliferation of XML dialects.

- The excessive use of attributes.

- The move from DTDs to XML Schemas.

A.7. Using XML Schema, one can define new data types by which one of the following?

- Restriction or extension.

- Inheritance.

- Grouping.

- Aggregation.

- Alliteration.

A.8. Given the following DTD (located at http://www.nscu.edu/address.dtd) for an address book,

```
<?xml encoding="UTF-8"?>
<!ELEMENT addressBook (person)+>
<!ELEMENT person (name, email*)>
<!ELEMENT name (family, first)>
<!ELEMENT family (#PCDATA)>
<!ELEMENT first (#PCDATA)>
<!ELEMENT email (#PCDATA)>
```

(a) Show how the DTD would have to be modified to add an attribute *gender* to person, where the *gender* attribute must be present, can have only the values male, female, and unknown, and has a default value of unknown.

(b) Construct an XML document for the addresses of two people that would be valid with respect to the DTD as modified in (a).

A.9. Consider a library that keeps track of books and borrowers in a database. The library decides to use XML to encode the reports it sends out. The database contains the following information: the name (family name and first name), address, gender, birth-year, and salary of each borrower, and the title, author (family name and first name), publisher's address, publication year, and ISBN of each book. Which of the above features should be elements and which should be attributes? Show your answer to this by constructing a DTD for this scenario.

A.10. Write an XML Schema that an employment service can use to advertise job and internship opportunities. The advertisements should include the following information: company name, location (such as city and state), annual salary, type (permanent or internship), degree requirements (major and BS, MS, or PhD), and years of experience required.

Then use your XML Schema to construct the single XML encoding for the following two job advertisements (i.e., both ads would be in the same XML structure):

(i) TopTech in Columbia, SC has openings for five interns. Each intern will be paid $2 000 per month. The positions require a master's degree in computer engineering.

(ii) New York, NY based Xcorp is looking for a permanent website developer. The successful candidate, who must have a PhD in computer engineering or five years of relevant experience, will earn $120 000 per year.

A.11. Construct a class diagram in UML (using, for example, Rational Rose or Microsoft Visio) for a domain of your choice. The diagram should have at least two classes with at least one relationship between them. For example, you might have entities ZooAnimals and Food, with the relationship Eats between them. Make sure that each class has several attributes and methods.

Construct an XML Schema document for data that might be transmitted to, from, or within applications in your domain.

Optionally, construct a DTD for data that might be transmitted to, from, or within applications in your domain.

Create two XML documents that are valid with respect to your XML Schema. Verify the validity by using Xerces and XSV (described in Section A.5), or one of the XML validation tools that you can download via the website for this book.

A.12. Write a program in a language of your choice that accepts an XML Schema and produces the code needed for a Web form. When a user enters data into fields on the form, the form converts it to XML that conforms to the schema and stores it in a file.

Appendix B

URI, URN, URL, and UUID

The Web architecture relies upon a way to identify resources uniquely. This leads to the concept of the *Uniform Resource Identifier*. URIs are defined via a number of the so-called *schemes*, such as HTTP, ftp, and so on. The various schemes correspond to different subspaces of the space of resource identifiers. The schemes must be registered with the Internet Assigned Numbers Authority (IANA) [IANA, 2004].

A *Uniform Resource Locator* is a kind of URI that gives an address or location of a resource. The most popular URI schemes, such as HTTP, HTTPS, and ftp, are all examples of URLs. URLs build on the *Domain Name Service* (DNS) to address hosts symbolically and use a file-path like syntax to identify specific resources at a given host. For this reason, mapping URLs to physical resources is straightforward and is implemented by the various Web browsers.

A *Uniform Resource Name* is a kind of URI that gives a name for a resource. Conceptually, URNs reside at a higher level than URLs. They name resources without regard to the host name and the host's physical resource hierarchy (i.e., file system). In principle, if resources are uniquely named, that means they are identified independently of where they might happen to reside on the Web. However, the conceptual flexibility comes at a price and URNs cannot be resolved to physical resources easily, and are therefore not supported by modern Web browsers or networking APIs. However, for the purposes of identification, URNs are just fine. Subspaces of URN must also be registered with IANA by the organization that wishes to serve as the authority for making URN assignments within that subspace.

As described in Section A.2.4, a namespace is identified via its URI. Two URI references to namespaces are considered identical only if they match literally, i.e., character by character. Nonidentical URI references may prove to be functionally equivalent if they resolve to the same address. In principle, because a namespace is identified via a URI, we can think of the namespace as a resource. An application that was designed to process XML documents might retrieve the resource corresponding to the namespace declaration, say, an XML Schema, and use that resource to validate a given document or to create an appropriate document. However, this need not be the case. A URI merely identifies a namespace uniquely. The URI might not refer to any specific physical resource that can be retrieved and reasoned about. In many cases, the applications that process XML documents might be hard-coded for the schemas that they can accommodate. Before such applications begin processing a document, they can check if the namespace URIs are the ones they are designed to accommodate, but not rely on the actual description of the namespace at run time.

The URIs used for identifying namespaces can be URLs or URNs. In general, they are URLs because it is easy to map them to physical resources. Further, although this is not required in principle, it is good practice to ensure that the URLs for a namespace do indeed physically contain the definition of that namespace. Even if automatic tools do not use such definitions, programmers can use them to understand how a document using a particular set of namespaces should be processed. However, URNs are conceptually cleaner than URLs because the location of a document is, after all, incidental to its meaning. Only a few of the namespaces are identified via URNs. One of the better known URNs for services is urn:uddi-org:api_v2, which identifies the UDDI Version 2 namespace.

As remarked above, unlike URLs, URNs do not map to physical locations in a trivial manner. Consequently, the location of the namespace definition needs to be encoded separately. This is the reason that the targetNamespace attribute, e.g., as used in XML Schema definitions, comes in handy. It is the same reason why the import element in WSDL has separate attributes for namespace (identifier) and schemaLocation. The document found in the specified schema location must have the given namespace as its target namespace. This approach is necessary for URIs that do not readily map to locations, but it provides an additional check even when URLs are being used and it allows copies of namespace definitions to be kept at multiple locations.

Often, we will see not a URI but a *URI Reference*. A URI Reference may include a so-called *optional fragment identifier*, i.e., the part of a URL including and after the # character, which points to a particular part within a page.

A *Universally Unique Identifier* (UUID) is a means to uniquely identify objects in a distributed system. Formally, a UUID is a 128-bit number, conventionally written as a hexadecimal string.

UUIDs are generated by executing a standard algorithm. This algorithm includes the hardware address of the given host, the current time, and a random component. The hardware address of a host corresponds to the MAC address of its network card, which is itself guaranteed to be unique because each network card vendor is assigned a unique part of the space and is charged with ensuring that the cards it manufactures are given unique numbers. Including the hardware address ensures that UUIDs generated by different hosts will not conflict. Using the current time means that the UUIDs generated by a given host will not conflict. However, this relies upon the host's system clock always having unique values. If a clock is reset, possibly due to a reboot, then the assumption is no longer valid. The UUID algorithm goes to great lengths to ensure that such problems would be avoided. The random component of the UUID further reduces the risk of nonuniqueness.

Consequently, UUIDs can be treated as unique for all practical purposes, although there may be a combination of extremely rare events under which this assumption is violated.

Appendix C

XML Namespace Abbreviations

Each XML namespace is defined via a URI. Namespaces are abbreviated within each document to provide qualified names. To reduce clutter and make it easier to understand the examples, this book uses the following conventions for abbreviated names:

bpel. Business Process Execution Language for Web Services

dc. Dublin Core metadata for documents, whose current version is at
http://purl.org/dc/elements/1.1/

owl. Web Ontology Language

owls. OWL for services

rdf. Resource Description Framework

rdfs. RDF Schema

soapenc. SOAP encoding

soapenv. SOAP envelope

tns. this namespace; refers to the current WSDL document. It is conventional and reasonable to define this abbreviation as the value of the targetNamespace attribute.

uddi. UDDI

wsdl. WSDL

wsse. WS-Security

xsd. XML Schema definition

xsi. XML Schema instance, which includes the standard types of the XML Schema, which can be used in instance documents.

Glossary

B2B. B2B is an abbreviation for business-to-business electronic commerce, which denotes the interactions among autonomous commercial organizations.

B2C. B2C is an abbreviation for business-to-consumer electronic commerce, which denotes the interactions between an individual and a commercial organization.

BPEL4WS. Business Process Execution Language for Web Services, also known as BPEL.

BPML. Business Process Management Language.

BPSS. Business Process Schema Specification, one of the ebXML suite of standards.

CORBA. Common Object Request Broker Architecture, a standard proposed by the Object Management Group. CORBA introduced a number of technical innovations for interoperation, but has generally been considered too cumbersome for widespread use.

Discovery. Web service discovery is the act of locating a machine-processable description of a Web service that may have been previously unknown and that meets certain functional criteria.

ebXML. Electronic Business eXtensible Markup Language.

EDI. Electronic Data Interchange.

IDL. The Interface Definition Language is a language for specifying operations (procedures or functions), parameters to these operations, and data types.

Process. A process is a series of actions or changes proceeding from one to the next over time.

Protocol. A protocol is a set of rules governing the format and order of messages exchanged among systems or components.

RPC. A Remote Procedure Call refers to the act of a local computation synchronously invoking functionality at a nonlocal site.

SAML. SAML, the Security Assertion Markup Language, targets the secure interchange of authentication and authorization assertions and supports single sign-on.

Schema. A schema is an outline for a document or a plan for a collection of activities.

Service. A service is the product of human, organizational, or computational activity meant to satisfy a need, but not constituting an item of goods. In WSDL, a service is a collection of ports.

SOAP. The Simple Object Access Protocol is the XML language and protocol for messages that enables clients and servers to communicate with each other and exchange XML data. It is built on common Web transport protocols, such as HTTP, SMTP, and FTP.

UDDI. The Universal Description, Discovery, and Integration (UDDI) protocol is used to describe Web services, so that providers can advertise their services in a registry and clients can locate and then use them.

Web service. A Web service is functionality that can be *engaged* over the Web.

WfMC. The Workflow Management Coalition is a group of organizations that defines standards and conventions for workflow management systems.

Workflow. The facilitation or automation of a business process by a computer system.

Workflow Management System. A system that defines, manages, and executes workflows through the execution of software whose order of execution is driven by a computer representation of the workflow logic.

WS-Chor. The W3C Web Service Choreography Working Group and its specifications for organizing the interactions of Web services. These are working drafts that go beyond WSCL and WSCI (see below).

WSCI. The Web Service Choreography Interface is a specification for organizing the interactions of Web services.

WSCL. The Web Service Conversation Language is a specification for stating constraints on the conversations in which a given Web service may participate.

WSDL. WSDL, the Web Services Description Language, is an XML format for describing how one software system can connect to and use the services of another software system over the Internet.

WS-Coordination. Web Service Coordination is a specification for a service whose job it is to coordinate the activities of the Web services that are part of a business process.

WS-Transaction. Web Service Transaction is a specification for two particular coordination types: a short-term atomic transaction and a long-duration business activity.

WS-Security. WS-Security, the most mature Web service security protocol, defines methods for embedding authentication, encryption, and security in SOAP messages, and provides a framework for the exchange of XML-based objects, such as X.509 certificates or SAML tokens.

XACML. The eXtensible Access Control Markup Language is a specification for expressing access-control policies in XML for information access over the Internet.

XPDL. The XML Process Definition Language can be used to describe a process using the XML syntax.

Bibliography

Achermann, Franz and Oscar Nierstrasz. Applications = components + scripts—a tour of Piccola. In Aksit, Mehmet, editor, *Software Architectures and Component Technology*, pages 261–292, Boston, 2001. Kluwer.

Arpinar, Sena, Asuman Dogac, and Nesime Tatbul. An open electronic marketplace through agent-based workflows: MOPPET. *International Journal on Digital Libraries*, 3(1):36–59, 2000.

Arthur, W. Brian, John H. Holland, Blake Lebaron, Richard G. Palmer, and Paul Tayler. Asset pricing under endogenous expectations in an artificial stock market. In *Proceedings on the Economy as an Evolving Complex System II*. Addison-Wesley, 1997.

Austin, John L. *How to Do Things with Words*. Clarendon Press, Oxford, 1962.

Bansal, Sharad. Matchmaking of Web services based on the DAML-S service model. Master's thesis, Department of Computer Science and Engineering, University of South Carolina, Columbia, December 2002.

Barringer, Howard, Ruurd Kuiper, and Amir Pnueli. Now you may compose temporal logic specifications. In *Proceedings of the ACM Symposium on Theory of Computing*, pages 51–63. ACM, 1984.

Bartelt, Andreas and Winfried Lamersdorf. A multi-criteria taxonomy of business models in electronic commerce. In *Proceedings of the IFIP/ACM International Conference on Distributed Systems Platforms (Middleware 2001), Workshop on Electronic Commerce*, volume 2232 of *LNCS*, pages 193–205, Berlin, 2001. Springer-Verlag.

Bauer, Bernhard, Jörg P. Muller, and James Odell. Agent UML: A formalism for specifying multiagent software systems. *International Journal of Software Engineering and Knowledge Engineering*, 11(3):207–230, April 2001.

Bichler, Martin, Arie Segev, and Carrie Beam. An electronic broker for business-to-business electronic commerce on the Internet. *International Journal of Cooperative Information Systems*, 7(4):315–330, 1998.

Boll, Susanne, Andreas Gruner, Armin Haaf, and Wolfgang Klas. EMP - a database-driven electronic market place for business-to-business commerce on the Internet. *Distributed and Parallel Databases*, 7(2):149–177, April 1999.

Bond, Alan and Les Gasser, editors. *Readings in Distributed Artificial Intelligence*. Morgan Kaufmann, San Francisco, 1988.

Box, Don, David Ehnebuske, Gopal Kakivaya, Andrew Layman, Noah Mendelsohn, Henrik Frystyk Nielsen, Satish Thatte, and Dave Winer. Simple object access protocol (SOAP) 1.1, 2000. http://www.w3.org/ TR/ SOAP.

Breese, John S., David Heckerman, and Carl Kadie. Empirical analysis of predictive algorithms for collaborative filtering. In *Proceedings of the 14th Annual Conference on Uncertainty in Artificial Intelligence*, pages 43–52, 1998.

Breitbart, Yuri, Hector Garcia-Molina, and Avi Silberschatz. Transaction management in multiadatabase systems. In *Kim [1994]*, chapter 28, pages 573–591. ACM Press and Addison-Wesley, 1994.

Castelfranchi, Cristiano and Rosaria Conte. Distributed artificial intelligence and social science: Critical issues. In Jennings, N. R. and G. M. P. O'Hare, editors, *Foundations of Distributed Artificial Intelligence*. John Wiley & Sons, Somerset, NJ, 1996.

Cesta, Amedeo, Maria Miceli, and Paola Rizzo. Help under risky conditions: Robustness of the social attitude and system performance. In *Proceedings of the International Conference on Multiagent Systems*, pages 18–25, 1996.

Christensen, Erik, Francisco Curbera, Greg Meredith, and Sanjiva Weerawarana. Web services description language (WSDL) 1.1, 2001. www.w3.org/TR/wsdl.

Curbera, Francisco, Rania Khalaf, Nirmal Mukhi, Stefan Tai, and Sanjiva Weerawarana. The next step in web services. *Communications of the ACM*, 46(10):29–34, October 2003.

Dalal, Sanjay, Sazi Temel, Mark Little, Mark Potts, and Jim Webber. Coordinating business transactions on the web. *IEEE Internet Computing*, 7(1):30–39, January-February 2003.

DAML. The DARPA agent markup language. http://www.daml.org, 2001.

Davenport, Thomas H. *Process Innovation: Reengineering Work through Information Technology*. Harvard Business School Press, Boston, MA, 1993.

Davies, Charles T.,Jr. Data processing spheres of control. *IBM Systems Journal*, 17(2):179–198, 1978.

Davis, Randall and Reid G. Smith. Negotiation as a metaphor for distributed problem solving. *Artificial Intelligence*, 20:63–109, 1983. Reprinted in Bond and Gasser [1988].

de Kleer, Johan. An assumption-based truth maintenance system. *Artificial Intelligence*, 28 (2):127–162, 1979.

Decker, Stefan, Sergey Melnik, Frank van Harmelen, Dieter Fensel, Michel Klein, Jeen Broekstra, Michael Erdmann, and Ian Horrocks. The semantic Web: The roles of XML and RDF. *IEEE Internet Computing*, 4(5):63–74, September 2000a.

Decker, Stefan, Prasenjit Mitra, and Sergey Melnik. Framework for the semantic Web: An RDF tutorial. *IEEE Internet Computing*, 4(6):68–73, November 2000b.

Delugach, Harry S. An exploration into semantic distance. In *Proceedings Seventh Annual Workshop on Conceptual Graphs*, volume 754 of *Lecture Notes in Artificial Intelligence*, pages 29–37. Springer, 1993.

Dennett, Daniel C. *The Intentional Stance*. MIT Press, Cambridge, MA, 1987.

Doyle, Jon. A truth maintenance system. *Artificial Intelligence*, 12(3):231–272, 1979.

Dumais, Susan T., George W. Furnas, Thomas K. Landauer, Scott Deerwester, and Richard Harshman. Using latent semantic analysis to improve access to textual information. In *Proceedings of the ACM SIGCHI Conference on Human Factors in Computing Systems*, pages 281–285. ACM Press, 1988.

Emerson, E. Allen. Temporal and modal logic. In Leeuwen, Janvan , editor, *Handbook of Theoretical Computer Science*, volume B, pages 995–1072. North-Holland, Amsterdam, 1990.

Escrow.com. Online escrow process, 2003. http://www.escrow.com /solutions /escrow /process.asp.

FCC. Comments on reverse search for electronic white pages, July 1996. Number DA 96-1069; available at www.fcc.gov/ Bureaus/ Common_Carrier/ Orders/ 1996/ da961069.txt.

Fielding, Roy T. and Richard N. Taylor. Principled design of the modern Web architecture. *ACM Transactions on Internet Technology*, 2(2):115–150, May 2002.

Fielding, Roy Thomas. *Architectural Styles and the Design of Network-based Software Architectures*. PhD thesis, University of California, Irvine, CA, 2000. Available from http://www.ics.uci.edu/ fielding/pubs/dissertation/top.htm.

FIPA. Foundation for intelligent physical agents (FIPA) specification, 1998. www.fipa.org.

Forgy, Charles. Rete: A fast algorithm for the many patterns/many objects pattern match problem. *Artificial Intelligence*, 19(1):17–37, 1982.

Foster, Ian. The Grid: A new infrastructure for 21st century science. *Physics Today*, 55(2): 42–47, February 2002.

Foster, Ian, Carl Kesselman, and Steven Tuecke. The anatomy of the Grid: Enabling scalable virtual organizations. *International Journal of Supercomputer Applications*, 15(3):200–222, Fall 2001.

Francez, Nissim and Ira R. Forman. *Interacting Processes: A Multiparty Approach to Coordinated Distributed Programming*. ACM Press and Addison-Wesley, New York, 1996.

Friedman, Eric and Paul Resnick. The social cost of cheap pseudonyms. *Journal of Economics and Management Strategy*, 10(2):173–199, 2001.

Garcia-Molina, Hector and Kenneth Salem. Sagas. In *Proceedings of ACM SIGMOD Conference on Management of Data*, pages 249–259, 1987.

Georgakopoulos, Dimitrios, Marek Rusinkiewicz, and Amit P. Sheth. Using tickets to enforce the serializability of multidatabase transactions. *IEEE Transactions on Knowledge and Data Engineering*, 6(1):166–180, 1994.

Ghenniwa, Hamada H. and Michael N. Huhns. Intelligent enterprise integration: eMarketplace model. In *Creating Knowledge Based Organizations*, pages 46–79. Idea Group Publishers, 2003.

Gomez-Perez, Asuncion and Oscar Corcho. Ontology languages for the semantic Web. *IEEE Intelligent Systems*, 17(1):54–60, January-February 2002.

Gray, Jim and Andreas Reuter. *Transaction Processing: Concepts and Techniques*. Morgan Kaufmann, San Mateo, 1993.

Grosof, Benjamin, Michael Gruninger, Michael Kifer, David Martin, Deborah McGuinness, Bijan Parsia, Terry Payne, and Austin Tate. Semantic Web services language requirements, 2004. http://www.daml.org/services/swsl/requirements/swsl-requirements.shtml.

Gruninger, Michael. Applications of PSL to semantic Web services. In Cruz, Isabel F., Vipul Kashyap, Stefan Decker, and Rainer Eckstein, editors, *Proceedings of the 1st International Workshop on Semantic Web and Databases*, pages 217–230, September 2003.

Haddadi, Afsaneh. Towards a pragmatic theory of interactions. In *Proceedings of the International Conference on Multiagent Systems*, pages 133–139, 1995.

Hammer, Michael and James Champy. *Reengineering the Corporation*. Harper Collins, New York, 1993.

Heflin, Jeff and James A. Hendler. Dynamic ontologies on the Web. In *Proceedings of the National Conference on Artificial Intelligence (AAAI)*, pages 443–449, 2000.

Hendler, James and Deborah L. McGuinness. DARPA agent markup language. *IEEE Intelligent Systems*, 15(6):72–73, 2001.

Hohfeld, Wesley Newcomb and Walter W. Cook. *Fundamental Legal Conceptions: As Applied in Judicial Reasoning*. Lawbook Exchange Ltd., Clark, NJ, 2001.

Horrocks, Ian, Ulrike Sattler, and Stephan Tobies. Practical reasoning for expressive description logics. In *Proceedings of the 6th International Conference on Logic Programming and Automated Reasoning (LPAR)*, volume 1705 of *LNCS*, pages 161–180. Springer-Verlag, September 1999.

Huhns, Michael and David M. Bridgeland. Multiagent truth maintenance. *IEEE Transactions on Systems, Man, and Cybernetics*, 21(6):1437–1445, 1991a.

Huhns, Michael N. and David M. Bridgeland. Multiagent truth maintenance. *IEEE Transactions on Systems, Man, and Cybernetics*, 21(6):1437–1445, December 1991b.

Huhns, Michael N. and Munindar P. Singh. Agents and multiagent systems: Themes, approaches, and challenges. In *Huhns and Singh [1998b]*, chapter 1, pages 1–23. Morgan Kaufmann, 1998a.

Huhns, Michael N. and Munindar P. Singh, editors. *Readings in Agents*. Morgan Kaufmann, San Francisco, 1998b.

Huhns, Michael N., Munindar P. Singh, and Tomasz Ksiezyk. Global information management via local autonomous agents. In *Huhns and Singh [1998b]*, pages 36–45. Morgan Kaufmann, 1998. (Reprinted from *Proceedings of the ICOT International Symposium on Fifth Generation Computer Systems: Workshop on Heterogeneous Cooperative Knowledge Bases, 1994*).

Huhns, Michael N., Larry M. Stephens, and Nenad Ivezic. Automating supply-chain management. In *Proceedings of the 1st International Joint Conference on Autonomous Agents and MultiAgent Systems (AAMAS)*, pages 1017–1024. ACM Press, July 2002.

IANA. Internet assigned numbers authority, 2004. http://www.iana.org/.

IOTP. Internet open trading protocol (IOTP), October 2003. IETF: Internet Engineering Task Force, http://www.ietf.org/html.charters/trade-charter.html.

Isenberg, David S. The dawn of the stupid network. *ACM netWorker*, 2(1):24–31, February 1998.

Jennings, Nicholas R. Coordination techniques for distributed artificial intelligence. In *Foundations of Distributed Artificial Intelligence*, pages 187–210. John Wiley & Sons, New York, 1996.

Joy, Bill. Keynote address at JavaOne, June 2000. http://java.sun.com/ javaone/ javaone00/ transcripts/ keynote2.html.

Jung, Jae-yoon, Wonchang Hur, Suk-Ho Kang, and Hoontae Kim. Business process choreography for B2B collaboration. *IEEE Internet Computing*, 8(1):37–45, January-February 2004.

Kagal, Lalana, Tim Finin, and Anupam Joshi. A policy based approach to security for the semantic Web. In *Proceedings of the 2nd International Semantic Web Conference (ISWC)*, October 2003.

Kagal, Lalana, Massimo Paolucci, Naveen Srinivasan, Grit Denker, Tim Finin, and Katia Sycara. Towards authorization, confidentiality and privacy for semantic web services. In *AAAI 2004 Spring Symposium on Semantic Web Services*. AAAI, March 2004.

Kephart, Jeffrey O., James E. Hanson, and Amy R. Greenwald. Dynamic pricing by software agents. *Computer Networks*, 36(6):731–752, May 2000.

Kim, Won, editor. *Modern Database Systems: The Object Model, Interoperability, and Beyond*. ACM Press and Addison-Wesley, New York, 1994. Reprinted with corrections, 1995.

Klein, Johannes. Advanced rule driven transaction management. In *Proceedings of the IEEE COMPCON*, 1991.

Klein, Mark and Abraham Bernstein. Searching for services on the semantic Web using process ontologies. In *Proceedings of the International Semantic Web Working Symposium (SWWS)*, July 2001.

Klein, Mark, Chrysanthos Dellarocas, and Juan A. Rodríguez-Aguilar. A knowledge-based methodology for designing robust multi-agent systems. In *Proceedings of the 1st International Joint Conference on Autonomous Agents and MultiAgent Systems (AAMAS)*, page 661. ACM Press, July 2002.

Kolp, Manuel, Paolo Giorgini, and John Mylopoulos. A goal-based organizational perspective on multiagent architectures. In *Intelligent Agents VIII: Agent Theories, Architectures, and Languages*, volume 2333, pages 128–140, New York, 2002. Springer-Verlag.

Kowalski, Robert and Marek J. Sergot. A logic-based calculus of events. *New Generation Computing*, 4(1):67–95, 1986.

Labbe, Joe. Web services: The new conventional "WSDM". *eAI Journal*, pages 25–27, January 2002.

LeBaron, Blake. Agent-based computational finance: Suggested readings and early research. *Journal of Economic Dynamics and Control*, 24(5–7):679–702, June 2000.

Lenat, Douglas. Cyc: A large-scale investment in knowledge infrastructure. *Communications of the ACM*, 38(11):32–38, November 1995.

Lenat, Douglas and R. V. Guha. *Building Large Knowledge-Based Systems: Representation and Inference in the Cyc project*. Addison-Wesley, Reading, MA, 1990.

Little, Mark. Transactions and Web services. *Communications of the ACM*, 46(10):49–54, October 2003.

Malone, Thomas W., Kevin Crowston, and George A. Herman, editors. *Organizing Business Knowledge: The MIT Process Handbook*. MIT Press, Cambridge, MA, 2003.

Mani, Anbazhagan and Arun Nagarajan. Understanding quality of service for web services, January 2002. http://www-106.ibm.com/developerworks/webservices/library/ws-quality.html.

Manocha, Nitish, Diane J. Cook, and Lawrence B. Holder. Structural web search using a graph-based discovery system. *ACM Intelligence*, 12(1):20–29, Spring 2001.

Marsh, Steven P. *Formalising Trust as a Computational Concept*. PhD thesis, Department of Computing Science and Mathematics, University of Stirling, April 1994.

Martin, Bruce and Bashar Jano. WBXML: WAP binary XML content format, 1999. http://www.w3.org/ TR/ wbxml.

Maximilien, E. Michael and Munindar P. Singh. Conceptual model of Web service reputation. *ACM SIGMOD Record*, 31(4):36–41, December 2002.

McBride, Brian. Jena: A semantic Web toolkit. *IEEE Internet Computing*, 6(6):55–59, November 2002.

McCarthy, John. Ascribing mental qualities to machines. In Ringle, Martin, editor, *Philosophical Perspectives in Artificial Intelligence*, pages 161–195. Harvester Press, Brighton, UK, 1979.

McFadzean, David, Leigh Tesfatsion, and Deron Stewart. A computational laboratory for evolutionary trade networks. *IEEE Transactions on Evolutionary Computation*, 5(5):546–560, October 2001.

Melnik, Sergey and Stefan Decker. A layered approach to information modeling and interoperability on the Web, 2000. www-db.stanford.edu/ ~melnik/ pub/ sw00/.

Metz, Cade. Testing the waters, November 2001. http://www.pcmag.com/ article2/ 0,4149,62217,00.asp.

Meyer, Bertrand. .NET is coming. *IEEE Computer*, 34(8):92–97, August 2001.

Miller, George A. WordNet: A lexical database for English. *Communications of the ACM*, 38(11):39–41, November 1995.

Noy, Natalya Fridman and Deborah L. McGuinness. Ontology development 101: A guide to creating your first ontology. TR KSL-01-05 and SMI-2001-0880, Stanford Knowledge Systems Laboratory and Stanford Medical Informatics, March 2001.

Paolucci, Massimo, Takahiro Kawamura, Terry R. Payne, and Katia Sycara. Semantic matching of Web services capabilities. In *Proceedings of the First International Semantic Web Conference (ISWC)*, June 2002.

Parunak, H. Van Dyke. Visualizing agent conversations: Using enhanced Dooley graphs for agent design and analysis. In *Proceedings of the 2nd International Conference on Multiagent Systems*, pages 275–282. AAAI Press, 1996.

Pearl, Judea. *Probabilistic Reasoning in Intelligent Systems: Network of Plausible Inference*. Morgan Kaufmann, San Mateo, California, 1988.

Pierre, John. Practical issues for automated categorization of Web sites. In *Proceedings of the ECDL Workshop of Semantic Web*, 2000.

Porter, Michael E. *Competitive Advantage*. Free Press, New York, 1985.

Python. CSV file reading and writing, February 2004. http://www.python.org /dev /doc /devel /lib /module-csv.html; versions 2.3 and 2.4.

Rachlevsky-Reich, Benny, Israel Ben-Shaul, Nicholas Tung Chan, Andrew W. Lo, and Tomaso Poggio. GEM: A global electronic market system. *Information Systems*, 24(6): 495–518, 1999.

Rosenschein, Jeffrey S. and Gilad Zlotkin. Designing conventions for automated negotiation. *AI Magazine*, pages 29–46, Fall 1994a.

Rosenschein, Jeffrey S. and Gilad Zlotkin. *Rules of Encounter*. MIT Press, Cambridge, MA, 1994b.

Russell, Stuart J. and Peter Norvig. *Artificial Intelligence: A Modern Approach, Second Edition*. Pearson Education, Inc., Upper Saddle River, NJ, 2003.

Saltzer, Jerome H., David P. Reed, and David D. Clark. End-to-end arguments in system design. *ACM Transactions on Computer Systems*, 2(4):277–288, November 1984.

Schulzrinne, Henning. Internet telephony. In *Singh [2004]*. Chapman Hall & CRC Press, 2004.

Schwarz, Reinhard and Friedemann Mattern. Detecting causal relationships in distributed computations: In search of the holy grail. *Distributed Computing*, 7(3):149–174, 1994.

Sen, Sandip. Reciprocity: a foundational principle for promoting cooperative behavior among self-interested agents. In *Proceedings of the 2nd International Conference on Multiagent Systems*, pages 322–329. AAAI Press, Menlo Park, 1996.

SET. Secure electronic transactions (SET) specifications, 2003. http://www.setco.org/ set_ specifications.html.

Shafer, Glenn. *A Mathematical Theory of Evidence*. Princeton University Press, Princeton, 1976.

Shapiro, Jonathan S., Jonathan M. Smith, and David J. Farber. EROS: A fast capability system. In *Proceedings of the 17th ACM Symposium on Operating Systems Principles (SOSP)*, pages 170–185, New York, 1999. ACM Press.

Singh, Munindar P. Agent communication languages: Rethinking the principles. *IEEE Computer*, 31(12):40–47, December 1998.

Singh, Munindar P. Synthesizing coordination requirements for heterogeneous autonomous agents. *Autonomous Agents and Multi-Agent Systems*, 3(2):107–132, June 2000.

Singh, Munindar P. Distributed enactment of multiagent flows. TR 2003-12, North Carolina State University, May 2003.

Singh, Munindar P., editor. *Practical Handbook of Internet Computing*. Chapman Hall & CRC Press, Baton Rouge, 2004.

Sirbu, Marvin A. Credits and debits on the Internet. *IEEE Spectrum*, 34(2):23–29, February 1997.

Sowa, John F. *Knowledge Representation: Logical, Philosophical, and Computational Foundations*. Brooks/Cole, Pacific Grove, CA, 2000.

Tate, Austin and Jeff Dalton. O-plan: a common lisp planning web service. In *Proceedings of the International Lisp Conference*, New York, October 2003.

Thesiger, Wilfred. *Arabian Sands*. Penguin, London, 1959.

Timmers, Paul. *Electronic Commerce: Strategies and Models for Business-to-Business Trading*. Wiley & Sons, 1999.

Traiger, Irving L. Trends in systems aspects of database management. In *Proceedings of the 2nd International Conference on Databases (ICOD-2*, pages 1–21. Wiley & Sons, 1983.

Trastour, David, Claudio Bartolini, and Javier Gonzalez-Castillo. A semantic Web approach to service description for matchmaking of services. In *Proceedings of the International Semantic Web Working Symposium (SWWS)*, July 2001.

Tsvetovatyy, Maksim, Maria L. Gini, Bamshad Mobasher, and Zbigniew Wieckowski. MAGMA: An agent-based virtual market for electronic commerce. *Journal of Applied Artificial Intelligence*, 11(6):501–523, 1997.

Tuecke, Steven, Karl Czajkowski, Ian Foster, Jeffrey Frey, Steve Graham, and Carl Kessel-man. Grid service specification. Draft 2/15/2002, University of Chicago, 2002. www.globus.org/research/papers/gsspec.pdf.

UDDI. UDDI technical white paper, 2000. www.uddi.org/pubs/Iru-UDDI-Technical-White-Paper.pdf.

Venkatraman, Mahadevan and Munindar P. Singh. Verifying compliance with commitment protocols: Enabling open Web-based multiagent systems. *Autonomous Agents and Multi-Agent Systems*, 2(3):217–236, September 1999.

Vinoski, Steve. Web services interaction models, part I: Current practice. *IEEE Internet Computing*, 6(3):89–91, May 2002.

WfMC. Workflow management coalition (WfMC) reference model. http:// www.aiai.ed.ac.uk /WfMC/, 1995.

Wiederhold, Gio. Ontology algebra. In *Proceedings of the ICOT International Symposium on Fifth Generation Computer Systems: Workshop on Heterogeneous Cooperative Knowledge Bases*, 1994.

Wiegold, C. Frederic, editor. *The Wall Street Journal Lifetime Guide to Money*. Hyperion, New York, 1997.

Wooldridge, Michael. A knowledge-theoretic semantics for concurrent METATEM. In *Intelligent Agents III: Agent Theories, Architectures, and Languages*, pages 357–374, 1997.

Wooldridge, Michael, Nicholas R. Jennings, and David Kinny. The Gaia methodology for agent-oriented analysis and design. *Autonomous Agents and Multi-Agent Systems*, 3(3): 285–312, 2000.

WSI. Web services interoperability organization, 2004. http://www.ws-i.org.

Yolum, Pınar and Munindar P. Singh. Emergent properties of referral systems. In *Proceedings of the 2nd International Joint Conference on Autonomous Agents and MultiAgent Systems (AAMAS)*, pages 592–599. ACM Press, July 2003.

Yu, Bin and Munindar P. Singh. An evidential model of distributed reputation management. In *Proceedings of the 1st International Joint Conference on Autonomous Agents and Multi-Agent Systems (AAMAS)*, pages 294–301. ACM Press, July 2002.

Index